GRE® BIOLOGY TEST

TestWare® Edition

Linda Gregory, Ph.D.
Lecturer
University of Victoria
Victoria, British Colombia, Canada

Larry "Zack" Florence, Ph.D.
Adjunct Professor
University of Alberta
Edmonton, Alberta, Canada

Marilyn Florence, M.A.
Freelance Writer

Kevin S. Reilley, B.S.
Researcher

Research & Education Association
Visit our website: www.rea.com
GRE Test Updates: www.rea.com/GRE

Planet Friendly Publishing
GREEN EDITION
✔ Made in the United States
✔ Printed on Recycled Paper
Text: 10%   Cover: 10%
Learn more: www.greenedition.org

At REA we're committed to producing books in an Earth-friendly manner and to helping our customers make greener choices.

Manufacturing books in the United States ensures compliance with strict environmental laws and eliminates the need for international freight shipping, a major contributor to global air pollution.

And printing on recycled paper helps minimize our consumption of trees, water and fossil fuels. This book was printed on paper made with **10% post-consumer waste**. According to Environmental Defense's Paper Calculator, by using this innovative paper instead of conventional papers, we achieved the following environmental benefits:

Trees Saved: 5 • Air Emissions Eliminated: 1,004 pounds
Water Saved: 936 gallons • Solid Waste Eliminated: 296 pounds

For more information on our environmental practices, please visit us online at **www.rea.com/green**

Research & Education Association
61 Ethel Road West
Piscataway, New Jersey 08854
E-mail: info@rea.com

GRE® Biology Test with TestWare® on CD-ROM

Printed in the United States of America

Library of Congress Control Number 2010920081

ISBN 13: 978-0-7386-0835-8
ISBN 10: 0-7386-0835-1

G10-0101

Contents

CHAPTER 1

Introduction   1

CHAPTER 2

Cellular Structure and Function   9

CHAPTER 3

Genetics and Molecular Biology   41

CHAPTER 4

Animal Structure, Function, and Organization   89

CHAPTER 5

Animal Reproduction and Development   133

Contents

CHAPTER 6

Plant Structure, Function, and Organization, with Emphasis on Flowering Plants   147

CHAPTER 7

Plant Reproduction, Growth, and Development with Emphasis on Flowering Plants   157

CHAPTER 8

Diversity of Life   165

CHAPTER 9

Ecology   171

CHAPTER 10

Evolution   189

About Research & Education Association

Founded in 1959, Research & Education Association is dedicated to publishing the finest and most effective educational materials—including software, study guides, and test preps—for students in middle school, high school, college, graduate school, and beyond.

REA's Test Preparation series includes books and software for all academic levels in almost all disciplines. Research & Education Association publishes test preps for students who have not yet entered high school as well as high-school students preparing to enter college. Students from countries around the world seeking to attend college in the United States will find the assistance they need in REA's publications. For college students seeking advanced degrees, REA publishes test preps for many major graduate school admission examinations in a wide variety of disciplines, including engineering, law, and medicine. Students at every level, in every field, and with every ambition can find what they are looking for among REA's publications.

REA's practice tests are always based on the most recently administered exams and include every type of question that can be expected on the actual exams.

REA's publications and educational materials are highly regarded and continually receive an unprecedented amount of praise from professionals, instructors, librarians, parents, and students. Our authors are as diverse as the subject matter represented in the books we publish. They are well known in their respective disciplines and serve on the faculties of prestigious high schools, colleges, and universities throughout the United States and Canada.

Today REA's wide-ranging catalog is a leading resource for teachers, students, and professionals.

We invite you to visit us at *www.rea.com* to find out how "REA is making the world smarter."

Staff Acknowledgments

We would like to thank REA's Larry B. Kling, Vice President, Editorial, for his overall direction; Pam Weston, Vice President, Publishing, for setting the quality standards for production integrity and managing the publication to completion; John Paul Cording, Vice President, Technology, for coordinating the design, development, and testing of REA's TestWare® software; Molly Solanki, Associate Editor, for her editorial contributions; Diane Goldschmidt, Senior Editor, for post-production coordination; Software Project Managers Heena Patel and Michelle Boykins-Smith for their software testing efforts; Christine Saul, Senior Graphic Artist, for designing the cover; Jeff LoBalbo, Senior Graphic Designer, for coordinating pre-press electronic file mapping; and Kathy Caratozzolo for typesetting the manuscript.

Installing REA's TestWare®

SYSTEM REQUIREMENTS

Pentium 75 MHz (300 MHz recommended) or a higher or compatible processor; Microsoft Windows 98 or later; 64 MB available RAM; Internet Explorer 5.5 or higher

INSTALLATION

1. Insert the GRE Biology TestWare® CD-ROM into the CD-ROM drive.
2. If the installation doesn't begin automatically, from the Start Menu choose the Run command. When the Run dialog box appears, type d:\setup (where d is the letter of your CD-ROM drive) at the prompt and click ok.
3. The installation process will begin. A dialog box proposing the directory "Program Files\REA\GRE_Biology" will appear. If the name and location are suitable, click ok. If you wish to specify a different name or location, type it in and click ok.
4. Start the GRE Biology TestWare® application by double-clicking on the icon.

REA's GRE Biology TestWare® is **EASY** to **LEARN AND USE**. To achieve maximum benefits, we recommend that you take a few minutes to go through the on-screen tutorial on your computer. The "screen buttons" are also explained there to familiarize you with the program.

TECHNICAL SUPPORT

REA's TestWare® is backed by customer and technical support. For questions about **installation or operation of your software**, contact us at:

Research & Education Association
Phone: (732) 819-8880 (9 a.m. to 5 p.m. ET, Monday–Friday)
Fax: (732) 819-8808
Website: www.rea.com
E-mail: info@rea.com

Note to Windows XP Users: In order for the TestWare® to function properly, please install and run the application under the same computer administrator-level user account. Installing the TestWare® as one user and running it as another could cause file-access path conflicts.

GRE Biology Pro Study Plan

Making the Most of REA's 12-Week Pro Study Plan

As a busy student or professional, you may be wondering how you will find time to work, have a social life, take care of family obligations, and prepare for the GRE Biology Subject Test. But don't worry, our GRE Pro Study Plan gives you a complete road map from now until test day. This flexible schedule allows you to work at your own pace and shows you how to prepare for the GRE Biology as efficiently as possible. You'll get organized, stay on track, and make the most of your valuable study time.

You will need to make time *every day* to study for the GRE Biology. How and when you study is up to you, but consistency is the key to completing your GRE preparation. Make studying a priority and consider it a "job" until you take the GRE exam.

Keep in mind that a high score on the GRE doesn't just help you get into a better grad school; it helps move your career forward. Therefore, a solid commitment to daily study is worth every minute of your time. The results will pay off in the long run!

This study schedule will help you become thoroughly prepared for the GRE Biology. Although the schedule is designed as a 12-week study program, it can be compressed into a 6-week plan by combining two weeks of study into one. If you choose the 12-week schedule, you should plan to study for *at least* one hour a day. GRE examinees who are following a 6-week program need to set aside *at least* two hours for studying every day.

Find a study routine that works for you and stick to it! Some people like to get up early and study for an hour or two before going to work. Others might choose to study while commuting, on their lunch hour, at the library, or at home after work. Whatever schedule you choose, make a commitment to study every day – even on weekends.

Remember, daily focused concentration on the GRE subject matter will help you retain more information, fully grasp the material tested, and improve your overall GRE score.

This GRE test prep with TestWare® offers you additional practice and reinforcement. We've included three practice tests in the book and two full-length practice tests on an interactive, easy-to-use CD. The TestWare® CD offers diagnostic feedback, on-screen detailed answer explanations, and timed testing conditions, so you can "practice for real" and know what to expect on the GRE Biology *before* exam day.

If you're new to computer-based tests, don't worry! The CD comes with step-by-step installation instructions and a tutorial that explains how to answer the practice test questions. No matter what your technological skill level, you will benefit from taking our exams on CD. It's a great way for you to become comfortable with computer-based tests while mastering the GRE subject matter.

Week	Activity
1	Read and study our introduction to the GRE Biology Subject Test on the following pages. Be sure you understand the format of the GRE Biology and know exactly what is tested on the exam. Now would be a good time to plan a study schedule. When and where will you study? How will you balance your responsibilities with studying? Decide when and where you study best, and get ready to tackle the GRE Biology!
2	Start your study routine by taking the GRE Biology Practice Test 1 on the TestWare® CD to determine your strengths and weaknesses. Make sure you give yourself at least 3 hours after work, on a weekend, or at another convenient time to take the exam. You will need to focus and concentrate, so take the practice test at a time and place where you will not be disturbed. When you take the test, try to do your best, even on sections where you may be confused. After you have finished the test, record your scores. This will help you track your progress as you study. Later in the week, study the detailed explanations for the questions you answered incorrectly. In the cases where you erred, find out why. Take notes and pay attention to sections where you missed a lot of questions. You will need to spend more time reviewing the related material.
3	Make a firm commitment to study for at least an hour a day, every day for the next few weeks. It may seem hard to find time in your busy schedule, but remember: the more you study, the better prepared you will be for the GRE Biology. This week, focus your study on chapter 2. Complete the practice questions and check your answers for each section of the review. Be sure you have a good grasp of the figures and formulas and that you are comfortable with the material before continuing to next week's study agenda.

4	Study chapter 3. Take your time and make sure you're familiar with all the vocabulary, figures, and topics presented in our review. Pay close attention to the problems and solutions in each section; they will show you how to solve questions you may encounter on the actual GRE Biology exam. If you find yourself in need of extra review or clarification on a topic, you may want to consult your biology textbook or ask a classmate or professor for additional help.
5	Keep working your way through the review chapters and practice problems. This week, study chapter 4. This chapter is filled with tables, figures, must-know vocabulary, and key concepts. Take your time and make sure you understand everything included in the chapter before continuing to next week's study plan. Give yourself more time to practice and review any areas in which you feel you need improvement.
6	Congratulations! You've reached the halfway point! Success is just around the corner, but keep going, there's more work to be done. Take Practice Test 3 in the book. Record your score and see how well you did. After you've evaluated your test results, go back through each chapter and brush up on the topics you need to review. Later this week, study the detailed explanations of answers for any questions you answered incorrectly. Make sure you understand why you answered the question wrong, so you can improve your test-taking skills.
7	Your hard work is paying off! By the time you've finished studying for the GRE Biology, you'll have an excellent understanding of all the topics tested on the exam – and that will help you get a high score. Take time this week to study chapter 5. If you feel you have a good grasp of the material, review your notes and restudy any areas of weakness from previous chapters.
8	Start off Week 8 by studying chapters 6 and 7. These two chapters contain tables, figures, and key biology terms that you will need to know for the GRE. After you've completed your study for the week, why not ask a friend or classmate to quiz you on some vocabulary terms?
9	This week, review any questions you answered incorrectly on the previous practice tests. Spend time studying the detailed answer explanations and re-read the relevant chapters for extra review. If you feel you need extra GRE Biology practice, why not review your notes during your lunch hour or on your commute?

(con't on next page)

(con't from previous page)

10	Are you ready for even more practice? It's time to finish the review chapters. This week study review chapters 8, 9, and 10. Although these are shorter chapters, comprehending the subject matter is vital to your success on the GRE Biology. Make sure you have a good grasp of these topics, so you're ready for any question on the GRE Biology exam.
11	You have just about reached your goal! After a day or two of additional study to reinforce any areas of weakness, take Practice Test 2 on CD. Allow yourself at least 3 hours to take the test in a quiet location. Don't rush! Remember what you've learned and answer every question to the best of your ability. How much has your score improved since you took the first TestWare® practice exam? After the test, thoroughly review all the explanations for the question you answered incorrectly. Remember, your primary goal right now is to get a high score on the GRE, so keep working!
12	Use this time to restudy any sections in which you need improvement. Now is a great time to re-read the review chapters or retake the practice tests to help you boost your skills. Just for fun, why not take Practice Test 1 on CD again? After you've completed the exam, compare your two scores and see how much you've improved in just a matter of weeks!

Congratulations! You've worked hard and you're ready for the GRE Biology!

Introduction

About This Book & TestWare®

This book, along with REA's exclusive TestWare® software, provides you with an accurate and complete representation of the GRE Biology Subject Test. REA's three full-length practice tests are based on the latest editions of the exam. Our topical reviews are designed to prepare you for the very kind of material you are most likely to encounter when taking the actual test. Our sample tests have been carefully calibrated to match the GRE Biology Subject Test's level of difficulty, its format, and, of course, the type and proportional representation of its content. Following each practice test you will find an answer key along with detailed step-by-step explanations designed to help you master the relevant material and score high.

The practice tests in this book and software package are included in two formats: in printed form in the book and in TestWare® format on the enclosed CD. We recommend that you begin your preparation by first taking the practice exams on your computer. The software provides timed conditions, and instantaneous, accurate scoring that makes it easier to pinpoint your strengths and weaknesses.

About REA's Test Experts

To aid us in meeting our objective of providing you with the best possible study guide for the GRE Biology Test, REA's test experts have carefully prepared our topical reviews and practice exams. Our authors come armed with specific knowledge of the GRE Biology Test. They have thoroughly examined and researched the mechanics of the GRE Biology Test to ensure that our model tests accurately depict the exam and appropriately challenge the student. Our experts are highly regarded in the educational community. They have taught and conducted scientific research at competitive institutions. They have an in-depth knowledge of the subjects presented in the book and provide accurate questions that will put you in a position to do your very best on the exam.

About the Test

The GRE Biology Test is taken by students applying to graduate programs in biology. Most programs require that applicants submit scores for both the GRE General Test and the GRE Biology Test; together with other undergraduate records, they are part of the highly competitive admission process to graduate school. Both tests are offered by Educational Testing Service (ETS) and administered throughout the United States and abroad. You can obtain a test registration booklet from your college or by contacting ETS directly. To determine if you should take the GRE Biology Test, contact the universities you are applying to for admission. For questions pertaining to GRE Biology policies, contact:

Graduate Record Examinations
Educational Testing Service
P.O. Box 6000
Princeton, NJ 08541-6000
Phone: (866) 473-4373
Fax: 1-610-290-8975
Website: *www.gre.org*

Format of the GRE Biology Test

The test is usually given three times a year and contains approximately 200 multiple-choice questions, which you must answer in 2 hours and 50 minutes. Items for the GRE Biology Test are composed by a committee of specialists who come recommended by the American Institute of Biological Sciences, the American Society of Zoologists, and the Botanical Society of America and are selected from various undergraduate and graduate faculties. Each of the 200 questions is worth one point. There is a penalty for wrong answers, which serves to correct for "guessing." For each wrong answer, one-quarter of a point is deducted from your score. Unanswered questions don't count for or against you. Some questions are grouped together, based on particular laboratory-experiment scenarios or descriptive paragraphs. Emphasis is placed on the major areas of biology.

These major areas are given equal importance on the test, but are covered in no particular order. In general, you should find that the test items are rooted in courses of study most commonly offered in the undergraduate curriculum.

SSD Accommodations for Students with Disabilities

Many students qualify for extra time to take GRE Biology, and our TEST*ware*® can be adapted to accomodate your time extension. This allows you to practice under the same extended-time accommodations that

THE GRE BIOLOGY SUBJECT TEST AT A GLANCE

Subject	Percent
Cellular and Molecular Biology:	**33–34%**
Cell Structure and Function	16–17%
Molecular Biology and Genetics	16–17%
Organismal Biology:	**33–34%**
Animal Structure, Function, and Organization	9–10%
Animal Reproduction, Growth, and Development	5–6%
Plant Structure, Function, and Organization	6–7%
Plant Reproduction, Growth, and Development	4–5%
Diversity of Life	6–7%
Ecology and Evolution:	**33–34%**
Ecosystems and General Ecology	16–17%
Evolutionary Processes and Consequences	16–17%

you will receive on the actual test day. To customize your TEST*ware*® to suit the most common extensions, visit our website at *www.rea.com/ssd*. For more information, contact:

GRE-ETS
PO Box 6000
Princeton, NJ 08541-6000
Phone: (609) 771-7670 Monday through Friday
          8 A.M. to 7:45 P.M. (Eastern time)
Fax: (609) 290-8975

About the Review

REA's Biology Review concisely and systematically summarizes the main areas tested on the GRE Biology Test. We have prepared it to help you better grasp concepts that your textbook explores in far greater detail.

By studying our review, your chances of scoring well on the actual exam will be greatly increased. It affords you a kind of master checklist for everything you need to know. After thoroughly studying the material presented in the Biology Review, you should go on to take the practice tests. Used in combination, the review and practice tests will enhance your test-taking skills and give you the confidence needed to obtain a high score.

How to Prepare for the GRE Biology Test

"Don't worry." Easier said than done, but rest assured that this book will help you assess yourself as well as the test. As with other GRE subject tests, the GRE Biology Test gauges knowledge that you have gained throughout your academic career. Most of what's tested on the GRE Biology Test will require you to make use of information you learned in your General Biology courses in college.

We at REA believe the best way to prep for the GRE Biology Test is to replicate the complete GRE test-taking experience. Toward that end, we provide three full-length exams that accurately reflect this subject test in terms of format, content, and degree of difficulty. Our practice exams mirror the latest GRE Biology Test forms and include every type of question that you can expect to encounter when you take the exam. Following each of our practice exams is an answer key complete with detailed explanations and solutions. Designed specifically to clarify the material for the student, the explanations not only provide the correct answers, but also explain why the answer to a particular question is indeed the best choice. By completing all three practice exams and studying the explanations that follow, you will isolate your strengths and weaknesses. This, in turn, will enable you to concentrate on attacking the sections of the exam you find to be toughest.

Participate in Study Groups

As a final word on how to study for this test, you may want to study with others. This will allow you to share knowledge and obtain feedback from other members of your study group. Study groups may make preparing for the exam more enjoyable.

Scoring the Test

Each correct response on both our practice tests and the actual exam earn you one "raw score" point, while each incorrect answer results in a 1/4-point deduction; omitted responses are not counted. Here is a formula for calculating your raw score:

# of questions correct		# of questions incorrect	

$$\underline{\hspace{2cm}} - (\underline{\hspace{2cm}} \times 1/4) = \underline{\hspace{2cm}}$$

Now use the Conversion Chart on the next page to determine your scaled score range.

GRE Biology Score Conversion Chart*

Total Score			
Raw Score	**Scaled Score**	**Raw Score**	**Scaled Score**
199–200	990	96–98	640
196–198	980	93–95	630
193–195	970	90–92	620
190–192	960	87–89	610
187–189	950	84–86	600
184–186	940		
181–183	930	81–83	590
179–180	920	78–80	580
176–178	910	75–77	570
173–175	900	72–74	560
		69–71	550
170–172	890	66–68	540
167–169	880	63–65	530
164–166	870	60–62	520
161–163	860	57–59	510
158–160	850	54–56	500
155–157	840		
152–154	830	51–53	490
149–151	820	48–50	480
146–148	810	45–47	470
143–145	800	42–44	460
		39–41	450
140–142	790	36–38	440
137–139	780	33–35	430
134–136	770	30–32	420
131–133	760	28–29	410
128–130	750	25–27	400
125–127	740		
122–124	730	22–24	390
119–121	720	19–21	380
116–118	710	16–18	370
113–115	700	13–15	360
		10–12	350
110–112	690	7–9	340
107–109	680	4–6	330
105–106	670	1–3	320
102–104	660	0	310
99–101	650		

*NOTE: Due to the statistical formulas used by Educational Testing Service, your performance on our practice tests can only *approximate* your performance on the actual GRE Biology Test; one will not directly equate with the other.

Scoring Worksheet

	Raw Score	Scaled Score
Practice Exam 1	_____	_____
Practice Exam 2	_____	_____
Practice Exam 3	_____	_____

Test-Taking Strategies

Although you will probably have to take both the GRE General Test and the Biology Subject Test, try to avoid taking them on the same day. Taking any test is stressful, and after sitting for one extremely long standardized test, you will hardly be at your best for a second.

Be sure to register for testing dates several months before the due date to ensure that the graduate schools you designate will receive your scores by the application deadlines. Most schools will not consider an incomplete application.

Because the test is not divided into sections, you are completely responsible for budgeting your own time. All the questions are worth the same number of points, so you should not spend too much time on any one item. The GRE Biology Test attempts to cover a broad range of topics. It is unlikely that you will have complete knowledge of all of them. It is important that you do not spend too much time on questions you find difficult at the expense of working on those that are easier for you.

The time constraints are such that, on average, a little less than a minute is allotted for each question. Thus, it is unlikely that you will have time to answer all 200 questions; however, you can still receive an excellent score without answering all of them. Because the questions are in no particular order, we recommend making a complete sweep through all the questions on the test. Answer the ones that are immediately easy for you and mark those that you want to revisit. Once you have answered all of the easier questions, you can use the remaining time to go back through the test and work on the harder questions, which require a greater amount of your time. In this way, you will ensure that you have the chance to answer all the questions you are likely to get correct, instead of spending valuable time on difficult questions near the beginning of the test and leaving easy questions at the end of the test unanswered.

The penalty for wrong answers should not deter you completely from guessing. If you have no clue what the answer might be, by all means press on. However, if you can eliminate one or two of the five choices, it is to your advantage to make an educated guess. Statistically, guessing randomly among the five choices would give you the possibility of guessing correctly 1/5 of the time. (This is what the quarter-point deduction for wrong answers is designed to balance.) Being able to eliminate three of the choices as wrong answers means that guessing between the two remaining choices would give you a far better chance of being correct.

The Day of the Test

On the day of the test, you should wake up early (after a decent night's rest, we hope) and enjoy a good breakfast. Make sure you dress comfortably—in layers—so that you are not distracted by being too hot or too cold while taking the exam. You should plan on arriving at the test center early. Doing so will spare you the needless anxiety that comes from racing the clock. It will also allow you to collect your thoughts, focus, and actually relax before taking the exam.

Before you leave for the test center, make sure that you have **two** forms of identification. You will not be admitted to the test center without proper identification. Acceptable forms of identification include a driver's license, Social Security card, birth certificate, passport, and green card.

Make sure you bring at least two sharpened No. 2 pencils, with erasers, to the exam. You may want to wear a watch to the test center; however, only ordinary watches will be permitted. Watches with alarms, calculator functions, flashing lights, beeping sounds, etc., will not be allowed. In addition, neither food nor calculators will be allowed into the examination room.

During the Test

When you arrive at the test center, try to sit in a seat where you feel you will be comfortable. No breaks are given during the exam. If you need to use the rest room, or if you become ill, you may leave the examination room, but you will not be allowed to make up any lost time.

Once you enter the test center, follow all of the rules and instructions given by the test supervisor. If you do not, you risk being dismissed from the examination or having your GRE Biology scores voided, meaning that they will not be scored.

When all of the test materials have been distributed, the test instructor will give you directions for filling out your answer sheet. You must complete this sheet carefully since the information on it will be printed on your score report. Write your name exactly as it appears on your identification documents and admission ticket, unless otherwise instructed.

Make sure you do not write in your test booklet or on your answer sheet, except to fill in the oval corresponding to the answer you choose. Scratch paper will be provided. You will be marking your answers on side two of your answer sheet. Each numbered row will contain five ovals corresponding to each answer choice for that question. Fill in the oval corresponding to your answer darkly, completely, and neatly. You can change your answer, but remember to completely erase your old answer. Only one answer should be marked. This is very important, as your answer sheet will be machine-scored, and stray lines or unnecessary marks may cause the machine to score your answers incorrectly.

Work only on the test section on which the test instructor has instructed you to proceed. You should begin only when instructed to do so, and stop immediately when instructed to stop. Do not turn to the next section of the test until you are told to do so. When all of the sections have been completed, you should remain seated until all of the test materials have been collected.

Good luck on the GRE Biology Subject Test!

GRE BIOLOGY

Review

Cellular Structure and Function

1. Biological Compounds

Inorganic and organic atoms and molecules combine into four classes of macromolecular compounds that are essential to living, cellular organisms. These compounds are **proteins**, **carbohydrates**, **lipids** and **nucleic acids**.

Macromolecular Structure and Bonding

Proteins

Structure and Bonding

A protein is a molecule composed of one or more **polypeptides**. A polypeptide is a polymer or chain of amino acids (monomeric units) connected by chemical bonds between the carboxyl group of one amino acid and the amino group of another. These are called peptide bonds; a molecule containing two bound amino acids is called a dipeptide, three a tripepetide, and many a polypeptide. The biologically active form of a protein may be composed of one or several polypeptide chains.

The structure of a protein assumes four levels of organization.

Amino acids, as the name implies, have properties of both amines and carboxylic acids. Isomers (mirror images) of amino acids are defined by their rotation about the alpha-carbon (chiral carbon) and are referred to as D (for dextro-) and L (for levo-); the prevalent form in protein molecules is the L isomer. The general formula for an amino acid is

$$+H_3N - CHR - COO^-,$$

where R can be any of a number of side chains. As can be seen from the structure, amino acids are zwitterions (from the German *zwitter*, "hybrid"), meaning that they have both negative and positive charges at neutral pH. Under acidic conditions they will react as:

$$+H_3N - CHR - COO^- + H_3O^+ \quad +H_3N - CHR - COOH + H_2O.$$

Under basic conditions, they will react as:

$$+H_3N - CHR - COO^- + OH^- \quad +H_2N - CHR - COO^- + H_2O.$$

There are 20 amino acids normally found in organisms, each with a unique three-letter abbreviation and a different side group called the R-group (Table 2.1). The different R-groups are important because they dictate the reactions (e.g., bonding by hydrogen, sulfur) possible for each amino acid, mainly solubility in water, which is based on the polarity of the R-group (e.g., polar amino acids on the surface make proteins soluble [hydrophilic]; nonpolar amino acids tend to aggregate to the interior of the helix and repulse water from the core of the molecule [hydrophobic]). Four amino acids predict the overall charge on a molecule at neutral pH: arginine and lysine carry positive charges; aspartate and glutamate are negatively charged. The simplest amino acid, glycine, has a single hydrogen for its R-group. Others, such as alanine and valine, have long aliphatic chains for their R-groups. Lysine, arginine, and histidine all have an additional ami-

no group and hence are basic. Serine and threonine have aliphatic hydroxyl side chains. Phenylalanine, tyrosine, and tryptophan have aromatic side chains. Aspartate and glutamate have an additional carboxylic acid group and hence are acidic. Asparagine and glutamine have amide side chains. Cysteine and methionine contain sulfur in their side chains. One- or three-letter symbols can be used to identify these amino acids (e.g., the primary sequence of a tripeptide could be displayed as A-L-T [alanine-leucine-threonine], or alternatively in 3-letter code, Ala-Leu-Thr). Amino acids that cannot be produced by the organism in question, and must be supplied by an external source, are called "essential amino" acids. Those marked with an asterisk in Table 2.1 are essential acids for *Homo sapiens*.

Table 2.1.
Symbols and Formulae of Amino Acids

Amino Acid	Three-Letter Abbreviation	Formula	Amino Acid	Three-Letter Abbreviation	Formula
Aliphatic Amino Acids					
Glycine	Gly	$^+H_3N - C - COO^-$ (H above, H below)	Isoleucine*	Ile	$^+H_3N - C - COO^-$ (H above; $H - C - CH_3$; CH_2; CH_3 below)
Alanine	Ala	$^+H_3N - C - COO^-$ (H above, CH_3 below)			
Valine*	Val	$^+H_3N - C - COO^-$ (H above, CH below branching to CH_3 CH_3)	**Aliphatic Hydroxyl Side Chains**		
			Serine	Ser	$^+H_3N - C - COO^-$ (H above; $H - C - OH$; H below)
Leucine*	Leu	$^+H_3N - C - COO^-$ (H above; CH_2; CH branching to CH_3 CH_3)	Threonine*	Thr	$^+H_3N - C - COO^-$ (H above; $H - C - OH$; CH_3 below)

* Essential for *Homo sapiens*.

(cont'd) →

Amino Acid	Three-Letter Abbreviation	Formula

Aromatic Side Chains

Phenylalanine* — Phe

$$^+H_3N - \overset{\overset{\displaystyle H}{|}}{C} - COO^-$$
$$|$$
$$CH_2$$

Tyrosine — Tyr

$$^+H_3N - \overset{\overset{\displaystyle H}{|}}{C} - COO^-$$
$$|$$
$$CH_2$$
$$OH$$

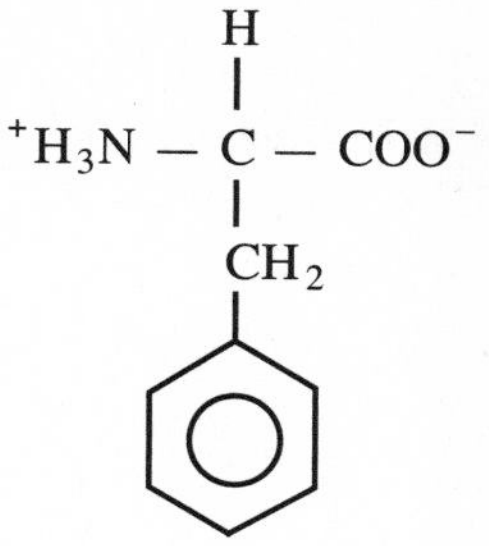

Tryptophan* — Trp

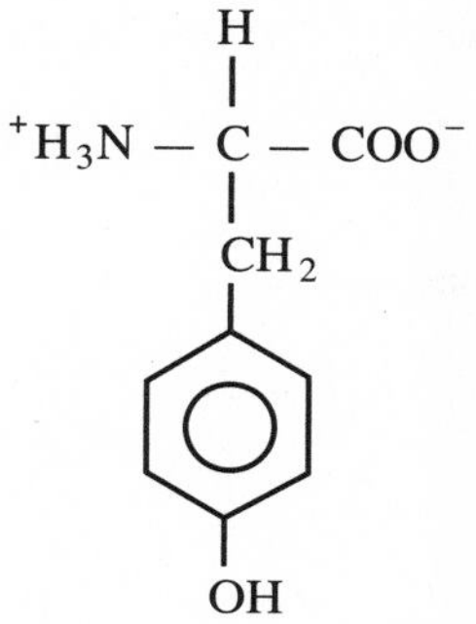

$$^+H_3N - \overset{\overset{\displaystyle H}{|}}{C} - COO^-$$
$$|$$
$$CH_2$$
$$\overset{\displaystyle C}{\underset{N}{\underset{H}{\|}} CH}$$

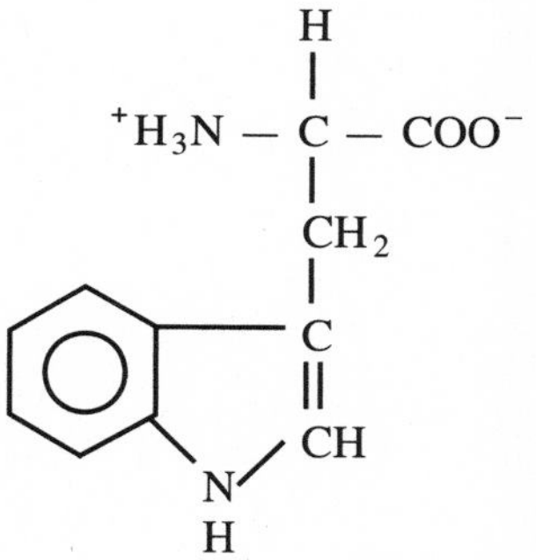

Basic Amino Acids

Lysine* — Lys

$$^+H_3N - \overset{\overset{\displaystyle H}{|}}{C} - COO^-$$
$$|$$
$$CH_2$$
$$|$$
$$CH_2$$
$$|$$
$$CH_2$$
$$|$$
$$CH_2$$
$$|$$
$$NH_3^+$$

Amino Acid	Three-Letter Abbreviation	Formula

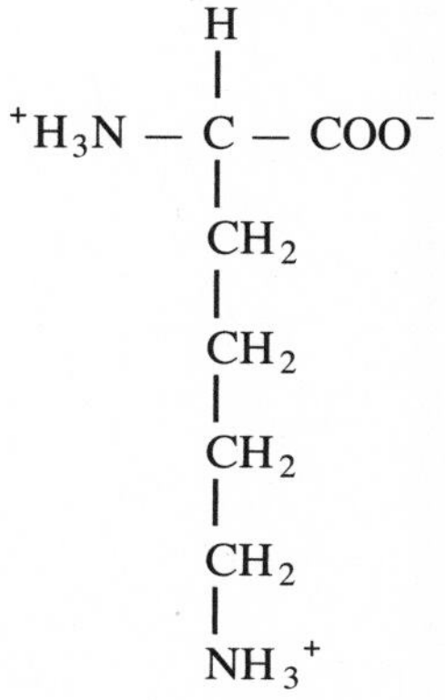

Arginine* — Arg

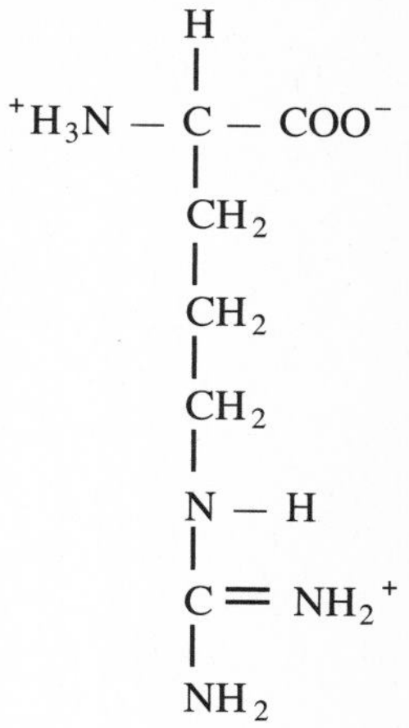

$$^+H_3N - \overset{\overset{\displaystyle H}{|}}{C} - COO^-$$
$$|$$
$$CH_2$$
$$|$$
$$CH_2$$
$$|$$
$$CH_2$$
$$|$$
$$N - H$$
$$|$$
$$C = NH_2^+$$
$$|$$
$$NH_2$$

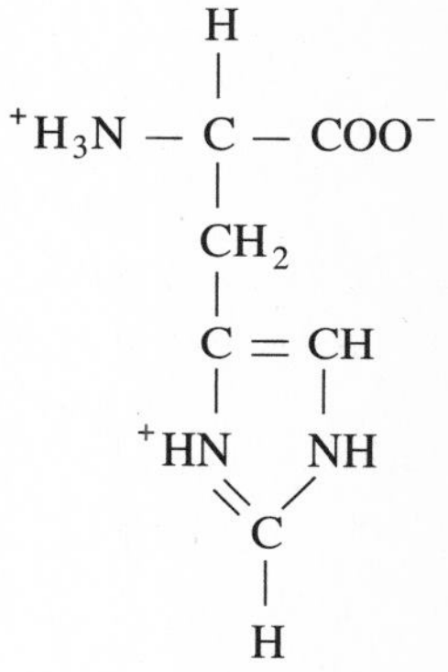

Histidine* — His

$$^+H_3N - \overset{\overset{\displaystyle H}{|}}{C} - COO^-$$
$$|$$
$$CH_2$$
$$|$$
$$C = CH$$
$$^+HN \quad NH$$
$$C$$
$$H$$

Acidic Amino Acids

Aspartate (or Aspartic Acid) — Asp

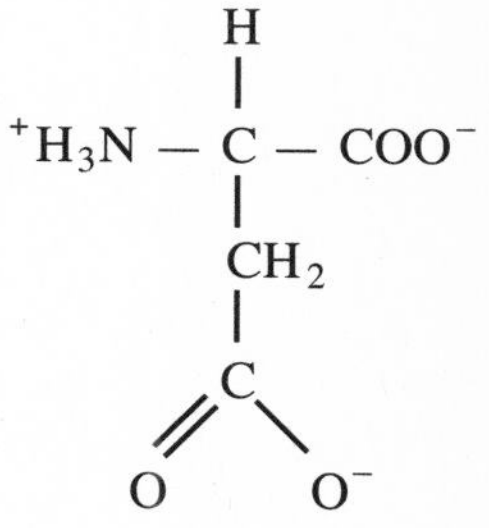

$$^+H_3N - \overset{\overset{\displaystyle H}{|}}{C} - COO^-$$
$$|$$
$$CH_2$$
$$|$$
$$C$$
$$O \quad O^-$$

Glutamate (or glutamic acid) — Glu

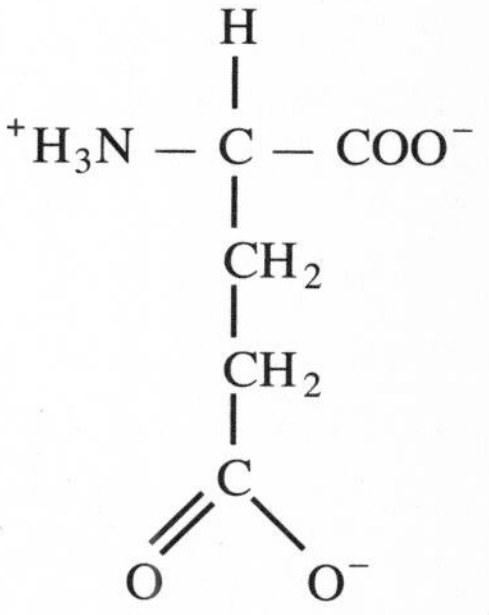

$$^+H_3N - \overset{\overset{\displaystyle H}{|}}{C} - COO^-$$
$$|$$
$$CH_2$$
$$|$$
$$CH_2$$
$$|$$
$$C$$
$$O \quad O^-$$

(cont'd) →

11

Amino Acid	Three-Letter Abbreviation	Formula	Amino Acid	Three-Letter Abbreviation	Formula

Amide-Containing Amino Acids

Asparagine — Asn

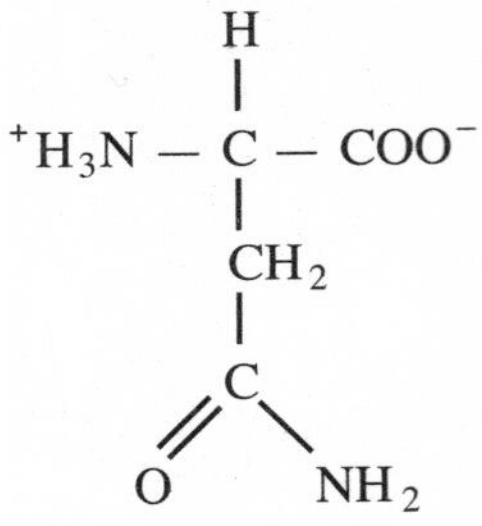

$$^+H_3N - C - COO^-$$ with H on top, CH_2, then C double-bonded to O and single-bonded to NH_2

Glutamine — Gln

$$CH_2 - C - COO^-$$ with H on top, CH_2, CH_2, then C double-bonded to O and single-bonded to NH_2

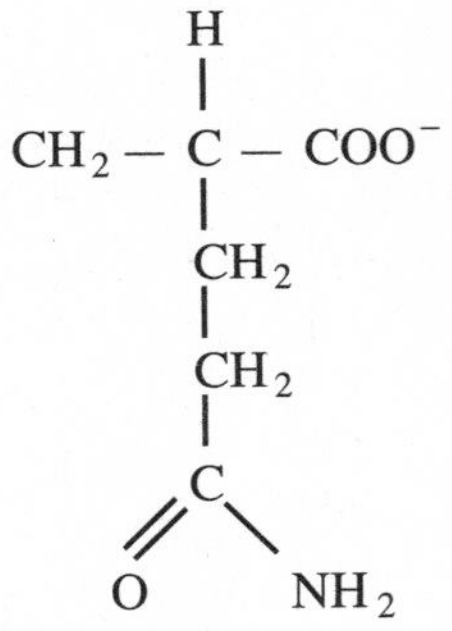

Sulfur-Containing Amino Acids

Cysteine — Cys

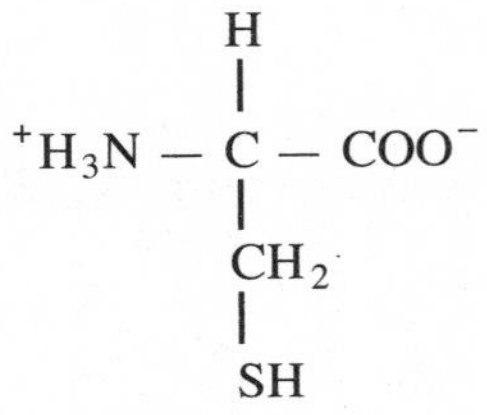

$$^+H_3N - C - COO^-$$ with H on top, CH_2, SH

Methionine* — Met

$$^+H_3N - C - COO^-$$ with H on top, CH_2, CH_2, S, CH_3

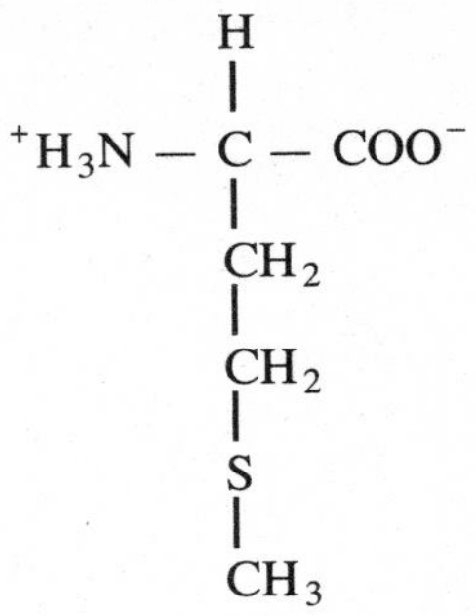

Other Amino Acids

Proline — Pro

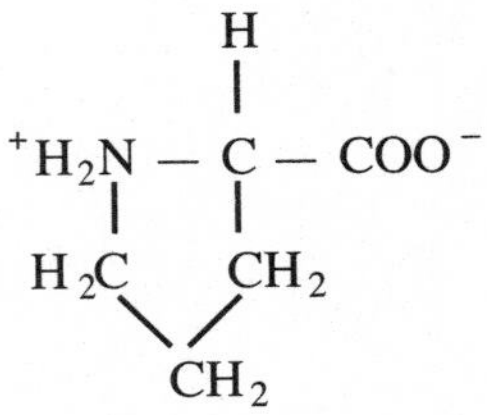

$$^+H_2N - C - COO^-$$ with H on top, ring: H_2C, CH_2, CH_2

The primary method of linking amino acids together to form peptides is by a peptide bond. The amino end of one amino acid forms a covalent bond with the carboxylic acid end of another amino acid (Figure 2.1).

This results in a dipeptide and water and is an example of dehydration. Breaking a peptide bond between amino acids requires water; hence it is known as hydrolysis. This is the reverse of dehydrogenation.

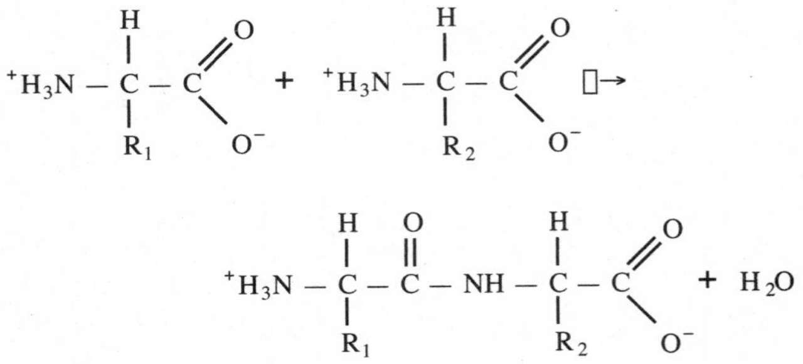

$$^+H_3N - C - C + {}^+H_3N - C - C \rightarrow$$

$$^+H_3N - C - C - NH - C - C + H_2O$$

Figure 2.1.
Formation of Peptide by Covalent Bonds

Amino acids with sulfur-containing side chains (e.g., cysteine and methionine) can also bond with one another by sulfide bonds (Figure 2.2). This is important because an amino acid in one peptide chain can bond with an amino acid in another by sulfide bonds.

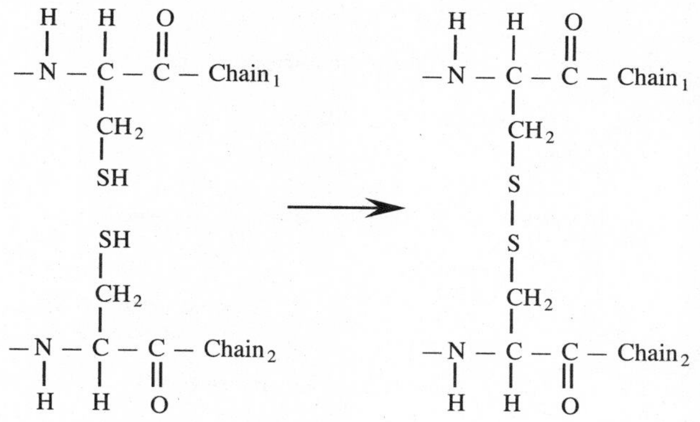

Figure 2.2.
Formation of Disulfide Bonds
Between Cysteine and Methionine

Proteins can have four levels of structure. The primary structure consists of the sequence of amino acids. Because there are 20 amino acids normally found in organisms, the sequence of amino acids can vary from protein to protein. This great variability enables proteins to assume many different functions. One important function is as a biological catalyst or **enzyme**.

The secondary structure is due to the formation of alpha-helices or beta-sheets. Both alpha-helices and beta-sheets arise from the periodic hydrogen bonding between peptides (Figure 2.3).

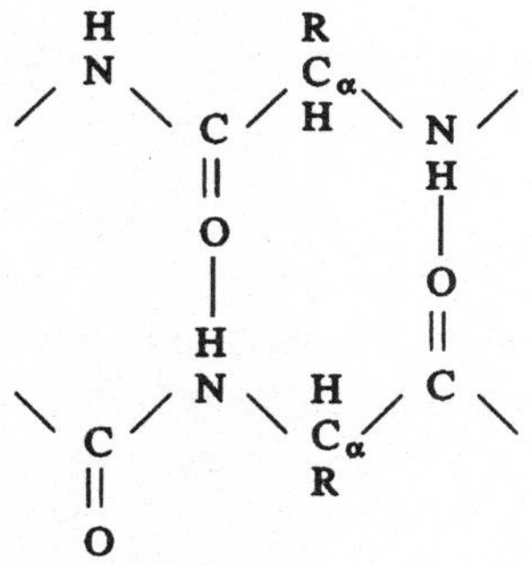

Figure 2.3.
Hydrogen Bonds Between Two Polypeptide Chains That Give the Secondary Structure

The tertiary structure of a protein is the complete three-dimensional description of each atom in the protein. Tertiary structure is built from units of secondary structure linked together by turns in the polypeptide backbone. Quarternary structure results when protein subunits combine to form a larger structure. Hemoglobin is an example of a protein with quarternary structure since it contains two alpha-subunits (α) and two beta-subunits (β) (i.e., $\alpha_2\beta_2$).

Proteins can be either water-soluble or bound to biological membranes. For water-soluble proteins, the hydrophilic amino acid side chains are generally found near the outer surface of the protein. The hydrophobic amino acid side chains are found in the interior of the protein, where they are not in contact with water. Hydrophobic amino acid side chains (Cys, Val, Ile, Leu, Met, Trp, and Phe) are relatively nonpolar and do not interact with water molecules. Hydrophilic amino acid residues (Lys, Arg, His, Asp, Glu, Asn, and Gln) are polar and interact favorably with water.

Membrane proteins are associated with the lipid bilayer of biological membranes. There are two general categories of membrane proteins: intrinsic membrane proteins and extrinsic membrane proteins. Intrinsic membrane proteins are strongly associated with the biological membrane and can only be removed with a denaturing detergent such as sodium dodecylsulfate (SDS). Intrinsic membrane proteins often have a sequence containing numerous hydrophobic amino acid side chains that are strongly associated with the hydrophobic domain of the lipid bilayer. Extrinsic membrane proteins are only loosely associated with biological membranes and can be removed by alterations in the ionic strength or by a chelator such as EDTA.

Proteins often contain a nonprotein moiety such as carbohydrate (e.g., gycoproteins), lipid (lipoprotein), or metal ion (e.g., metalloprotein).

Carbohydrates

Structure and Bonding

Carbohydrates and lipids (fats) are important sources of energy in biological systems. A carbohydrate is a molecule composed of three types of atoms: carbon, hydrogen, and oxygen (i.e., *carbo-* and *hydrate-*, signifying atoms in a water molecule). Carbohydrates, or monosaccharides (sugar), all have the empirical formula: $(CH_2O)_n$, where $n=3,4,5,6$ or 7. All contain a hydroxyl (-OH) group and either a keto group (C-C:O-C) or an aldehyde group (C-C:O-H). Viewing these molecules from the complex to simple, a carbohydrate that cannot be hydrolyzed to a simpler compound is called a *monosaccharide*. If a carbohydrate contains an aldehyde group, it is known as an *aldose*; if it contains a ketone group, it is known as a *ketose*.

When two or more monosaccharides are linked (glycoside linkage), they are called disaccharides, trisaccharides, and so forth, with longer chains being called polysaccharides. When these bonds are broken during hydrolysis, energy is released for use by the cell. Figure 2.4 shows the synthesis of the Krebs Cycle starting material. Note that since water is involved, it is a type of hydrolysis.

Monosaccharides are also classified by the number of carbon atoms they contain. For example, triose, tetrose, and pentose carbohydrates would have three, four, and five carbon atoms, respectively. An aldopentose would be a five-carbon monosaccharide with an aldehyde group. A ketohexose would be a six-carbon monosaccharide containing a ketone group. All monosaccharides are reducing sugars. Most disaccharides are also reducing sugars (sucrose is one of the exceptions). (A reducing sugar means that the carbonyl group of a sugar is not attached to another structure.) Carbohydrates can exist either in an open chain form or a closed-ring form (hemiacetal forms). It is easy to change from one to the other. Figure 2.5 shows the two forms for D-glucose, the first a linear chain and the second a six-member ring (the most common form), which results when the aldehyde group on the 1 carbon reacts with the hydroxyl group on the 5 carbon. A five-member hemiacetal ring forms when the 4 carbon and 1 carbon react. These variations are important to the recognition by enzymes during metabolism of carbohydrates. The carbon atoms are numbered in each form for reference. The figure is actually β-D-glucose. Such diaste-

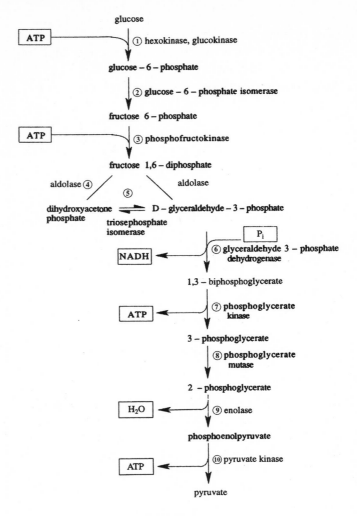

Figure 2.4.
Hydrolysis of a Polysaccharide

reoisomers that differ only about carbon 1 are called *anomers*.

Among the aldoses, any diastereoisomers that differ only about the configuration of carbon 2 are called *epimers*. Note that any carbohydrate can have an anomer, but only an aldose can have an epimer. Aldoses can be oxidized by Fehling's or Tollen's reagent, bromine water, nitric acid, and periodic acid. Ketoses can also be oxidized by Fehling's or Tollen's reagent, but not any of the other listed oxidants. The oxidation of

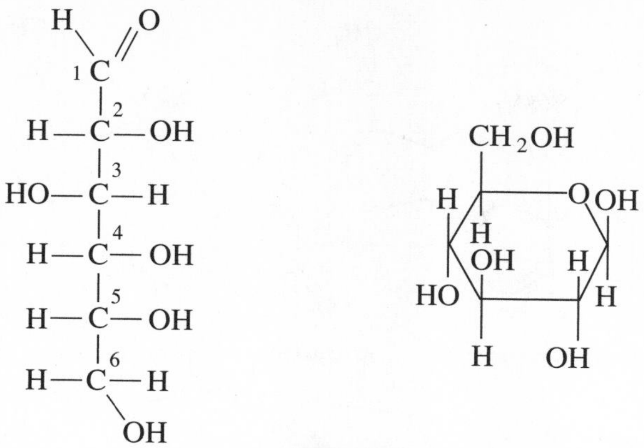

$$\begin{array}{c}
\overset{H}{\underset{1}{C}} \diagdown^{O} \\
\end{array}$$

Figure 2.5.
The Open Chain and a Hemiacetal, Six-member Closed-Ring Form of D-Glucose

aldoses by bromine water and nitric acid is shown in Figure 2.6.

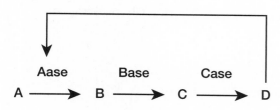

Figure 2.6.
Oxidation of an Aldose by Bromine Water and by Nitric Acid

Lipids

Structure and Bonding

Lipids are water-insoluble biological molecules. They include fats, sterols, and phospholipids.

Chemically, fats are carboxylic esters derived from glycerol and are known as glycerides. Fatty acids are generally straight-chained compounds from 3 to 18 carbons long. Living organisms tend to produce fatty acids containing an even number of carbons: 16, 18, or 20 with up to three double bonds. Fatty acids having at least one double bond are called *unsaturated* fats, and those with none are called *saturated* fats. See Figure 2.7.

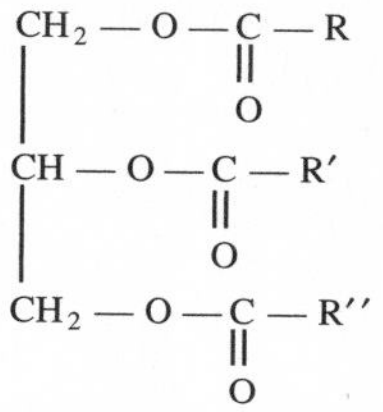

Figure 2.7.
The General Formula for a Triglyceride

Sterols

Saturated fatty acids can be converted to sterols, the second group of lipids. See Figure 2.8. One important sterol is cholesterol, an important constituent of plasma membranes and a precursor of the steroid hormones (e.g., glucocorticoids, mineralocorticoids, estrogens, androgens).

Phospholipids

The final group of lipids is the *phospholipids* in which one of the fatty acid chains of a triglyceride is replaced by a phosphate-containing group. These compounds are the most common amphipathic lipids (i.e., one end of the molecule interacts with water [hydrophilic], the other does not [hydrophobic]). Phospholipids are important in biomembranes of cells and membrane-bound organelles (e.g., mitochondrion). See Figure 2.9.

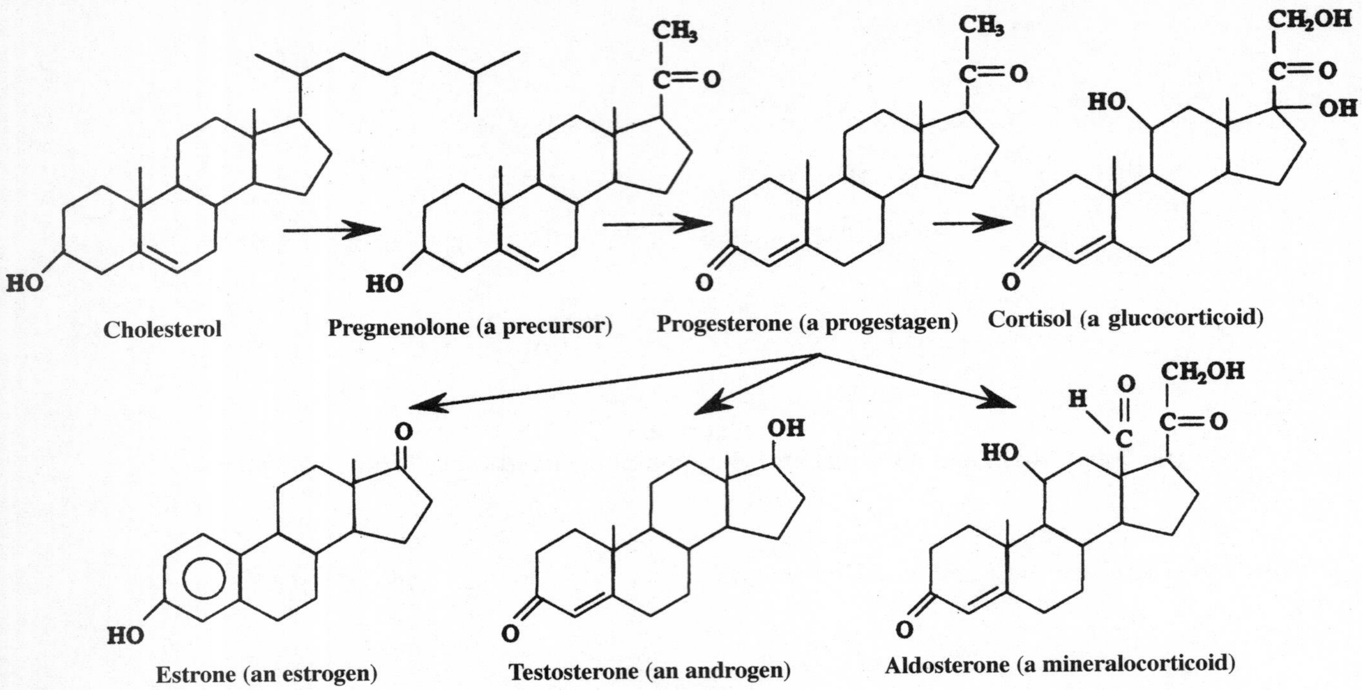

Figure 2.8.
Cholesterol and Some of Its Hormone Derivatives

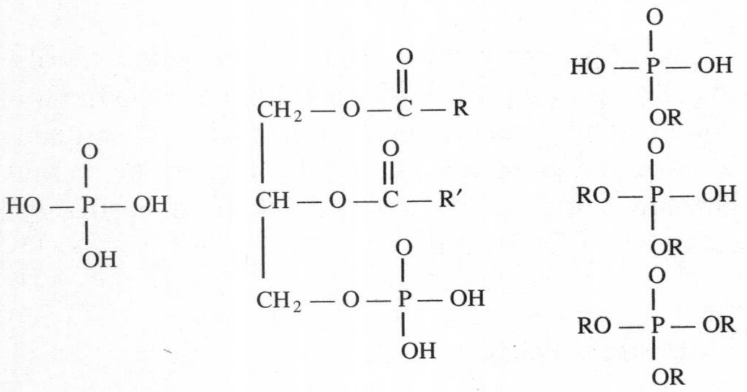

Nucleic Acids: Structure and Bonding

The fourth and final group of biological molecules is the nucleic acids. They are polymers; the monomers that compose them are nucleotides. See Figure 2.10. Nucleotides consist of a five-carbon sugar, a phosphate group and a nitrogenous base. There are two important nucleic acids: deoxyribonucleic acid (**DNA**) and ribonucleic acid (**RNA**).

Figure 2.9.
Phosphatidic Acid,
Simplest of the Phospholipids,
and General Form
(R-Groups Can Be Any Fatty Acid)

Phosphoric Acid   Phosphatidic Acid   Various phosphate esters

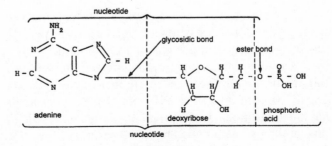

Figure 2.10.
Nucleotides

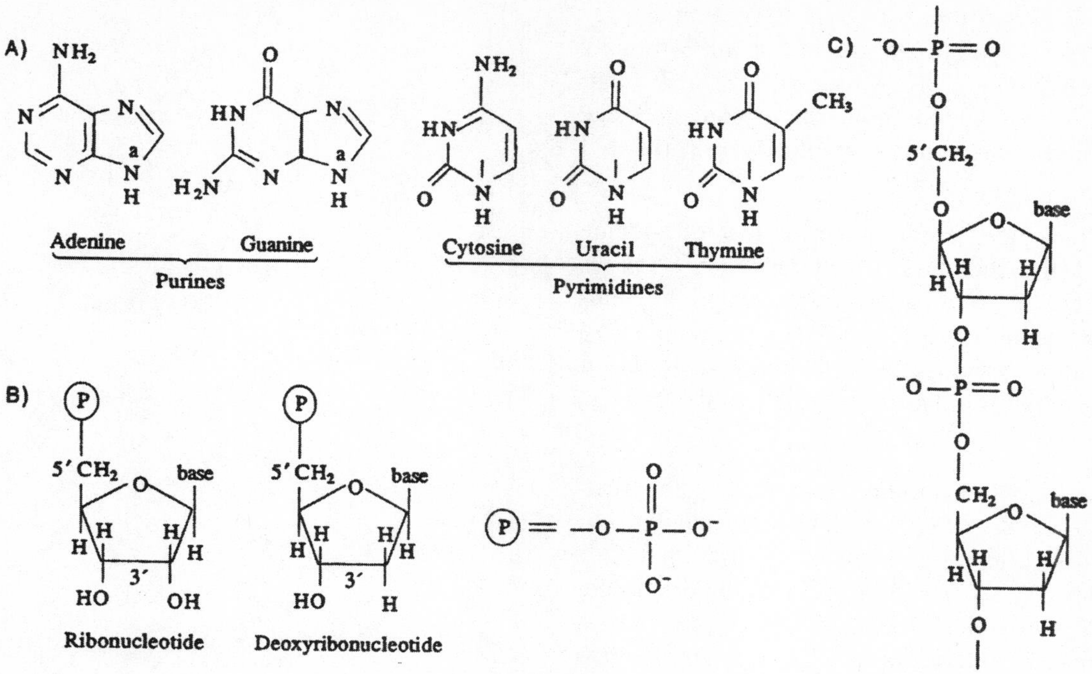

Fig 2.11 A, B and C
Nucleic Acid Structure
(A) Structure of bases in nucleic acids. Purines attach to ribose (or deoxyribose) at the 9-position, and pyrimidines at the 1-position. (B) Structure of a ribonucleotide and deoxyribonucleotide. (C) Structure of single chain deoxyribonucleic acid.

Deoxyribonucleic Acid (DNA)

Nucleotides contain either a purine or a pyrimidine base that is attached to a five-carbon sugar-phosphate (see Figure 2.11A). In DNA the sugar is 2-deoxyribose (note the loss of an oxygen on carbon 2, Figure 2.11B); in ribonucleic acids (RNA) the sugar is ribose. Four bases are found in both DNA and RNA (Figure 2.11, A, B). Adenine and guanine are the two purine bases present in DNA. Thymidine and cytosine are the two pyrimidines. RNA also contains adenine, guanine, and cytosine, but uracil substitutes for thymine. Nucleotides are linked to each other by phosphodiester bonds between the 3-prime (3′) hydroxyl group of one nucleotide and the 5-prime (5′) end of the next (see Figure 2.11C) to form two antiparallel strands (i.e., the direction of one sugar-phosphate "backbone" chain runs from the 5-carbon of the sugar to the 3-carbon end of the next

sugar; the complementary strand runs from 3-carbon to 5-carbon).

DNA contains the cell's genetic information: the information (encoded sequence) determining the ordering of amino acids in a polypeptide. The composition of DNA provides important clues with regard to both its function and structure. Chargaff's rules state that:

(1) The proportions (A/T, G/C, A+T/G+C, or A+G/T+C, etc.), or base compositions, of the four bases remained constant among different tissues from the same species. This supported the idea that the genetic composition of an organism remains constant during mitotic cell divisions during growth and differentiation, whereas, if genetic material were a protein, such would not be the case among tissues.

17

(2) The base composition of DNA from similar species was also similar, and the base composition of DNA from widely divergent species was dissimilar. Organisms having common phylogenic ancestry shared similar molecular traits that were correlated with DNA composition.

(3) In DNA from all species, the number of adenine bases equaled the number of thymine (A=T) bases and the number of guanine bases equaled the number of cytosine bases (G=C). DNA was implicated to have an underlying, universal code giving rise to predictable ratios of bases, regardless of the species.

The Watson-Crick model (published in the British journal *Nature* in 1953) of DNA provides an immediate and simple explanation for the fact that A=T and G=C. The Watson-Crick model was the first to describe the way in which DNA could be structured as a double-stranded, antiparallel helix containing two sugar-phosphate backbones bound by the four complementary bases. The purines and pyrimidine bases are stacked on top of each other, forming the inside of the double helix. The planes of the bases are essentially parallel to one another and perpendicular to the long axis of the DNA molecule. Adenine, on one strand, forms a specific base pair with thymine on the other, antiparallel strand. The AT base pair is stabilized by two hydrogen bonds. Guanine and cytosine also form a specific base pair (GC), but it is stabilized by three hydrogen bonds (this explains why it requires more energy to break apart DNA with high proportions of GC pairs than it does AT-rich sequences). The complementary base pairing for double stranded DNA is illustrated in Figure 2.12. In addition to hydrogen bonding and charge separation between phosphates along the helix, the structure of DNA is stabilized by hydrophobic interactions. The stacked bases are removed from contact with water. Further, with complementarity being consistent with the Watson-Crick model, earlier findings were substantiated. For example, if the proportion of one base was known, the other three could be determined by difference because if it was determined that A=15%, then T was also 15%, leaving 70% remaining evenly split between the GC pairs.

The purine (G = Guanine, A = Adenine) and pyrimidine (T = Thymine, C = Cytosine) base pairs are

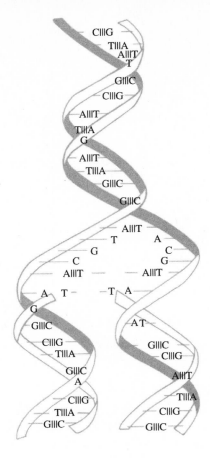

Figure 2.12.
The DNA Double Helix

connected by hydrogen bonds. When these bonds are broken, the DNA can unwind and replicate (as in mitosis and meiosis) or act as a template for mRNA synthesis. The deoxyribose sugar and phosphate groups act as a backbone of DNA, helping to maintain the sequential order of the nitrogenous bases.

Ribonucleic Acid (RNA)

The operational difference between DNA and RNA is that DNA's molecule stores information and RNA utilizes that information to synthesize a protein or functional polypeptide. The *central dogma* in molecular genetics is as follows:

DNA *(replication to maintain the information within and between generations)* → **transcription to**

mRNA *(messenger RNA)* → **translation** *(involving rRNA, tRNA, and amino acids) to a functional protein or polypeptide*

We now know that this is not always true (e.g., retroviruses), but this will be covered in Chapter 3.

Other differences, as noted above, are that RNA uses ribose as the sugar and is usually single-stranded, not double-stranded like DNA. The RNA molecule contains uracil, which replaces the thymine found in DNA. The functional forms of RNA are described in Table 2.2.

Table 2.2.
Types of RNA and Their Functions

Type of RNA	Size (number of nucleo-tides)	Function
mRNA (messenger RNA)	500–3000 +	Encodes (via a 3-base codon) an amino acid sequence specified in the DNA template.
rRNA (ribosomal RNA)	100–3000	Associates with protein to form ribosomes, which support and catalyze protein synthesis.
tRNA (transfer RNA)	75–80	Anticodon binds an mRNA codon on one end, an amino acid on the other, thus linking a gene's message to the amino acid sequence it encodes.

Briefly, the relationships between the DNA template and the RNA molecules and synthesis of a protein are the following:

DNA template sequence : 3′ – GGGTATAAA – 5′

mRNA transcript   : 5′ – **CCCAUAUUU** – 3′

tRNA (anticodon)   : 3′ – **GGG**U**AU**AAA –5′

amino acid sequence   :   pro tyr lys

proline-tyrosine-lysine

Abiotic Origin of Biological Molecules

What, when, where, and how are some of the questions that will likely continue for a long time regarding the emergence of biological molecules from abiotic precursors. Candidates like CO_2, CO, NH_3, H_2, CH_4, S, P, and H_2O combined with a minimum amount of energy are believed to have given rise to some of the earliest biomolecules. It is widely accepted that these earliest molecules would have to evolve the capacity to self-replicate and self-organize into more complex pre-cell structures. It is believed that RNA, rather than DNA, is a better genetic candidate because it is single stranded, and a virus-based enzyme like reverse transcriptase could synthesize the complementary DNA in which the information for the earliest dipeptides were encoded.

2. Enzyme Activity, Receptor Binding, and Regulation

Enzymes alter the rate of chemical reactions without undergoing change themselves. They do so by reducing the activation energy of the reaction. The enzyme (E) forms a complex with the substrate (S) at a specific active site. The resulting enzyme—substrate complex (ES) is then converted to the product (P), and the enzyme is released:

$$E + S \rightarrow ES \rightarrow P + E$$

Temperature, pH, cofactors, substrate concentration, and inhibitors affect the activity of individual enzymes. Because enzymes are proteins, the effect of temperature is similar for most enzymes, but individual enzymes have different optimal pH values. Cofactors or coenzymes are small molecules that work with enzymes. They can be metals such as zinc or magnesium and vitamins.

A **Lineweaver-Burke plot** (Figure 2.13) shows the relationship between the reciprocal of the reaction rate ($1/v_0$) and the reciprocal of substrate concentration ($1/[S]$). The Michaelis-Menten constant (K_m) is the substrate concentration at which the initial reaction rate is one-half of the maximum (V_{max}). As the reciprocal of substrate concentration increases, the reciprocal of reaction rate increases. Increased substrate concentration increases the reaction rate until the enzyme becomes limiting.

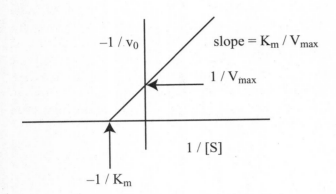

Figure 2.13.
Lineweaver-Burke Plot

Inhibitors affect the slope and intercepts of the Lineweaver-Burke plot. Competitive inhibitors bond to the active site of an enzyme and thus compete with the substrate (Figure 2.14). Therefore, the enzyme activity and reaction rate depends on the concentration of substrate and inhibitor. In a Lineweaver-Burke plot, there is an increase in the slope (K_m/V_{max}) due to an increase

in the x-intercept ($-1/K_m$) but no change in the y-intercept ($1/V_{max}$). That is, there is no change in V_{max}, but a higher substrate concentration is required to displace the inhibitor which increases the K_m. Noncompetitive inhibitors bind to a part of the enzyme outside the active site (Figure 2.14). The result is a change in shape of the enzyme such that the active site is altered. In a Lineweaver-Burke plot there is an increase in the slope due to an increase in the y-intercept ($1/v_0$) with no change in the x-intercept. That is, there is no change in K_m but an increase in V_{max} because the substrate does not displace the inhibitor. These effects are shown in Figure 2.14.

PROBLEM

Enzymes are

a) proteins.   b) catalysts.

c) carbohydrates.   d) both a) and b).

Solution

d) Enzymes are proteins that act as catalysts. Carbohydrates are not proteins, and therefore cannot be enzymes.

Many enzymes have two (or more) alternative conformations, and the binding of ligands (substrates or other molecules) can influence which conformation the

Figure 2.14.
(a) Competitive Inhibition; (b) Noncompetitive Inhibition

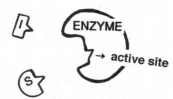

(a) Competitive inhibition
Both substrate (S) and inhibitor (I)
compete for the same active site.

(b) Noncompetitive inhibition
The inhibitor (I) binds at a site different from
the active site and does not prevent binding
of the substrate. The inhibitor does decrease
enzymatic activity.

enzyme assumes. For allosteric proteins, one conformation is enzymatically active, and the other conformation is inactive. Allosteric enzymes are very important in the regulation of metabolic reactions.

Cells are required to synthesize an enormous number of essential compounds for their survival. Bacteria, although structurally simple compared to eukaryotic cells, can use glucose to provide for their energy needs and to synthesize necessary organic components. The synthesis of these organic compounds is accomplished by specific metabolic pathways in which a precursor molecule is converted to a product by a series of enzyme-catalyzed reactions. The flow of metabolites in a metabolic pathway is often regulated by controlling the activity of key enzymes (genetic control via transcription) in the pathway. Usually, the first enzyme in a metabolic pathway is regulated by the end product of the pathway. This type of regulation is called feedback inhibition. In negative feedback inhibition (see Figure 2.15) the end product of a pathway inhibits the first enzyme in the pathway. Negative feedback is one of the most important mechanisms for maintaining organismal homeostasis (i.e., relative *internal constancy* that is essential to survival).

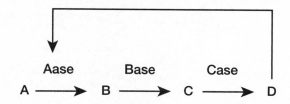

Figure 2.15.
Negative Feedback Inhibition of a Metabolic Pathway (Metabolite D Inhibits Enzyme 1 [E$_1$])

The first enzyme in a metabolic pathway is usually an allosteric enzyme. The end product of the pathway binds to a regulatory site on the enzyme, causing it to assume an inactive conformation. The regulatory site is different from the active site. In the case of negative feedback inhibition, the pathway end product is usually a noncompetitive inhibitor. A noncompetitive inhibitor binds to the enzyme at a locus different from the active site. Furthermore, for noncompetitive inhibition, the rate of the enzymatically catalyzed reaction depends only on the concentration of the inhibitor and not on the concentration of substrate. In summary, feedback regulation:

(a) usually involves an allosteric enzyme,

(b) is very rapid, and

(c) can involve enzymatic inhibition or enzymatic activators (positive feedback inhibition).

Feedback regulation also provides an efficient method of conserving cellular energy and preventing the build-up of metabolic intermediates which, at high levels, could be toxic.

PROBLEM

In feedback inhibition of metabolic pathways, which are controlled directly?

a) End products
b) Metabolites
c) Enzymes
d) Precursor molecules

Solution

b) In feedback inhibition, the flow of metabolites is often regulated by controlling the activity of key enzymes in a pathway. Usually, the first enzyme in a metabolic pathway is controlled by the end product of the pathway.

3. Major Metabolic Pathways and Regulation

Respiration, Fermentation, and Photosynthesis

Respiration

The body converts food energy (stored in the bonds of glucose) into high-energy phosphate bonds of **adenosine triphosphate (ATP)**, the carrier of energy used by the body in a catabolic process called cellular respiration (i.e., intake and/or release of O_2 and expiration of CO_2). Although cellular respiration is a continuous process, it is divided into three stages: *glycolysis, citric acid (Krebs) cycle,* and *electron gluconeogenesis transport.* Some yeasts and bacteria use only glycolysis to obtain ATP. However, most organisms have higher ATP requirements and produce additional ATP via the Krebs cycle and electron transport.

Glycolysis

The sequence of energy-producing catabolic reactions called glycolysis takes place in the cytosol (cytoplasmic fluid) of all living cells. Glycolysis results in the production of adenosine triphosphate (ATP), which provides cells with an efficient source of chemical energy. Catabolism is the chemical breakdown of food molecules to provide energy and building blocks for the synthesis of macromolecules. The first step in catabolism is the breakdown of macromolecular polymers, such as carbohydrates and lipids, to their monomeric units. Carbohydrates, as polysaccharides, for example, are broken down into sugars such as glucose. The process of glycolysis further catabolizes glucose.

Anaerobic environment. Glycolysis does not require the presence of oxygen. This metabolic pathway evolved when the Earth's atmosphere contained very little oxygen. In eukaryotes (organisms having cells with a nucleus) and many prokaryotes (organisms having cells without a nucleus), glycolysis results in the net production of two molecules of ATP, two molecules of NADH, and two molecules of pyruvate per molecule of glucose:

$$\text{D-glucose} + 2HPO_4^{2-} + 2ADP + 2NAD \rightarrow$$
$$2CH_3C{-}CO_2 + 2ATP + 2NADH + 2H^+.$$

The ten steps in the glycolytic pathway are detailed below (see Figure 2.16). The first three steps of glycolysis convert glucose to fructose 1,6-diphosphate (FDP) at the cost of two ATPs. Fructose and glucose are both six-carbon sugars (hexoses). The second stage results in the splitting (by aldolase) of FDP into two three-carbon sugars (trioses) (i.e., dihydroxyacetone

phosphate [DHAP] and D-glyceraldehyde 3-phosphate). DHAP is converted to D-glyceraldehyde 3-phosphate by triose phosphate isomerase. The third phase of glycolysis produces ATP by converting D-glyceraldehyde 3-phosphate into metabolites that can transfer phosphoryl groups to ADP. The pyruvate produced by glycolysis is a key branch-point metabolite. Under anaerobic conditions, yeast converts pyruvate to ethanol and carbon dioxide, and animals convert pyruvate to lactic acid. Some prokaryotes utilize different pathways than the one described above. All function to generate ATP, $NADH_2^+$, and pyruvate, but some are less efficient than glycolysis.

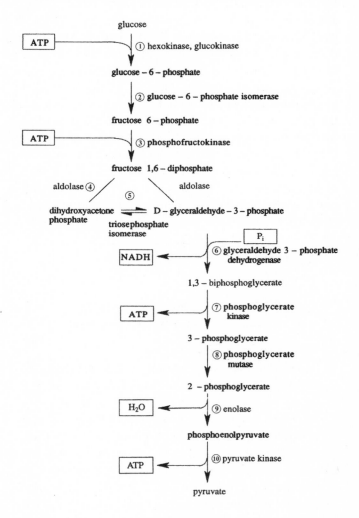

Figure 2.16.
Gycolysis

The key regulatory enzyme in glycolysis is phosphofructokinase (see Figure 2.16). Its activity is inhibited by ATP, citrate, and fatty acids and is activated by ADP, AMP, cyclic AMP, and FDP.

Aerobic environment. When cells that are undergoing anaerobic glycolysis are switched to aerobic conditions, the rate of glycolysis rapidly drops. This is called the **Pasteur effect**. The effect is explained by the fact that under aerobic conditions the pyruvate produced by glycolysis can undergo further oxidation, via the Krebs cycle (also called the citric acid cycle; see Figure 2.17). This results in the production of 18 ATP molecules per pyruvate. Thus, the energy needs of the cell are met with a considerably reduced rate of glycolysis. The decreased rate of glycolysis with higher levels of ATP is consistent with the fact that phosphofructokinase is inhibited by ATP.

PROBLEM

Glycolysis does NOT

a) occur in the cytoplasm.   b) require oxygen.

c) produce ATP.   d) break down glucose.

Solution

b) Glycolysis is the series of metabolic reactions by which glucose is converted to pyruvate (a 3-carbon sugar) with the concurrent formation of ATP. Glycolysis occurs in the cytoplasm of the cell and for this process the presence of oxygen is unnecessary.

Citric Acid Cycle (Also Known as the Krebs Cycle or the Tricarboxylic Cycle)

For most eukaryotic cells and aerobic bacteria, the pyruvate produced by glycolysis, after being transitioned into acetyl CoA, enters mitochondria and is completely oxidized to CO_2 and H_2O. The citric acid (or Krebs) cycle occurs in the mitochondrial matrix (i.e., nonmembrane-bound reactions). This is where the final oxidation of carbohydrates and lipids occurs. (Note that lipids are hydrolyzed to glycerol and fatty acids: the glycerol is catabolized in glycolysis, and the fatty acids are beta-oxidized prior to entering the citric acid cycle

as acetyl CoA.) This process produces reducing power in the form of NADH and $FADH_2$. NADH and $FADH_2$ are then utilized by the electron transport system to produce ATP. Electron transport occurs in the mitochondria of eukaryotic cells and in the plasma membrane of aerobic bacteria.

In eukaryotic cells, the pyruvate produced by anaerobic glycolysis enters the mitochondrion and is decarboxylated to yield acetyl-CoA and NADH by pyruvate dehydrogenase. The primary function of the citric acid pathway is to oxidize acetyl groups to CO_2 and H_2O. The overall reaction is:

$$\text{acetyl CoA} + 2H_2O + 3NAD^+ + FAD + GDP + HPO_4^- \rightarrow 2CO_2 + 3NADH + FADH_2 + GTP + H^+ + CoA$$

Steps catalyzed by enzymes are shown in Figure 2.17. It is noteworthy that molecular oxygen (O_2) does not enter the citric acid cycle. The additional oxygen atoms required for CO_2 production come from H_2O. The one GTP produced by step 7 is easily converted to ATP (GTP + ADP = GDP + ATP). The oxidation of one NADH molecule by the electron transport system produces three ATPs, and similarly, the oxidation of one $FADH_2$ produces two ATPs. The complete oxidation of glucose yields 38 ATPs.

PROBLEM

In the citric acid cycle, all of the following occur EXCEPT

a) oxidation of succinate.

b) formation of $FADH_2$.

c) formation of NADH.

d) transformation of NADH to NAD.

Solution

c) NADH is converted to NAD during oxidative phosphorylation, which yields 3 ATPs; in the citric acid cycle, NAD is reduced to NADH.

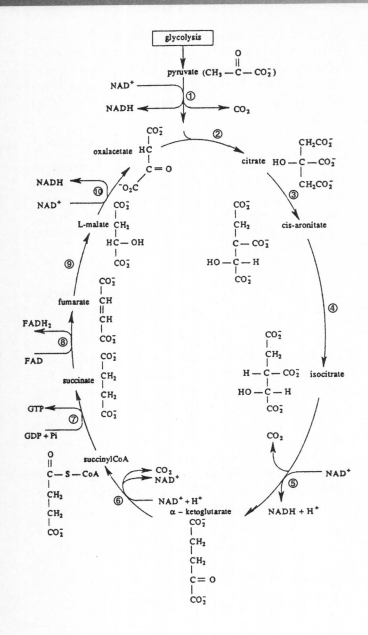

**Figure 2.17.
Citric Acid (Krebs) Cycle**

(1) pyruvate
  carboxylase

(2) citrate synthetase

(3) aconitase

(4) aconitase

(5) isocitrate
  dehydrogenase

(6) alpha-ketoglutarate
  dehydrogenase

(7) succinyl CoA synthetase

(8) succinate
  dehydrogenase

(9) fumarase

(10) malate
  dehydrogenase

Electron Transport Chain and Oxidative Phosphorylation

The last steps in catabolism, called oxidative phosphorylation, result in the efficient production of ATP. In these steps, electrons (e^-) are ultimately transferred to oxygen (see Figure 2.18) with the generation of ATP. Oxidative phosphorylation is dependent upon the structure of mitochondria (see Figure 2.20) because: (1) pyruvate and/or fatty acids are oxidized to CO_2 in the matrix or on inner-membrane proteins (within the cristae), (2) electron transfer form NADH to FADH to O_2 occur in the inner membrane, and (3) energy moves as a proton gradient for ATP synthesis in the inner membrane.

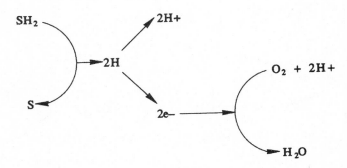

**Figure 2.18.
Oxidative Phosphorylation: Substrates (SH_2) contain H-atoms consisting of a proton (H^+) and an electron (e^-). The energy locked in H^+ and e^- is harnessed during oxidative phosphorylation to produce ATP.**

ATP can be generated in two ways:

(1) by substrate level phosphorylation as indicated in glycolysis and the citric acid cycle (i.e., transfer of a high energy phosphoryl group to ADP to make ATP), and

(2) by adding a phosphate group to ADP (i.e., the reverse hydrolysis of ATP; ATP = ADP + Pi).

Oxidative phosphorylation generates ATP by the second mechanism. The enzyme catalyzing this reaction is a proton (H^+) driven adenosine triphosphatase (H^+-ATPase). H^+-ATPase is a transmembrane protein embedded in the inner mitochondrial membrane. The

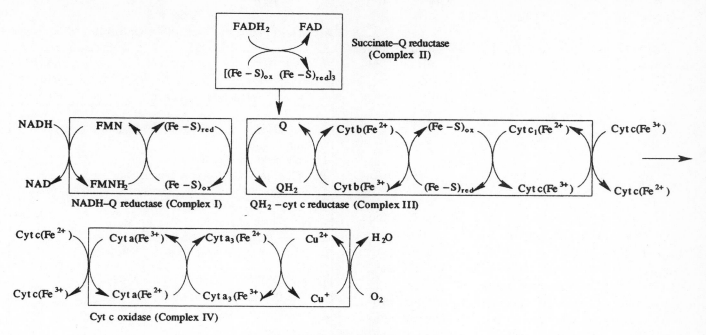

**Figure 2.19.
Respiratory Chain**

energy to drive ATP formation by H^+-ATPase comes from a proton gradient across the inner mitochondrial membrane. This proton gradient is generated by the movement of electrons down the respiratory chain (Figure 2.19).

The respiratory chain is a series of membrane-bound redox carriers with cytochrome oxidase being the terminal electron acceptor (see Figure 2.19).

The electrons that ultimately reach O_2 (and form H_2O) are initially carried by the hydrogen atoms of NADH and $FADH_2$ (from glycolysis and the citric acid cycle). These hydrogen atoms can be dissociated into an electron (e^-) and a proton (H^+). The electrons are transported by the respiratory chain, and the protons are released into the aqueous medium. The released H^+ ions are translocated from the matrix space to the intermembrane space, and a pH gradient is established. The energy created by this gradient is trapped by the H^+-ATPase when the H^+ ions flow back into the matrix (see Figure 2.20). This process is known as chemiosmosis.

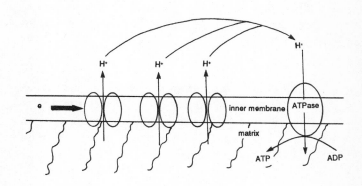

**Figure 2.20.
Oxidative Phosphorylation**

The process of oxidative phosphorylation illustrates a milestone in biochemistry, because it is an example of factorial metabolism—the coupling of metabolism with transport across a membrane (i.e., in the mitochondrion).

PROBLEM

The ratio of ATP produced aerobically to an-aerobically by the oxidation of one molecule of glucose is

a) 2:1. b) 1:2.

c) 1:18. d) 18:1.

Solution

d) The aerobic production of ATP involves the Krebs (citric acid) cycle and the oxidation of glucose. The anaerobic production of ATP takes place during glycolysis. The citric acid cycle produces 34 ATPs and the oxidation of glucose produces two. This makes the total number of ATPs produced during aerobic processes 36. Glycolysis yields two ATPs. The net ratio of aerobic ATP to anerobic ATP is 36:2, which is equal to 18:1.

As high energy electrons move down the respiratory chain, a proton gradient is created. H^+-ATPase uses this gradient to make ATP.

The energy required for phosphorylation, described above, may be obtained three ways: oxidative phosphorylation, photosynthetic phosphorylation, and substrate-level phosphorylation.

Oxidative Phosphorylation: Energy is released as electrons; it passes through a series of electron acceptors (the electron transport system) to either oxygen or some other inorganic compound. Coenzymes (e.g., FAD, NAD, NADP) are required.

Photosynthetic Phosphorylation: Electrons that are released as light are absorbed by chlorophyll; they subsequently pass through the electron transport system.

Substrate-level Phosphorylation: Energy, in the form of a high-energy phosphate, is released from a substrate (e.g., a metabolic intermediate) through enzyme activity.

Table 2.3.
ATP Yield From the Complete Oxidation of Glucose

Reaction sequence	ATP yield per glucose
GLYCOLYSIS: GLUCOSE TO PYRUVATE (*in the cytoplasm*)	
Phosphorylation of glucose	1
Phosphorylation of fructose 6-phosphate	1
Dephosphorylation of 2 molecules of 1, 3 DPG	+ 2
Dephosphorylation of 2 molecules of phosphoenolpyruvate	+ 2
2 NADH are formed in the oxidation of 2 molecules of glyceraldehyde 3 phosphate	
CONVERSION OF PYRUVATE TO ACETYL CoA (*inside mitochondria*)	
2 NADH are formed	
CITRIC ACID CYCLE (*inside mitochondria*)	
Formation of 2 molecules of guanosine triphosphate from 2 molecules of succinyl CoA	+2
6 NADH are formed in the oxidation of 2 molecules of succinate	
OXIDATION PHOSPHORYLATION (*inside mitochondria*)	
2 NADH formed in glycolysis; each yields 2 ATP (not 3 ATP each, because of the cost of the shuttle)	+ 4
2 NADH formed in the oxidative decarboxylation of pyruvate; each yields 3 ATP	+ 6
2 FADH formed in the citric acid cycle; each yields 2 ATP	+ 4
6 NADH formed in the citric acid cycle; each yields 3 ATP	+ 18

Fermentation

During fermentation, reduction-oxidation (redox) occurs in the absence of any added electron acceptor; glucose is only partially broken down during catabolism. Organic molecules act as both electron acceptors and electron donors. Energy is released from sugars (e.g., glucose) or other organic molecules in the absence of an added electron acceptor (i.e., an organic molecule serves as the final electron acceptor). This is in contrast to **aerobic respiration** where glucose is completely broken down with molecular oxygen serving as the final electron acceptor. In anaerobic respiration, an inorganic ion other than molecular oxygen serves as the final electron acceptor. As shown in Table 2.3, anaerobic respiration yields fewer ATP molecules than aerobic respiration because only part of the Krebs cycle is operative under an aerobic conditions.

There are three kinds of fermentation: *alcoholic* fermentation, when ethanol and CO_2 are produced; *heterolactic* fermentation, when the pentose phosphate pathway is used to produce lactic acid and ethanol; and *lactic acid* fermentation, when pyruvic acid is reduced to lactic acid.

Photosynthesis

Photosynthesis is metabolism in chlorophyll-containing organisms (e.g., higher plants, cyanobacteria, or blue-green algae), whereby light energy from the sun is used to synthesize carbohydrates from CO_2 and water. These reactions are centered in an organelle called the chloroplast. Photosynthesis is anabolic metabolism in which carbohydrates are created; by contrast, fermentation is catabolism, in which carbohydrates are broken down to supply energy.

Photosynthesis (there are two systems in plants, PS-I and PS-II) occurs via a metabolic pathway called the Calvin cycle in which the gas, CO_2, is "fixed" as starch (polymer of alpha 1-4 glucose; 1-4 indicates the carbons connecting the monomers) and as sugars (the disaccharide, sucrose, which is glucose alpha 1-2 fructose; 1-2 indicates the carbons connecting the monomers) using light, chlorophylls *a* and *b*, ATP and NADH with O_2 as a by-product (oxygen is not released in bacteria systems). The pathway to fixation in C3-plants (e.g., garden peas) is catalyzed by the enzyme ribulose 1,5-biphospahate carboxylase, but in C4 plants (e.g., corn), it is by phosphoenolpyruvate.

Photosynthesis involves conversion of the light energy into chemical energy that can be used for carbon fixation (e.g., α-D-glucose is synthesized and bound into polymers of cellulose, which are layered in criss-cross orientation to form plant cell walls). There are two different sets of reactions in photosynthesis—the light reactions and the dark reactions (also know as the Calvin-Benson cycle).

$$6CO_2 + 12 H_2O \rightarrow C_6H_2O_6 + 6O2 + 6H_2O$$

In the *light* reactions, electrons from chlorophyll travel through an electron transport chain, producing ATP via chemiosmosis. The light reactions can involve cyclic photophosphorylation, wherein the electrons return to chlorophyll, or photolysis accompanied by non-cyclic photophosphorylation, wherein the electrons are used to reduce NADP; electrons from H_2O or H_2S return to chlorophyll.

In the *dark* reactions, CO_2 is reduced to synthesize carbohydrates.

4. Membrane Dynamics and Cell Surfaces

Cellular Structures, Functions, and Dynamics

Structure

All living matter is composed of cells. Cells are the basic units of life and are self-contained units. All cells are divided into two categories: prokaryotes and eukaryotes (refer to Table 2.4).

Prokaryotes

Prokaryotic cells are more simplified and primitive and generally are circumscribed by only a plasma membrane. **Eubacteria** and **archaebacteria** are the two primary groups of prokaryotes, the former being the ones most commonly referred to in this and other texts. A subgroup of eubacteria, the gram-negative (which include the ubiquitous intestinal bacterium, *Escherichia coli* or *E. coli*), contain two surface mem-

Table 2.4.
Characteristics of Eukaryotes and Prokaryotes

Characteristic	Eukaryotic cells	Prokaryotic cells
Chromosomes	multiple, composed of nucleic acids and protein	single, composed only nucleic acids
Nuclear membrane	present	absent
Mitochondria	present	absent
Golgi apparatus, endoplasmic reticulum, lysosomes, peroxisomes	present	absent
Photosynthetic apparatus	chlorophyll, when present, is contained in chloroplasts	may contain chlorophyll
Microtubules	present	rarely present
Ribosomes	large	small
Flagella	have 9-2 tubular structure	lack 9-2 tubular structure
Cell wall	when present, does not contain muramic acid	contains muramic acid

branes, the inner, plasma membrane (see below), an outer containing porin proteins, and a cell wall (not to be confused with the highly lignified wall of many plants) between, composed of peptidoglycan (polysaccharide and protein), which provides rigidity to the cell. Gram-positive bacteria have only a cell wall and plasma membrane.

Eukaryotes

Unlike prokaryotes, eukaryotes have a nucleus and complex organelles in the cytoplasm. Cells having a rigid, cell wall exterior to the plasma membrane are typical of higher plants (note that some authors in the plant literature may refer to the plasma membrane as the *plasmalemma*); plant cell walls are primarily cellulose and hemicelluloses. Generally, animal cells do not have cell walls. Most commonly, eukaryotes are thought of as multicellular organisms; however, yeasts, which are unicellular, are eukaryotes.

Cytoplasm refers to *all* of the cell solute including an organized complex of organic and inorganic substances (i.e., proteins, water, and elements) in a cell *outside* of the nucleus, if a nucleus is present. Different cell types have cytoplasm of differing compositions. In eukaryotes, the *cytosol* includes all the solute of the cell cytoplasm that lies outside of the organelles.

The plasma membrane surrounds the cell and all its contents. It separates the intracellular fluid from the extracellular environment, responds to signals from other cells, responds to foreign material/organisms and chemicals such as hormones, and controls the movement of materials in and out of the cell. Movement in and out of the cell is by diffusion, active transport, osmosis, and exocytosis and endocytosis.

The plasma membrane is composed of a lipid bilayer and membrane-bound protein globules (Figure 2.21). Lipid bilayers are also present in organelles of eukaryotic cells (e.g., nuclear membrane, mitochondria, and endoplasmic reticulum). Phospholipid (PL) molecules are the primary lipid constituents of most lipid bilayers (Figure 2.22). Cholesterol and glycolipids are also present in many biological membranes. PL molecules are amphipathic molecules (i.e., they have a polar head group and two nonpolar hydrocarbon "tails"). PL molecules self-aggregate to form a lipid bilayer because in this molecular arrangement, their head groups remain in contact with water, and their tails are removed from contact with water.

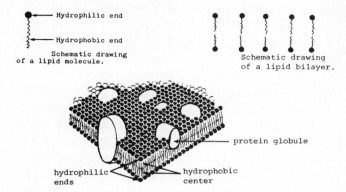

Figure 2.21.
Lipid Bilayer of the Plasma Membrane

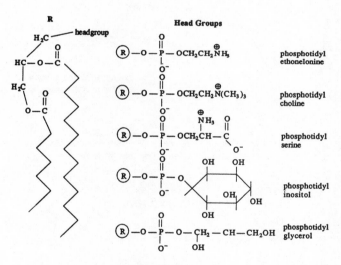

Figure 2.22.
The Structure of Phospholipid (PL) Molecules

The lipid bilayer structure has two important properties:

1. Charged, hydrophilic molecules cannot move through the lipid bilayer because they would have to give up thermodynamically favorable interactions with polar water molecules. Water molecules, although they can permeate the bilayer, have an extremely low concentration in the hydrophobic domain of the bilayer. For charged molecules to pass through the bilayer, specific transport proteins must

be present. Hydrophobic molecules such as O_2 as well as small uncharged polar molecules (H_2O, CO_2, and urea) are, however, membrane permeable.

2. The individual PL molecules move rapidly in the plane of the lipid bilayer. Proteins associated with the lipid bilayer also can have rapid lateral motion. The bilayer, therefore, acts as two-dimensional fluid and this fluidity is necessary for diffusion of membrane-bound enzymes and receptor molecules.

The plasma membrane and the membranes of other subcellular organelles are asymmetric with respect to the head groups found on the inner (toward cytoplasm) and outer (extracellular, facing interstitial space between adjoining cells) monolayers. In addition, the proteins associated with biological membranes are also embedded in the bilayer in an asymmetric manner. For example, glycoproteins (as well as glycolipids) in the plasma membrane usually have their carbohydrate moieties facing the extracellular space.

Although plant and animal cells may be many times the size of prokaryotic cells, their chemical compositions can be quite similar. A "typical cell" has been estimated to be approximated by the following: 70% water, 7% small molecules and ions, 20% protein, 2% RNA (ribonucleic acid), and ~1%DNA (deoxyribonucleic acid).

Transport Across the Plasma Membrane

Passive and Active Transport Across the Bilayer of the Plasma Membrane. The hydrophobic core of the bilayer is selective against salts, sugars, and other smaller hydrophilic molecules. This is not true for lipids, of course, because the monolayers are composed of lipids. Proper functioning of cellular processes (i.e., maintaining homeostasis) at the cellular level is largely dependent upon specialized transport proteins; these may differ among tissues and organs. Transport proteins can achieve great specificity and permit only one class of molecules to be transported (e.g., only sugars or only amino acids) or one specific molecule in a class. Some transport molecules simply permit a solute to reversibly diffuse from one side of the membrane to the other. This process is called *passive transport*. The direction of transport for a solute will be influenced by a concentration gradient across the membrane (i.e., from high to low concentration), as well as the electric charge across the membrane (i.e., the membrane potential).

Cell membranes are generally composed of a double layer of:

a) phospholipids with proteins dispersed throughout the membrane.

b) phosphoproteins with glucose dispersed throughout the membrane.

c) nucleic acids.

d) proteins with phospholipids dispersed throughout the membrane.

Solution

a) The plasma membrane (see Figure 2.21) contains about 40 percent lipid and 60 percent protein by weight, although there is considerable variation between different cell types. The lipid molecules of the plasma membrane are polar, "lollypop-shaped" molecules. One end is hydrophobic (interior of the bilayer); the other end is hydrophilic (interior and exterior membrane surfaces). The lipid molecules are arranged in two layers so that the hydrophobic ends, between the two monolayers, are near each other, and the hydrophilic polar ends face outward: one surface to the intracellular space, the other to the extracellular space (refer again to Figure 2.21). The individual lipid molecules can move laterally, so the bilayer is fluid and flexible. Protein molecules of the plasma membrane may be arranged at various sites and imbedded to different degrees. The highly selective permeability of the plasma membrane is dependent upon the specific types and amounts of proteins and lipids present.

The combination of the chemical and electrical gradient is called the **electrochemical gradient**. Plasma membranes are more negatively charged on the cytoplasmic side than the extracellular side, and this hinders the passive transport of positively charged ions. In some cases, the **passive transport** of a solute is through an aqueous pore created by the transport protein. This type of transport protein is called a channel protein or porin. Passive transport of a solute via a channel protein cannot become saturated, which means that the rate of input increases/decreases in proportion to the concentration in the extracellular fluid (see upper linear curve, Figure 2.23). In facilitated diffusion, a process similar to a substrate binding to the active site of an enzyme, the solute molecule binds to a transport protein, which then facilitates its translocation to the other side of the membrane. This process can become saturated (see lower nonlinear curve, Figure 2.23). In order to transport a solute *against* an unfavorable electrochemical gradient, it is necessary to expend energy, usually ATP. This type of transport is called **active transport**.

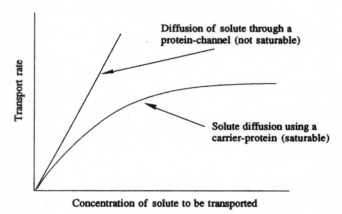

Figure 2.23.
Passive Transport

Porins are important in

a) facilitated diffusion.   b) substrate binding.

c) processes that can   d) passive transport.
be saturated.

Solution

c) Passive transport may occur when a solute passes through an aqueous pore created by a channel protein, or porin. These are not subject to saturation like answers a, b, and c.

The membrane potential of plasma membranes is generated by two important transport proteins: the Na^+, K^+-ATPase and the K^+ channel. Na^+, K^+-ATPase uses ATP to pump Na^+ ions *out* of the cell and K^+ ions *into* the cell (see Figure 2.24). This is an example of *active transport* because the concentration of Na^+ outside the cell is higher than inside. The reverse is true for K^+ ions. The K^+-channel permits K^+ ions to diffuse out of the cell; this loss of positive ions causes the inside of the cell to become more negative than the outside. Eventually, the increasing negative charge inside the cell retards the outflow of K^+ ions (i.e., the negative charge inside the cell attracts the positively charged K^+ ions), and equilibrium is achieved when the inflow of K^+ ions equals the outflow. The end result is a plasma membrane potential between -20 and -70 mV depending on the cell type.

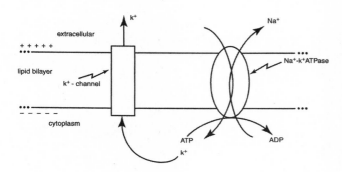

Fig 2.24.
Membrane Potential and Active Transport

PROBLEM

In most cells, the concentration of Na^+ is _____ in the cell than outside it, because of _____.

a) higher . . . active transport

b) higher . . . passive transport

c) lower . . . active transport

d) lower . . . passive transport

Solution

c) The Na^+–K^+ pump is a form of active transport where energy is used to pump Na^+ ions out of the cell and K^+ ions into the cell.

Maintenance of cell volume and osmotic pressure. The Na^+, K^+-ATPase and K^+ channel control the movement of ions and thus the level of ions inside the cell. These elements are also the main controllers of intracellular osmotic pressure and cellular volume. Inside the cell the charged macromolecules (e.g., proteins) require counterbalancing ions like Na^+, K^+, and Cl^-. This creates an osmotic pressure causing cellular swelling from the influx of water. Counterbalancing this intracellular osmotic pressure is the osmotic pressure caused by the ions in the extracellular fluid, which are primarily Na^+ and Cl^-. These ions tend to move down their concentration gradient and into the cell. Were it not for the Na^+, K^+-ATPase pumping Na^+ out and consequently preventing Cl^- from leaking in (by maintaining a negative membrane potential), the cell would swell and burst.

PROBLEM

Which ions must be pumped out of the cell to prevent its rupture?

a) Na+

b) K+

c) Cl–

d) All of the above.

Solution

d) The macromolecules inside the cell require counterbalancing ions outside the cell, such as Na^+, K^+, and Cl^-. These counterbalancing ions create an osmotic balance that prevents the influx of water into the cell, thus preventing cell rupture.

Exocytosis and Endocytosis. Macromolecules are too large to be transported through the plasma membrane by specific transport proteins. The processes of **exocytosis** and **endocytosis** accomplish the transport of these macromolecules. In exocytosis, an intracellular vesicle, containing for example hormones, is transported to the plasma membrane where it fuses with the plasma membrane. The fusion process releases the contents of the vesicle outside the cell to the extracellular space. Endocytosis is essentially the reverse of this process. For example, very large molecules, such as low density lipoprotein (LDL), can be brought into the cell via endocytosis (refer to Figure 2.28).

PROBLEM

Which is most likely to be transported by exocytosis?

a) Urea   b) Lipoprotein

c) Na+   d) Hormones

Solution

b) Exocytosis is used for the transport of the largest molecules, macromolecules. Lipoproteins have larger molecules than any of the other materials listed.

Cell Matrix and Interactions: Recognition, Junctions, and Plasmodesmata

The plasma membrane contains a wide variety of protein receptors to which ligands can bind. A major function of these receptors is to recognize and accept signals from the extracellular environment. Neurotransmitters and hormones are examples of ligands that bind to protein receptors on target cells and influence the behavior of the cell.

Receptor proteins (see Figure 2.25) are usually transmembrane proteins that have an extracellular domain where signals are received, a hydrophobic domain going through the lipid bilayer, and a cytoplasmic (intracellular) signal-transducing domain. The initial binding of the signal molecule (i.e., the first message) alters the conformation of the protein receptor, and this activates an intracellular signal pathway. The intracellular signal is often transmitted by a second class of small and rapidly transportable molecules (by diffusion) called second messengers. Calcium and cyclic AMP are two important second messengers. Alternately, the cytoplasmic domain of the receptor may have protein kinase activity, which is activated upon ligand binding. Thus, the receptor molecule itself can activate or inactivate certain intracellular substrates via phosphorylation.

The second messengers can then regulate a wide variety of biochemical and physiological processes. For

example, the release of fatty acids from adipocytes (fat cells) is regulated by catecholamines.

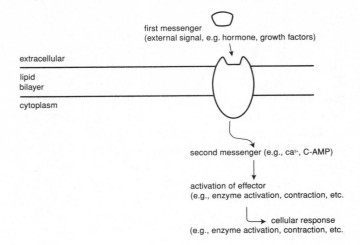

Fig. 2.25.
Cell Receptors and Signals

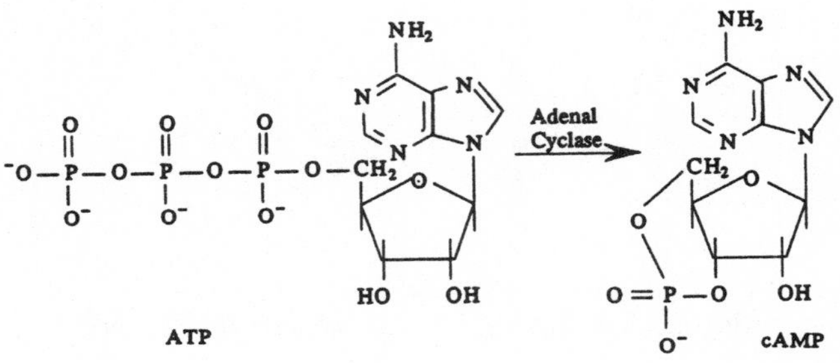

Fig 2.26.
Structure of Cyclic AMP, a Secondary Messenger

When catecholamines (the first message) bind to a surface receptor on the adipocyte plasma membrane, this causes the receptor molecule to activate an adenylate cyclase enzyme, which catalyzes the production of intracellular cAMP from ATP (Figure 2.26). The increased cAMP (the second messenger) activates a protein kinase which, in turn, phosphorylates the hormone-sensitive lipase enzyme. The phosphorylation activates hormone-sensitive lipase, and it then hydrolyzes triglyceride into fatty acids (see Figure 2.27).

Some receptors serve to bind very large molecules that are brought into the cell as a source of nutrients.

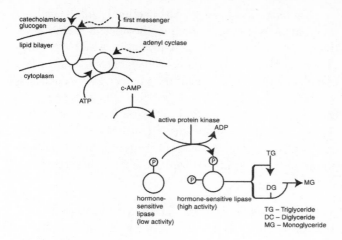

Figure 2.27.
Hormone-Sensitive Lipase

Most cells, for example, have receptors for low density lipoprotein (LDL), which is a very large lipid-protein complex. The LDL receptor is called the apoB,E receptor, and it recognizes the apoB protein moiety of LDL. After binding to the apoB,E receptor, LDL is internalized by endocytosis (see Figure 2.28) and provides the cell with an external source of cholesterol and other lipids.

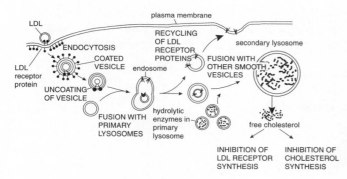

Fig 2.28.
LDL Uptake by Receptor-Mediated Endocytosis

PROBLEM

Ligands bind to

a) target cells.

b) neurotransmitters.

c) protein receptors.

d) hormones.

Solution

c) Neurotransmitters and hormones are examples of ligands that bind to protein receptors on target cells.

Cellular Adhesion. Cells in a tissue are in contact with a network of molecules called the extracellular matrix. This matrix plays a major role in promoting cell–cell adhesion. In addition, cells that are in direct contact with each other can form cell junctions between specialized regions of their plasma membranes.

Surface proteins provide recognition among like-cells so that communication and exchange between cells is possible. Cell adhesive proteins are mainly these: *cadherins, selectins,* and the *immunoglobulin (Ig) superfamily*. Early experiments demonstrated that when cells of, for example, tissues of liver and kidney were disassociated, then mixed, that liver cells aggregated with liver, and kidney with kidney. These preferential cell–cell interactions also permit communication and the movement of ions and molecules between cells in a tissue via *gap junctions*.

Plasmodesmata.   Intercellular communication among plant cells is facilitated by plasmodesmata. These are round, physical channels through the cell wall and through which the plasma membranes of adjacent cells exchange nutrients and molecules.

5. Organelles: Structure, Function, and Targeting

The endomembrane system of a cell, contained within the cell plasma membrane, is a continuous system of organelles that work together in synthesis, storage, and secretion of important molecules. It includes the nuclear envelope around the nucleus, rough endoplasmic recticulum, smooth endoplasmic recticulum, Golgi apparatus, transport vesicles, lysosomes, and vacuoles. The nuclear envelope and rough and smooth endoplasmic recticulum are connected structurally.

Transport vesicles (recall endo- and exocytosis) move materials between the endoplasmic reticulum and the Golgi apparatus and from the Golgi apparatus to lysosomes, vacuoles, or the plasma membrane.

The cellular organelles that are commonly identified in eukaryotic cells are described below. The locations of these structures are shown for a typical animal cell (Figure 2.29) and plant cell (Figure 2.30).

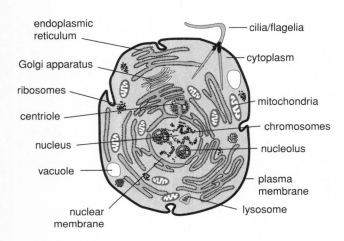

Figure 2.29.
The Components of a Typical Animal Cell

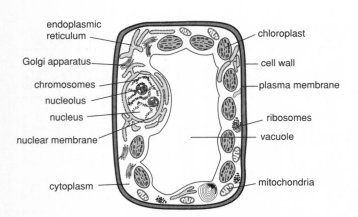

Figure 2.30.
The Components of a Typical Plant Cell

Nucleus. The nucleus is contained within a phospholipid, bilayer membrane (i.e., the nuclear membrane). It contains DNA stored as chromatin, which becomes visible at the time of cell division as chromosomes. The nucleus of the cell contains the nuclear DNA encoding the genetic information required for cellular replication, differentiation, and functions. DNA replication and RNA synthesis occur in the nucleus. The mRNA in the nucleus can be processed (e.g., RNA splicing) before being transported to the cytoplasm where protein synthesis occurs.

Nucleosomes. Nuclear DNA associates with histone proteins to form nucleosomes, the unit particles of chromatin. **Chromatin**, in turn, is packaged to form very compact structures called chromosomes. Other proteins, called nonhistone proteins, are also associated with nuclear DNA. Most of the DNA in the nucleus does not code for proteins.

Nuclear Envelope and Nuclear Pores. The nucleus is bound by an envelope that is made up of two layers of phospholipid (i.e., an inner and an outer bilayer forming the nuclear membrane). The inner and outer nuclear membranes are fused at points called nuclear pores. The nuclear pores contain a nuclear pore complex believed to permit the selective transport (in and out) of macromolecules. For example, DNA and RNA polymerases, which are synthesized in the cytoplasm, must be transported into the nucleus through nuclear pore complexes.

PROBLEM

Nuclear pore complexes

a) transport micromolecules.

b) transport the nucleus.

c) fuse the nuclear and cell membranes.

d) fuse the inner and outer nuclear membranes.

Solution

d) Nuclear pore complexes fuse the inner and outer nuclear membranes, to permit the transport of macromolecules to and from the nucleus.

Nucleolus. This is a dense region of the nucleus that is composed of protein and RNA. This is also the site where rRNA is made. The nucleolus is a highly ordered structure specially designed to produce the rRNA required for ribosomes. The synthesized rRNA immediately complexes with ribosomal proteins made in the cytoplasm and is transported through the nuclear pores into the nucleolus. The final maturation of ribosomes occurs, however, in the cytoplasm.

PROBLEM

rRNA is made in the

a) endoplasmic reticulum.   b) nucleolus.

c) ribosomes.   d) cytoplasm.

Solution

b) rRNA is synthesized in the nucleolus and combined with ribosomal proteins.

Mitochondrion. This organelle has a bilayer membrane and is the site of chemical reactions that provide the cell with energy via respiration. The mitochondria are responsible for 95 percent of all ATP produced in the cell. For this reason, the mitochondria are commonly referred to as the "powerhouse" of the cell. Mitochondria are the primary sites for the production of ATP. Mitochondria appear to be associated with the microtubules of the cytoskeleton. Mitochondria contain their own genome. Proteins from both the mitochondrial genome and the nuclear DNA are required for mitochondrial DNA replication.

Chloroplasts. These are found in plants, algae, and certain single-cell organisms like bacteria (e.g., *Rhodopseudomonas viridis*). The organelle is enclosed by a bilayer of phospholipids, and internally contains flattened thylakoid membranes filling the stroma (analogous to the cell cytoplasm). The thylakoid membranes are where photosynthesis occurs (within stacks of thylakoids called grana).

Lysosomes. Lysosomes are membrane-bound vesicles also found in the cytoplasm. Unlike most organelles, these are surrounded by a single layer membrane (only one layer of lipid molecules rather than a bilayer found in the plasma membrane) that contain acid hydrolase, an enzyme that digests substances within the cell. Lysosomes are responsible for the intracellular digestion of macromolecules. A primary lysosome is a newly synthesized vesicle and contains a wide variety of hydrolytic enzymes (all are acid hydrolases) such as proteases, phospholipases, and nucleases. These hydrolytic enzymes are almost all glycoproteins and have optimal enzymatic activities at pH 5.0, the pH inside the lysosomes. The primary lysosome arises from budding of specialized regions of the Golgi apparatus. A secondary lysosome is a lysosome that is actively digesting a substrate (see Figure 2.28). The substrate can be a foreign pathogen such as a bacterium or an endogenous macromolecule such as LDL (low-density lipoprotein).

PROBLEM

Lysosomes contain

a) glycogen stores.   b) lipids.

c) acid hydrolases.   d) ATP.

Solution

c) Lysosomes are cell organelles found in the cytoplasm. They are vesicles surrounded by a single membrane and contain enzymes, mostly acid hydrolases. These hydrolases are released when the membrane bursts, permitting the digestion of cellular structures and macromolecules. During the normal metabolism of the cell, enzyme release is carefully controlled by mechanisms that are still very poorly understood.

Ribosomes. These are small structures composed of ribosomal RNA and are the sites of protein synthesis; they may be associated with the endoplasmic reticulum or free in the cytoplasm. Messenger RNA (mRNA), which carries genetic information from the nucleus to the ribosomes, associates with the small ribosomal subunit first and then binds to the large subunit as a prelude to protein synthesis. This association of mRNA to ribosomes makes the system of protein synthesis more efficient than if the complex was dispersed freely into the cytoplasm. The mRNA then pairs with complementary molecules of transfer RNA (tRNA), each carrying a specific amino acid, which bind with each other to form a highly specific protein molecule. Thus, ribosomes are the sites where proteins are synthesized under genetic control.

Endoplasmic Reticulum. This is a double unit membrane, which may have ribosomes present. If ribosomes are present, the endoplasmic reticulum is termed "rough." Rough endoplasmic reticulum (ER) synthesizes proteins. The "smooth" endoplasmic reticulum lacks ribosomes. The rough and smooth endoplasmic reticulum transport substances. Also, the endoplasmic reticulum (present only in eukaryotic cells) accounts for about one-half the cellular membrane.

Eukaryotic cells contain a smooth and rough endoplasmic reticulum, which represents about one-half of all the cellular membrane. Prokaryotic cells do not contain an ER. Structurally, the ER is a single, highly convoluted membrane sheet enclosing a single space called the ER lumen. The cytoplasm is separated from the ER lumen by a single membrane (the ER membrane). The ER membrane is continuous with the outer nuclear membrane. The rough endoplasmic reticulum has ribosomes along the endoplasmic membrane, which gives it a rough appearance in an electron microscope. The two main functions of the rough endoplasmic recticulum are to make membrane proteins and secretory proteins.

Proteins that are to be secreted by the cell or sent to other intracellular organelles are delivered to the lumen of the rough ER. The polypeptides being translated on these ribosomes are transported from the cytoplasmic side of the ER membrane into the ER lumen. The ribosomes attached to the rough ER are identical to ribosomes that are not attached to the rough ER. Attachment of some ribosomes to the ER is directed by a small sequence of amino acids at the amino end of the polypeptide being translated (i.e., the signal sequence).

The signal sequence is removed once the polypeptide has been delivered to the ER lumen. Many polypeptides undergo "core glycosylation" in the ER. The smooth ER has no attached ribosomes. Transport vesicles carrying newly synthesized lipids and proteins bud off the smooth ER for transport to the Golgi apparatus. Many important detoxification and lipid metabolism reactions take place on the smooth ER.

PROBLEM

Prokaryotic cells contain

a)   an endoplasmic reticulum.

b)   ribosomes.

c)   a nuclear membrane.

d)   both an endoplasmic reticulum and ribosomes.

Solution

b) Prokaryotic cells contain ribosomes, but they do not have an endoplasmic reticulum or nuclear membrane.

Golgi Apparatus. This organelle is a collection of membranes that packages and secretes proteins and polysaccharides; in plants, it synthesizes and transports polysaccharides for the cell wall. The Golgi apparatus (also referred to as the *Golgi body* by some authors) is composed of a collection of flattened membrane-bound sacs (dictyosome, a group of 5–10 Golgi cisternae [i.e., folded membranes]) enveloped by smaller membrane-bound vesicles called "coated vesicles." Proteins associated with the ER are transported to the Golgi apparatus by these small vesicles, which are coated with a protein called clathrin. Glycoproteins are received by the convex side of Golgi apparatus and undergo "terminal glycosylation" in the Golgi apparatus. The sugar component of glycoproteins are extensively modified by enzymes in the Golgi apparatus, and the modified glycoproteins are sorted and delivered to either other organelles or to the plasma membrane, where they can be secreted into the extracellular fluid.

PROBLEM

The Golgi apparatus primarily functions in

a) packaging protein for secretion.

b) synthesizing protein for secretion.

c) packaging protein for hydrolysis.

d) synthesizing protein for hydrolysis.

Solution

a) The Golgi apparatus is an organelle that is responsible only for the packaging of protein for secretion.

Proplastid. A specialized, precursor organelle found in plant tissues, which depending on the tissue where it is located and its exposure to light, can differentiate into several different kinds of functional plastid organelles. These can include: chloroplasts, amyloplasts, elaioplasts, chromoplasts, and etioplasts.

Peroxisomes or Microbodies. These single-membrane (only one layer of lipid molecules) organelles are contained in most all eukaryotes, are high in protein and are the sites where enzymes are found that form H_2O_2. The enzyme, catalase, is in the peroxisome and degrades H_2O_2 to reduce that molecule's potentially harmful oxidative effects.

Vacuoles. These are membrane-enclosed fluid-filled spaces in plant cells that are involved in the storage of substances and chemical breakdown of macromolecules. Some refer to it as a lysing compartment (i.e., it assumes the roles of the lysosome in animal cells). In mature plant cells, the pH is 4–5 and the vacuole contains lytic enzymes that have optimum activity in that acidic environment. They include proteinases, glycosidases, phosphatases and nucleases. The vacuole also serves as a reservoir for soluble materials such as sucrose, amino acids and various plant pigments, and toxic by-products that must be detoxified.

Cell Wall. The cell wall is present in plant cells for support and protection. While it must be tear resistant, it must also be flexible enough to adjust to osmotic pressure change and yield to plant growth. The cell wall lies outside of the plasma membrane, is invaginated with plasmodesmata that facilitate communication among adjacent cells, and is part of what some refer to as the "nonliving" part of the cell having arisen from secretions via the cytoplasm or protoplasm. The chemical composition of plant cell walls varies among species. Generally, there are four types of polymers, which are present in varying proportions among species. These are cellulose, hemicelluloses, pectin, and glycoproteins. Pectins are also the major constituents of the middle lamellae that "cement" adjacent cells to one another, and woody species contain lignin in their cell walls.

Centrosome. The centrosome is an organelle in animal cells located near the nucleus; it forms the microtubule-organizing center that contains a pair of centrioles. This complex, along with the centromere, is important to the proper separation of chromosomes during cell division.

Centriole and Centromere. See below.

Microtubules, Microfilaments, and Intermediate Filaments. These are fibrous and/or rodlike organelles (solid or hollow) in eukaryotic cells that are important for the integrity of the cell cytoskeleton (see section below for more) and act in many specific functions related to cell motility and cell division (spindle fibers).

6. Cytoskeleton, Motility, and Shape

The cytoskeleton of eukaryotic cells plays a key role in maintaining the cellular structure and facilitating cell motility. Microfilaments and microtubules are composed of cytoskeletal filaments. These filaments are polymers of soluble subunits and can rapidly assemble and dissemble. The assembly process is energy dependent and requires energy rich molecules like ATP. A third type of filament is designated as intermediate filament because it has a diameter in between that of microfilaments and microtubules. Intermediate filaments are found in most animal cells. They are of a more permanent nature than either microfilaments or microtubules.

Actin-Based Systems

These are solid, rod-shaped structural organelles composed of *actin protein* found in the cytoplasm of most eukaryotic cells and which have several functions in eukaryotic cells. Those include, but are not limited to, muscle contraction, cytoplasmic streaming, and cytokinesis (i.e., cell division).

Microtubule-Based Systems

These are hollow, rod-shaped structural organelles composed of *tubulin protein* and found in the cytoplasm, cilia, and flagella of eukaryotic cells. They, along with microfilaments and intermediate filaments, are part of the cell cytoskeleton. In plant cells they are constructed with 13 rows (protofilaments) of tubulin subunits (α- and β-tubulin), staggered in such a way as to appear in a helical structure. Microtubules are continuously being assembled and disassembled. They, in general, act as a means of stabilization within the cell's cytoplasm. In growing plant cells, they can influence the orientation of cellulose being layered for the cell wall. During cell division (mitosis), microtubules have been shown to compose the mitotic spindle fibers.

Intermediate Filaments

These fiber organelles are also part of the cytoskeleton and are formed by the polymerization of several different cell proteins. Among other functions, intermediate filaments form the major structural protein of hair and skin and act as support to myofibrils and Z-disks in muscle.

Cilia

Cilia are hairlike projections (0.25 µm long) that extend from the surface of many animal cells. These structures are used for cell movement (as in protozoa) or to move fluid (such as mucus) at the surface of the cell. Ciliated epithelial cells are found in the respiratory tract. Ciliary movement is dependent on movement of the axoneme, which is a relatively permanent structure primarily composed of microtubules. Ciliary motion requires ATP hydrolysis, which generates a sliding movement of microtubules.

In order for microtubules to perform their functions, they must be attached to other parts of the cell. Cilia end in a structural unit, the basal body, located at the base of the ciliary axoneme.

PROBLEM

Intermediate filaments are

a) found in animal cells.

b) more permanent than microfilaments.

c) intermediate in diameter between microfilaments and microtubules.

d) all of the above.

Solution

d) Intermediate filaments, found in animal cells, are more permanent than, and intermediate in diameter between, microfilaments and microtubules.

Bacterial Flagella and Movement

Flagella in eukaryotic cells have a structure very similar to cilia and generate movement using the same principle detailed for cilia. Sperm cells and protozoa are examples of flagellated eukaryotic cells. The flagella of bacteria differ markedly from those of eukaryotic cells; they are smaller, lack internal fibrils, and do not flex.

7. Cell Cycle, Growth, Division, and Regulation

Phases of the Cell Cycle. Cells are continuously subjected to various kinds of stress that can result in cell death. For an organism to grow, survive, and repair/maintain itself, cells must reproduce themselves. For cellular division to occur, a cell (mother cell) must first *double* its contents, nuclear and cytoplasmic, divide its nucleus (called mitosis), and divide its cytoplasm (called cytokinesis). Some authors will refer to the complete cell division process as cytokinesis. Mitosis (Figure 2.31) results in two genetically identical (barring mutation) daughter cells. A cell that is not undergoing active division is said to be in interphase. The interphase period has been further delineated on the basis of when DNA synthesis occurs. The period of active DNA synthesis and replication (sister chromatids [i.e., identical copies of each chromosome joined by a centromere]) is called S phase. The gap period before S phase is called G1 phase, and the gap period after S phase is called G2 phase. The mitotic phase, designated M phase, begins after the G2 phase. The M phase includes prophase, metaphase, anaphase, and telophase.

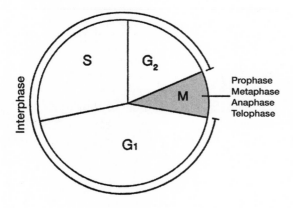

**Figure 2.31.
Phases in Cell Mitosis**

Prophase—The chromatids shorten, thicken, and become visible as chromosomes. The nucleolus and nuclear membrane break down and disappear. The centrioles divide and move to opposite poles of the nucleus, and spindle fibers begin to form.

Metaphase—The chromosomes move to the equator of the spindle fibers (the middle). The paired chromosomes attach to the equator of the spindle at the centromeres.

Anaphase—The sister chromatids separate and move to opposite poles of the spindle fibers. Cytokinesis begins so that the two new daughter cells will contain approximately equal contributions of cytoplasmic contents of the original mother cell (recall that all cell contents were duplicated during interphase).

Telophase—Cytokinesis and mitosis are complete, having formed two daughter cells. The chromosomes in each new cell uncoil, and the nucleus and the nuclear membrane reform. In plant cells, a cell plate appears, dividing the parent cell into two daughter cells. In animal cells, the plasma membrane invaginates, dividing the plasma membrane.

PROBLEM

A mitotic cell produces

a) two cells with half of the chromosomes of the first cell.

b) two cells each with the full chromosome complement of the original cell.

c) four cells with half of the chromosome complement of the original cell.

d) four cells with the full chromosome complement of the first cell.

Solution

b) Mitosis refers to the process by which a cell divides to form two daughter cells, each with exactly the same number and kind of chromosomes as the parent cell.

A number of cellular organelles/structures are visible during plant and animal cell division. This is largely due to the staining and microscopy methods used, but also, because as the cell progresses into M phase, chromosomes begin to organize and condense into thicker, darker structures, indicating that nucleic acids are complexed with histone and nonhistone proteins (referred to as chromatin). Among the more obvious structures are the following.

Chromatids and Centromere. Most of the synthetic events necessary for cell division occur during interphase. In animal cells, the initiation of DNA replication (i.e., the start of S phase) is closely associated with the replication of the cell's pair of centrioles. The centrioles are the microtubule organizing centers of the cell. Each centriole pair forms the spindle pole during mitosis. Once DNA synthesis is initiated, it continues until the entire DNA molecule is replicated. As the DNA replicates, new histones are attached and chromatin is formed. Each chromosome is duplicated in S phase, forming two sister chromatids joined by a centromere.

Centrioles, Asters, Spindles. G2 phase begins at the end of the S phase. During the G2 phase, cells prepare for mitosis by constructing much of the macromolecular machinery used in the mitotic spindle. In addition, the mitotic spindle is formed. The mitotic apparatus consists of the two centromeres, a set of microtubules, and two pairs of centrioles. The cytoplasmic microtubules observed in interphase cells are attached to the centrioles. Basal bodies and centrioles have very similar structures, each having a nine-fold array of triplet microtubules. The centrosome, which is present in most animal cells, has a centriole pair at its center. Higher plants do not have centrosomes. The centrosome, also called the cell center, is adjacent to the cell nucleus. It serves to organize microtubules and plays a major role in cell division. The microtubules form a radial array called the aster around each pair of centrioles. Some of the microtubules eventually connect each pair of centrioles. These microtubules comprise the spindle. The microtubules are responsible for the movement of chromosomes during mitosis.

Kinetochore. Early metaphase starts with the dissolution of the nuclear envelope. Microtubules subsequently become attached to chromosomes at a site called the kinetochore. During metaphase, the chromosomes become aligned at a plate halfway between the spindle poles. At anaphase, the chromosomes are broken apart by the microtubules attached to the kinetochores. During telophase, the daughter chromosomes arrive at opposite spindle poles, the kinetochore microtubules dissociate, a nuclear envelope appears around each set of new chromosomes, and nucleoli reappear. This completes the process of mitosis. The subsequent division of the cytoplasm is called cytokinesis.

PROBLEM

Kinetochore microtubules dissociate during

a) prophase.   b) metaphase.

c) telophase.   d) anaphase.

Solution

c) Kinetochore microtubules attach during prometaphase. At anaphase, they break apart the chromosomes. During telophase, the kinetochore microtubules dissociate.

Genetics and Molecular Biology

1. Genetic Foundations

Genetics is the study of genes, what they are, and how they work, including how they store, express, and replicate information, and how that information is transmitted to subsequent generations (*heredity*). *Molecular genetics* is a subdiscipline that emphasizes the study of DNA, RNA, proteins, and the biochemistry of gene action and expression. *Medical genetics* looks at how mutations express themselves as diseases and/or disabilities, how they might be alleviated, and how they are transmitted in family pedigrees. *Transmission genetics* is another subdiscipline that focuses on how genes are transmitted or inherited from one generation to the next. *Population/evolutionary genetics* concentrates on estimating allele frequencies within and among populations; it often asks questions about genetic changes over time and space. Today, there are even more new subcategories that were not named even a few years ago (e.g., genomics, bioinformatics, forensic genetics, to name a few). They are now even popularized in movies, on the Internet, and in television and radio news and dramas, in which basic genetic principles and DNA technology like PCR (*polymerase chain reaction*) are used to solve crimes, identify diseases, determine kinships, and much more.

Genetic Terms and Definitions

Allele—an allele is a variant, or alternative form, of a gene. Alleles are located at the same location (the locus, see below) on homologous chromosomes (see below) and code for the same protein, although the functional molecule *may* differ (e.g., petal color in snapdragons is controlled at a gene that produces a protein in the biosynthetic pathway that produces petal color; one allele results in red pigment, the other in the absence of color [white]).

Chromatin—Complex of constituents of a eukaryote chromosome, made up of DNA, histone proteins, nonhistone binding proteins, and RNA.

Chromosome—Structure not usually visible except when it condenses (shortens and gets thicker) and is complexed with dark staining histone and nonhistone proteins during cell division (i.e., either mitosis or meiosis). Chromosomes contain DNA, RNA, histones, and other proteins (see chromatin).

Codon—This is a sequence of three nucleotides (a triplet) in DNA, and mRNA, that codes for an amino acid. A nucleotide is composed of a 5-carbon sugar, phosphate, and a nitrogenous base.

Euchromatin and heterochromatin—Lighter and darker staining regions, respectively, within a chromosome.

Gene—The basic unit of inheritance, is a segment of double-stranded DNA (deoxyribonucleic acid), composed of a sequence of nucleotides (may include thousands of codons) that specify the structure of a functional product, usually a polypeptide or protein.

Locus—Is the position (location) of a gene within the double-stranded DNA molecule.

sex chromosomes—In humans there are two sex chromosomes, the X and Y. Gender is defined by the Y chromosome (i.e., a female zygote has no Y, a male zygote contains a Y chromosome). Diploid normal females have two X's (homogametic sex), and males have one X and one Y (heterogametic sex); typical notation for these *karyotypes* (see below) is 46XX and 46XY, respectively. Not all species are alike in sex determination (e.g., in birds, the males are ZZ [homogametic] and females are ZW [heterogametic]). An *autosome* is any chromosome that is not a sex chromosome, i.e., in human there are 22 pairs of autosomes and one pair of sex chromosomes.

Sex-limited and sex-influenced traits—A sex-limited trait affects only one of the sexes (e.g., in humans, the alleles at the gene for heavy beard occurs in both sexes but are only expressed normally in males). A sex-influenced trait is one that that is dominant in one sex but recessive in the other. The best-known example in humans is the gene for pattern baldness. In males, the baldness allele (B) is dominant, causing baldness in the heterozygote; in women, the same allele is recessive and does not cause baldness in the heterozygote. While in very low frequency, the BB genotype in women can cause baldness but not to the same extreme as in men. Obviously, this gene's expression is correlated with sex hormone differences, qualitatively and quantitatively, during fetal development.

Ploidy—The ploidy level or number of chromosomes in a species indicates the number of chromosomes that are commonly found. Many eukaryotes are *diploid* (2N). For example, humans have two copies (sets), or a diploid number of chromosomes, 2N= 46; in shorthand, and denoting gender, we can write this as 46 XX (female) and 46XY (male). Normal gametes ordinarily have half the number of chromosomes (called the *haploid* number, N) that comprise the nuclear genome (e.g., in humans, the sperm and egg each should contain N=23 chromosomes). Animals seldom are *polyploid* (having more than two sets of chromosomes), but many plant species are polyploid (e.g., wheat varieties can be *hexaploid*, having six copies of each chromosome, and each gamete would contain 3N number of chromosomes). *Aneuploidy* occurs when there is an extra (trisomy) or missing chromosome (monosomy) (e.g., 47 XXY, is the shorthand notation for a human male [Kleinfelter syndrome] who is trisomic for the X chromosome). *Nondisjunction* causes aneuploidy, that is, when chromosomes fail to separate properly during meiosis.

Homologue—A diploid organism has two sets of chromosomes. During division in meiosis I, each set of chromosomes individually align or pair (synapse) at respective colinear gene loci (i.e., allelic nucleotide sequences). They are said to be homologues, or homologous pairs, during meiosis. Because each set of homologues has $2 \times 2 = 4$ chromatids, they may be referred to as *tetrads*.

Chiasma (chiasmata, plural)—During meiosis (prophase I) homologous chromosomes pair or synapse. They are said to form a tetrad or bivalent structure; recall there are four chromatids following DNA replication during interphase. During synapsis, breaks occur in the chromosomes and crossovers (exchanges) occur between nonsister chromatids. These are called chiasmata.

Karyotype—Eukaryote species have a characteristic number of chromosomes. The karyotype is the pictorial/graphical display of the chromosomes of a species ranked by size, shape, and banding patterns.

Genetic mutation—This is a change in an allele or segment of a chromosome that causes genetic variation, a prerequisite for evolution; not all mutations are "bad," although those kinds get the most attention. Severe deleterious mutations can cause either partial or total loss of function in the protein affected by the genetic change. Even minor nucleotide changes, if they cause a change in amino acid sequences at an active site in a polypeptide, can have large consequences.

Genome—All of the genes contained within the cells of an organism are referred to as the genome. Animal cells have two genomes, the chromosomes in the nucleus and the much smaller genome in the mitochondrion (cytoplasmic). Plant cells have three genomes, the nuclear, the mitochondrion (cytoplasmic), and another small one in the chloroplast (cytoplasmic).

Polymorphic gene locus—When there is more than one detectable variant sequence at a gene locus, it is called an *allele*, and the gene is said to be *polymorphic* in a population if there are two or more alleles and the least frequent allele at a locus has a frequency of 1% or greater.

Mendelian inheritance—Many gene products are known that are simply inherited (Mendelian inheritance), that is, the expression of the functional protein is controlled by one or a few genes (e.g., the three ABO alleles for blood antigens controlled at one gene locus in humans).

Homozygote—This is an organism possessing an *identical* pair of alleles on homologous chromosomes for a given trait, said to be homozygous. For example, a polymorphic diploid gene having alleles C and c in a population will have two homozygous genotypes among individuals in the population. They will be (CC) or (cc); the third possible genotype will be heterozygous (Cc), that is, having two dissimilar alleles.

Heterozygote—Represents, for example, in a diploid organism, a genotype with *different* alleles on ho-

mologous chromosomes for any given character (e.g., Rr); see also *Homozygote.*

Phenotype vs. genotype—The phenotype is the outward expression of the genotype in an organism. For simply inherited, Mendelian traits, the phenotype is affected by the relationship between the alleles at a gene locus (and, possibly often, other nonallelic genes interacting, called *epistasis*). Using the snapdragon example for petal color (above), and using symbols for the three possible genotypes (RR, Rr, rr), if we write the homozygous R-allele genotype as *RR*, then the petals are red. The homozygous *rr* genotype has white petals. Because in this particular species the petal color locus exhibits *incomplete dominance*, the heterozygote (Rr) is pink, thus an intermediate color to the two homozygotes. In these cases, the phenotype distribution of progeny should be the same as the genotype (e.g., a monohybrid cross of *Rr x Rr* would yield a 1:2:1 ratio distribution represented by *RR, Rr, rr*, which is the same outcome for the phenotypic ratio as for the genotypic ratio: 1/4 red, 2/4 pink, and 1/4 white). If a locus demonstrates having complete dominance, the phenotypic ratio is different than the genotypic. This can be shown using the three alleles at the human ABO blood antigen gene locus; by the way, there are many different kinds of blood antigen genes. It will also serve to demonstrate *codominance* between alleles at a locus. These are the possible genotypes at this locus: AA, AO, AB, BB, BO, and OO. The dominance relationships are such that A and B are dominant in expression to the O allele, but they are codominant when in the AB genotype (i.e., an assay for an AB person's blood type yields results indicating both antigens present). However, due to complete dominance of the A and B alleles to the O allele, AO is seen as A_ and BO is seen as B_; we underscore to indicate that without parental information, *and* with complete dominance between alleles, you might not be able to identify the person's genotype (i.e., an A_ phenotype could be either genotype AA or AO, as for the B_ phenotype). Thus, with full dominance at a locus, the genotype and phenotype are not perfectly related in terms of gene expression, only when they are codominant. These relationships get even more complicated when *multifactorial or polygenic* traits are studied (below).

Mutifactorial or polygenic trait—These are controlled by the actions of alleles at many genes (e.g., height in humans, yield of corn). The phenotype (the appearance) of a multifactorial trait is influenced by both the genotype (alleles that are present) and the environmental background in which the genes exist. These traits, like animal growth and crop yields, have been studied a lot by plant and animal breeders, but for ethical reasons, we do not choose to do controlled breeding experiments using humans. We can, however, study identical twins because, barring mutations, the differences we observe are due mainly to the environmental differences experienced by the twins, especially after they are separated upon leaving a common home environment. Today, with the tremendous explosion in biochemical and recombinant genetic technology, many species, including humans, are being described entirely by identifying the single changes in nucleotides that differentiate individuals within and among populations (e.g., the Human Genome Project).

Gregor Mendel's three laws (published 1865)— Mendel's experiments laid the foundations for future scientists to unravel the ways in which traits (factors, later genes) were transmitted between generations. **(1)** *Law of Dominance:* when crosses were made between true breeding garden peas that differed for any of the seven traits that he studied, Mendel observed that all of the progeny expressed the trait of only one of the two parental types (see Table 3.1, F1 generation). **(2)** *Law of Segregation*: Segregation refers to the *separation of alleles* at a gene, each randomly allocating to a gamete (e.g., given the "J" gene and genotype Jj, there will be four *chromatids* after DNA replication, thus JJjj copies; when these separate at the second division, the four meiotic products [gametes] will contain J, J, j, j in four different, new cells). This is possible because during Interphase prior to meiosis I (gamete formation), DNA molecules replicate and form what are called *chromatids,* which are later seen to be connected at the centromere of the chromosome; note that replication also occurs prior to mitosis, but there is only one cell division in mitosis but two (meiosis I and II) occur when gametes are formed (i.e., diploid germ cells are reduced from 2N to 1N [haploid] in the second division). **(3)** *Law of Independent Assortment:* this rule was discovered when Mendel made crosses and tracked the outcomes for two or more traits simultaneously. Independent assortment refers to the

relationships of alleles at *two or more different* genes that are located on different chromosomes; recall from the Law of Segregation that, by definition, the alleles at a single gene separate independently. The Law of Independent Assortment recognized that the action of inherited factors of one trait had no effect on that of another. Bateson and Punnett (between 1905–17, linkage studies at Cambridge University) further characterized this conclusion, and Punnett (you've likely heard of the "Punnett square" that we use in this text), in particular, combined the mathematical/statistical and biological components. For example, consider two genes in a heterozygous genotype, each having two alleles: RrJj. There are four possible gamete genotypes: RJ, Rj, rJ, rj. If these genes are independent (on different chromosomes or very far apart on the same so as to appear on different chromosomes), they will segregate and assort independently so that the outcome in the gametes is ¼, ¼, ¼, ¼, or 1:1:1:1. This rule permits simple tests to be done to determine if genes are on the same chromosome (*linked*, see below) by comparing the observed phenotype ratios to the expected under independent assortment. A testcross is a classical way of determining these relationships when dominance is present. For example, if *dihybrid* genotype (RrJj) is crossed with the homozygous recessive (rrjj), and if the two genes are independent, the progeny will exhibit four phenotypes (RJ, R_, _J, rj) in equal proportions, the same proportions as the parental gametes: 1:1:1:1 (refer back to Mendel's two laws which predict this outcome). This result also permits us to define both the genotype and the phenotype of the parental generation, which is not apparent with dominance present until progeny are observed (this is why breeders call these *progeny tests*, which are really tests to better characterize the parents).

Linkage—Genes that do not assort independently and thus are close together on the same chromosome and are inherited together, are said to be linked (*synteny* is also a word used to describe being on the same chromosome). Their phenotype ratio will significantly deviate from the expected 1:1:1:1 under independent assortment of genes (see above). When genes are mapped to a chromosome, they may be referred to as a block or linkage group. Some suggest that linkage groups are important in keeping advantageous combinations of alleles together (i.e., crossing-over is rare and restricted when genes are located very tightly together).

Crossing-over (see also *Chiasma*)—Paired chromosomes may break and their "sticky" fragments reunite in new combinations. This occurs in prophase of meiosis I when the homologous chromosomes are aligned and paired. In simple two-stranded crossovers, the gametes will results in two parental types (non-crossovers) and two crossover types. This is a major means whereby new combinations of alleles at different genes come together in the next generation resulting from recombination (see below). For example, consider two genes with a parental genotype (AB, ab). The gametes, with crossing over, can exhibit four combinations of alleles at the two genes—parental (AB and ab) and the crossovers (Ab and aB)—thus increasing the genotypic variation that may result in the next generation.

Genetic recombination—Genes may be rearranged from separate groups so that genes from two chromosomes are recombined into one chromosome containing some genes from each of the original chromosomes. This contributes to genetic diversity. In eukaryotes, the process is associated with sexual reproduction, during which haploid gametes, produced by meiosis, fuse to form a diploid zygote. Portions of chromosomes may be exchanged during meiosis by a process known as crossing-over.

Translocations—The shifting of gene positions in chromosomes that may result in a change in the serial arrangement of genes. In general, it is the transfer of a chromosome fragment to a nonhomologous chromosome. For example, in humans 1 in 20 cases of Down's syndrome results from translocation between the long arms of chromosomes 14 and 21; the carrier has 45 chromosomes and may or may not show phenotypic evidence, depending on which genes are affected near the translocation. Translocations may or may not be reciprocal (i.e., equal exchange between nonhomologues).

Inversion—These occur when a broken segment of a chromosome flips around and reinserts itself in re-

verse order. The actual genotype composition has not changed (such as in translocations, deletions, and additions), but the order of the genes has changed. For example, consider the group of genes abcdEfG and invert cdE, which now reads abEdcfG. It has been estimated in human populations that among carriers of inversions, about 5–10% may exhibit some medical problems thought to do with breaking up linkage blocks of genes or breaks within genes coding for critical proteins.

Meiosis—This is the cell division process that occurs in the germ line of cells and leads to the formation of haploid sex gametes in eukaryotes. Like mitosis, DNA in chromosomes replicates during interphase (there are then 4N chromosomes), then homologues form pairs (this does not happen in mitosis) during prophase I, these tetrads align at the cell's equator (metaphase plate) during metaphase I, the tetrads separate, and spindle fibers connected to the centromere pull the newly recombined chromosomes to opposite poles in anaphase I (chromatids are still connected, but contain new combinations due to crossovers). See Figure 3.1A. The diad cell (not two independent cells) then goes through a second series of prophase II, metaphase II (align in perpendicular plane to metaphase I), anaphase II (centromeres divide, thus chromatids become "new" chromosomes), and finally telophase, which may or may not be concordant with cytokinesis (cell separation). With the second division, each cell that may become a gamete contains the haploid (1N) number of chromosomes: 4N → 2N → 1N. If, or when, a sex gamete is involved in a fertilization event, the diploid number is restored in the zygote for the next sexual generation. See Figure 3.1B.

Figure 3.1. (A)

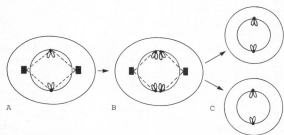

■ = centriole

● = kinetochore

- - - = spindle fiber

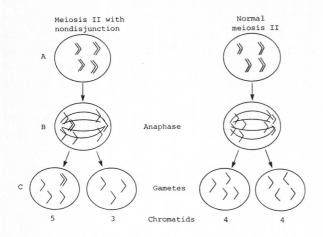

Figure 3.1. (B)

Mendelian Inheritance

In 1865, Gregor Mendel, an Augustinian monk in what is now a part of modern-day Germany, published his results of breeding experiments using garden peas. We now know that he had discovered the basic laws of genetics. It was not until 1900 when the scientific community revisited his research that the significance of his findings was realized. Table 3.1 describes his results for seven different traits: seed color, seed shape, pod color, height or stem length, flower position in the plant, pod shape, and flower color. These data represent the outward appearances (phenotype) of monohybrid crosses among three generations: the original parents (P_1), the first hybrid generation (called F1, for first filial), and the F2 generation (second filial) resulting from intercrossing the plants grown from the F1 seeds. Luckily for Mendel, and the science to later become genetics, we now know that these seven traits happened to all behave *independently*, thus he did not have to contend with distortions among his ratios due to linkage.

The very nearly 3:1 ratio resulting from this data enabled Mendel to recognize that the offspring of each plant had two factors instead of a single factor for any given characteristic. Each of the crosses among F1 to produce the F2 are now referred to as *monohybrid crosses* because we know that the proteins encoded in the DNA for each of the seven traits is controlled by different alleles at a single gene locus with complete dominance present.

Dominant/Recessive Monohybrid Inheritance: Two Alleles at a Single Gene

There are many different types of genetic inheritance patterns. The simplest is that of dominant and recessive inheritance of two alleles at one gene. Recall that dominance is not a physical "masking" of effects but stems from changes in alleles at the DNA or RNA level whereby the gene's allelic proteins function in slightly

Table 3.1.
A Summary of the Data Obtained by Mendel from His Breeding Experiments with Garden Peas

Parental Cross Phenotypes	F_1 Phenotype	F_2 Phenotpe (counts)	F_2 Phenotype Ratio
Yellow seeds x green seeds	all yellow	6,022 yellow:2,001 green	3.01:1
Round seeds x wrinkled seeds	all round	5,474 round:1,850 wrinkled	2.96:1
Green pods x yellow pods	all green	428 green:152 yellow	2.82:1
Long stems x short stems	all long	787 long:277 short	2.84:1
Axial flowers x terminal flowers	all axial	651 axial:207 terminal	3.14:1
Inflated pods x constricted pods	all inflated	882 inflated:299 constricted	2.95:1
Red flowers x white flowers	all red	705 red:224 white	3.15:1
		Average	2.98:1

different ways (e.g., no function, excessive function, interrupted biosynthetic pathway, and so forth). The Punnett square below (Table 3.2; about 1905 W. Bateson and R.C. Punnett [Cambridge] published the first account of gene linkage in sweet peas, and Punnett developed the "Punnett square" to depict the number and variety of genetic combinations) helps by displaying gametes and by showing the probabilities of outcomes from the mating between the two heterozygotes: Yy x Yy. These are Mendel's crosses shown in Table 3.1 among all yellow seeds in the F1. Note that the probabilities for each allele in the gametes are shown to be 0.50, which we expect if normal and random segregation of alleles occurred in meiosis. If gametes randomly combine, then genotypes should result in the frequencies indicated.

Table 3.2.
Monohybrid Crosses Testing Complete Dominance at the Seed Color Gene in Garden Peas. Numbers in Parentheses Are the Probabilities of Each Outcome

Gametes	Y (0.50)	y (0.50)
Y (0.50)	YY (0.25) yellow	Yy (0.25) yellow
y (0.50)	Yy (0.25) yellow	Yy (0.25) green

Phenotypically, 3:1, or 0.75 or 75%, of the F2 offspring will have yellow seeds due to dominance of the yellow *Y* allele over the green *y* allele (i.e., 0.25+2(0.25)=0.75). Note that the genotype frequencies differ from the phenotype (i.e., 0.25, 0.50, 0.25 [1:2:1] vs. 3:1). This is the typical relationship when one allele is completely dominant at a locus.

Dominant/Recessive Dihybrid Inheritance: Two Alleles at Two Independent Genes

For discussion sake, consider two genes in Mendel's peas (Table 3.1) at the locus controlling seed shape (round or wrinkled) and seed color (yellow or green). Each gene has two alleles, Rr and Yy; the R allele is dominant to the r, and the Y allele is dominant to y. This is the original cross: RRYY x rryy (i.e., round yellow seeded plant x wrinkled green seeded plant). The

F1 has the genotype RrYy, and because of dominance, all progeny are this phenotype, round and yellow. Table 3.3 describes the dihybrid cross among the F2 plants and the frequencies of the gametes expected following segregation and independent assortment of the chromosomes during meiosis.

Table 3.3.
Dihybrid Cross Involving Round and Yellow Seeded Pea Plants

Gametes	RY (0.25)[1]	Ry (0.25)	rY (0.25)	ry (0.25)
RY (0.25)	RRYY	RRYy	RrYY	RrYy
Ry (0.25)	RRYy	RRyy	RrYy	Rryy
rY (0.25)	RrYY	RrYy	rrYY	rrYy
ry (0.25)	RrYy	Rryy	rrYy	rryy

[1] Note that each cell represents (1/4)(1/4)=1/16 or (0.25)(0.25) = 0.0625 of the total progeny.

After tabulating the data and collecting like genotypes and phenotypes, the summary is in Table 3.4.

Table 3.4.
Dihybrid Genotypes and Phenotypes

Genotype (frequency)	Phenotype	Phenotype description
RRYY (1/16)	R_ Y_	round, yellow
RRYy (2/16)	R_ Y_	round, yellow
RRyy (1/16)	R_ yy	round, green
RrYY (2/16)	R_ Y_	round, yellow
RrYy (4/16)	R_ Y_	round, yellow
Rryy (2/16)	R_ yy	round, green
rrYY (1/16)	rr Y_	wrinkled, yellow
rrYy (2/16)	rr Y_	wrinkled, yellow
rryy (1/16)	rr yy	wrinkled, green

Note that there are 9 genotype classes in Table 3.4, and because of complete dominance at each gene, there are only 4 phenotype classes. When we sum the frequencies among the phenotypes, there is this distribution: 9:3:3:1—round, yellow: round, green: wrinkled, yellow: wrinkled, green. Further note, because the alleles at the two genes behave randomly, round:wrinkled indicates a 12:4 or 3:1 phenotypic ratio as we would expect for only one of the genes. This also holds true for the other gene (i.e., 12:4 or 3:1 yellow:green).

Dominant/Recessive Dihybrid Inheritance: Two Alleles at Two Linked Genes

Incomplete Dominance. Some Mendelian traits are inherited, but the expression of neither allele in the genotype is fully dominant to the other. Refer again to th e common ornamental snapdragon. Let *R* represent the allele for red petal pigment and *r* for no pigment (i.e., white). To determine if the alleles are dominant, we first cross two "true breeding" lines and observe that all the F1 progeny phenotypes are pink. Next, we make crosses among those offspring to produce the F2 generation. That cross is shown in Table 3.5.

Table 3.5.
Monohybrid Test to Determine
Incomplete Dominance for Petal Color

Gametes	R (0.50)	r (0.50)
R (0.50)	RR (0.25) red	Rr (0.25) pink
r (0.50)	Rr (0.25) pink	rr (0.25) white

The results demonstrate that petal color segregates to produce a phenotypic *and* genotypic ratio of 1:2:1, or 0.25, 0.50, 0.25, red:pink:white. This outcome is what we would predict if the two alleles exhibited incomplete dominance.

Codominance. Codominant expression by two alleles in a diploid occurs when a heterozygote has two dominant alleles that are equally expressed. Codominance can be demonstrated by the inheritance of human blood antigens at the ABO locus; this is an autosomal gene located on chromosome 9. There are multiple alleles, I^A, I^B, and I^O (note: "O" used here for discussion, often is designated as *i*), possible at the locus that codes for ABO blood type; this locus exhibits both complete dominance and codominance among the three alleles. I^A and I^B are dominant to I^O but are codominant with each other.

Type A blood (phenotype) is expressed by the genotypes $I^A I^A$ and $I^A I^O$. Such a person has only A antigens on his/her red blood cells. A phenotype that is type B has the genotype $I^B I^B$ or $I^B I^O$. This person has only B antigens on his/her red blood cells. The AB phenotype is expressed by the single genotype $I^A I^B$. These alleles are codominant, and the person has both A *and* B antigens on his/her red blood cells. The homozygous recessive genotype $I^O I^O$ is expressed phenotypically by type O blood (recessive to A and b). This person has neither A nor B antigens on his/her red blood cells.

Punnett squares can be used here as well to predict blood type probabilities for offspring. Suppose a heterozygous type-A man ($I^A I^O$) mates with a heterozygous type-B female ($I^B I^O$). The Punnett square below (Table 3.6) shows that there will be an equal probability (25%) of each blood type in the offspring.

Table 3.6.
Mating Showing Both Complete
Dominance and Codominance at the
Human ABO Blood Antigen Gene

Male (sperm)	Female (egg)	
	I^B (0.50)	I^O (0.50)
I^A (0.50)	$I^A I^B$ (0.25)	$I^A I^O$ (0.25)
I^O (0.50)	$I^B I^O$ (0.25)	$I^O I^O$ (0.25)

The outcomes show that there is a 0.25 (25%) probability for each of four different genotypes *and* phenotypes. These are codominance type AB ($I^A I^B$), dominance type B ($I^B I^O$), dominance type A ($I^A I^O$), and recessive type O ($I^O I^O$).

Sex-Linked Inheritance. Humans have two sex chromosomes, called X and Y, and 22 pairs of autosomal chromosomes. A female has two X chromosomes, one inherited from each parent; commonly, 46XX is the notation for the normal female karyotype and is referred to as the homogametic sex. A male has an X chromosome inherited from his mother and a Y chromosome inherited from his father; commonly, 46XY is the notation for the normal male karyotype and is referred to as the heterogametic sex. Genes on the X chromosome are called X-linked, and those on the Y are Y-linked. Gender is determined by the presence/absence of the Y-chromosome rather than the female requiring two copies of the X. Interestingly, this was determined by finding males who were XX and females who were XY. Karyotype analyses led to the realization that a small region of the Y chromosome was contained in cells of the XX males and the same region was missing in the XY females. This led to the discovery of the spy codes for a protein called *transcription factor*; this gene turns out to be the one that puts into action genetic-biochemical processes that determine maleness. The Y chromosome is relatively small compared to the X chromosome and has only two small homologous regions (the pseudoautosomal region, PAR1 and PAR2) with the X, one at the tip of each arm of the Y. The disparity in size and homology between the X and Y chromosomes also means that recombination due to crossing-over has much smaller influence, and for practical purposes, it is usually assumed that crossing-over in the X chromosome only occurs in the female and little to none in the Y, except for the pseudoautosomal regions. For most genes that are sex-linked, only the X expresses the gene product in males, and for this reason the male is said to be hemizygous (i.e., only one copy of a gene), and this leads to the important differences in the way that sex-linked genes are expressed in males relative to females, who have two alleles per gene rather than only one. Genetically and functionally, the pair of X chromosomes in females acts the same as the other 22 autosomes during inheritance. An important evolutionary difference in humans is that even though all normal human female somatic cells (any cell except germ cells) contain two X chromosomes, only one is expressed because each female cell has a Barr-body (one of her X chromosomes is inactivated), and the inactivation is random between the two X chromosomes; deleterious recessives (below) in heterozygotes are usually not fully expressed because half the body's cells are synthesizing normal protein.

Sex-linked traits that result in either abnormal or deleterious expression (inborn errors of metabolism or genetic disease) are often caused by deleterious recessive alleles. A dominant allele will be expressed in both sexes, and if deleterious (selected against), can be lost from the sexually mature population very quickly. However, recessives are maintained in the population in carrier females who are heterozygous and may pass the recessive to their sons, who being hemizygous, express the recessive trait that resides on their one X chromosome. A daughter will not express the disease unless she has also acquired the recessive gene from an affected father. The probability that a woman will be homozygous is a probability based upon the frequency of the allele in the population and the gametic/transmission frequencies possible within a pedigree containing individual genotypes. For example, *hemophilia A* is X-linked. In North America the incidence is about 1 in 10,000 male births, or an allele frequency of 0.0001, due to males having only one X chromosome. The probability of two X chromosomes coming together in a female genotype and being expressed is $(0.0001)^2$ or 1 in 100 million. The frequency of female carriers is ~1 in 5000. Table 3.7, below, shows a Punnett square in which the chances of affected sons and carrier daughters are much greater at the level of a single mating between one of the possible 5,000 carriers in the population and an unaffected man.

Table 3.7.
Mating Between a Woman ($X^H X^h$)
Who Is a Carrier of the X-Linked Disease,
Hemophilia A, and an Unaffected Man ($X^H Y$)

		Female (egg)	
Male (sperm)		X^H **(0.50)**	X^h **(0.50)**
Daughters	X^H **(0.50)**	$X^H X^H$ (0.25)	$X^H X^h$ (0.25)
Sons	Y **(0.50)**	X^H Y(0.25)	X^hY(0.25)

Observations we can draw from the above information are as follows:

1. One-half of the daughters may be carriers.

2. One-half of the sons may be affected.

3. The probability of this couple having an affected *child* is 0.25 or 1 in 4 because the probability of having a son is 0.50 and the probability the son will be affected is 0.50. Therefore (0.50)(0.50)= 0.25 or 1 in 4; obviously, the chance that none will be affected is 0.75 or 75%.

4. The Y chromosome has no hemophilia gene locus, and barring mutations, none of the daughters will be affected with hemophilia.

Pedigree Analysis

Genetic relationships among family members can be studied by making a pedigree chart (Figure 3.2). Commonly used symbols in human pedigrees are shown in Table 3.8. Having some knowledge about Mendelian inheritance will help a lot toward interpreting pedigree charts. Some useful guidelines are listed below.

When referring to Table 3.8, note that males are always represented by square symbols and females with circular symbols. A line drawn between a square and a circle represents a mating of that male and female. Two lines drawn between a square and a circle would indicate a consanguineous mating (between relatives).

Figure 3.2.

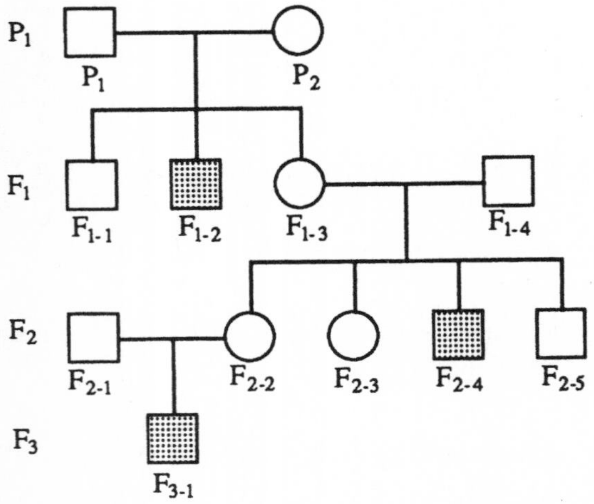

Generations are connected by a vertical line extending down from the mating line to the next generation. Children of a mating are connected to a horizontal line, called the sibship line, by short vertical lines. The children of a sibship are always listed in order of birth, the oldest being on the left. Sometimes, to simplify a pedigree only one parent is shown. Normal individuals are represented by an open square or circle, depending on the gender, and affected individuals by a solid square or circle. Each generation is numbered to the left of the

Table 3.8.
Common Pedigree Symbols, Definitions and Abbreviations[1]

	Male	Female	Sex unknown
Individual	b. 1925	30 y	4 mo
Affected individual (Define shading in key/legend)			
Affected individual (more than one condition)			
Multiple individuals, number known	5	5	5
Multiple individuals, number unknown	5	5	5
Deceased individual	d. 35y	d. 4 mo	
Stillbirth (SB)	SB 28 wk	SB 30 wk	SB 34 wk
Pregnancy (P)	P LMP: 7/1/94	P 20 wk	P
Spontaneous abortion (SAB)	male	female	ECT
Affected SAB	male	female	16 wk
Termination of pregnancy (TOP)	male	female	
Affected TOP	male	female	

[1]**Adapted from:** Bennett, R.L., Steinhaus, K.A., Uhrich, S.B., O'Sullivan, C., Resta, R.G., Doyle, D., Markel, D.S., Vincent, V., Hamanishi, J., 1995, Recommendations for standardized human pedigree nomenclature. *Am J Hum Genet* 56(3):745–752.

sibship line with roman numerals. Individuals in each generation are numbered sequentially, beginning on the left, with Arabic numerals. For example, the third individual in the second generation would be identified as individual II-3.

Identifying autosomal recessive inheritance in a pedigree:

— Both males and females may be affected.

— The trait may skip generations.

— Consanguinity (inbreeding or matings between relatives) increases the chance of bearing affected children.

— Deleterious recessives are often more severe than a dominant allele.

— Unaffected parents can have affected children.

Identifying autosomal dominant inheritance in a pedigree:

— Both males and females may be affected.

— Affected individuals may occur every generation.

— Each affected individual has at least one affected parent.

Identifying X-linked recessive inheritance in a pedigree:

— Most always expressed in the male.

— Homozygous females can express the trait but not a heterozygote.

— Always transmitted from mother to son.

— Affected children can have unaffected parents.

— An affected female must have an affected father and a mother who either is homozygous or heterozygous.

Identifying X-linked dominant inheritance in a pedigree:

— The trait will be expressed in females when they are heterozygous.

— Usually expresses more severely in males because they have only one allele at X-linked gene loci (i.e., no alternate allele to counter some of the effects).

— There usually is a high rate of miscarriages of male fetuses because of early lethality.

PROBLEM

Analyze this pedigree, then answer the questions below:

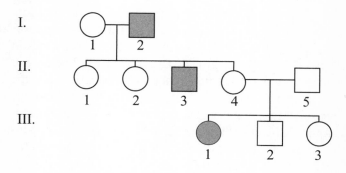

(i) What do you think is the most likely mode of inheritance of this trait? Give at least two reasons to support your answer.

(ii) Calculate the probability that individual III-3 is a carrier.

Solution

(i) Autosomal recessive. Reason: both sexes are affected, and unaffected parents can have affected children (e.g., II-4 x II-5 are unaffected, but III-1 (daughter) is affected).

(ii) II-4 & -5 must be *Aa* because III-1 is affected, therefore:

	Gametes from Parent II-4 (mother)		
Gametes from Parent II-5 (father)		½ A	½ a
	½ A	AA	Aa
	½ a	Aa	aa

Thus, the probability that III-3 is carrier = ¼(=Aa) + ¼(=Aa) = ²⁄₄ = ½.

Prokaryotic Genetics (Transformation, Transduction, and Conjugation)

This section on prokaryote genetics introduces concepts that deviate from the Mendelian modes of transmission (ones requiring meiosis and sex gametes) that were previously reviewed.

Bacterial growth is the result of cellular division. For bacteria the process of cell division is straightforward (i.e., the cell doubles in size and then divides in two). This type of growth is exponential. The time it takes for a doubling of a number of bacteria is called the mean generation time, which is typically less than one hour. During cellular division, the DNA of the chromosome replicates (Figure 3.3), the resulting homologous chromosomes separate, a cross wall forms between the chromosomes, the cell divides and separates.

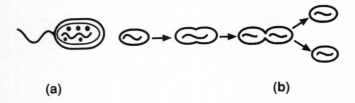

(a)   **(b)**

Figure 3.3.

Bacterial cells are noted for their metabolic versatility and their highly efficient regulation of metabolic and catabolic activities; you will recall that in bacteria DNA transcription and translation of proteins occur simultaneously (see section 5, "Gene Expression and Regulation" later in this chapter). Under adverse conditions some bacteria shift from their normal vegetative state to a dormant state (i.e., they undergo *sporogenesis*). Sporogenesis is a form of cellular differentiation resulting in a metabolically dormant structure such as an endospore, which is formed by the gram-positive bacteria of the genera *Bacillus* and *Clostridium*. Under favorable conditions the spore can undergo germination to return the cells to a vegetative state.

A considerable store of fundamental information about molecular biology has come from microbiologi-cal studies. Strong evidence for DNA being the genetic material comes from studies in which DNA from one bacterial strain is transferred to another. The transfer of the donor DNA is accompanied by the transfer of some donor phenotype(s) (such as virulence) to the recipient strain. DNA (genes) can be transferred from one bacterium to another.

Gene transfer is the movement of genetic information from one bacterium to another. There are three processes for genetic transfer in bacteria: *transformation*, *transduction*, and *conjugation* (refer to Figure 3.3). All of these are significant in that they bring about an increase in the amount of genetic variation within a population.

Transformation. "Naked" DNA is transferred from one bacterium to another in solution. DNA fragments are released as double-stranded DNA into the medium. *Endonucleases* (enzymes that cut at very specific base sequences) cut the double-stranded DNA in solution and the resulting fragments separate—only single-stranded molecules are transferred. The transferred DNA is spliced into the recipient cell's DNA. Uptake is dependent upon the presence of a protein known as competence factor. *Bacterial competence* is the ability of a bacterium to take up DNA from the extracellular environment. Transformation occurs naturally among some bacteria. It is used in the laboratory to create recombinant DNA and is also used to study the effects of introducing DNA into a cell, and in mapping gene locations.

The Experiments of Griffith and Avery. The following experiment of Frederick Griffith (1928) demonstrated transformation. There are two kinds of pneumococcus cells—rough and smooth. Only the smooth form is *virulent* (i.e., capable of infecting and killing mice). Griffith showed that neither heat-killed smooth cells nor live rough cells alone were capable of causing infection. However, when live rough cells were mixed with heat-killed smooth cells, the mice died and live smooth cells could be recovered. Oswald T. Avery (1940) showed that only the smooth cells contained the capsular polysaccharide that is responsible for virulence, and that the substance involved in the transfer was neither a protein nor the polysaccharide itself. Instead, it was the DNA containing the gene for the capsular polysaccharide that was being transferred from the dead smooth cells to the live rough cells, transforming them into the virulent smooth pneumococci.

Transduction. Transduction occurs when DNA is transferred from one bacterium to another via a bacteriophage (virus) and then incorporated into the recipient's DNA. The bacterial virus may be *virulent or temperate* (i.e., a prophage).

Conjugation. Conjugation occurs when DNA is transferred from one live bacterium to another through direct contact; large quantities of DNA can be transferred in this way.

F factors are *plasmids* transferred from a donor cell (an F^+ cell) to a recipient cell (an F^- cell) during conjugation. Plasmids are circular, extrachromosomal DNA molecules that are generally not essential for cell survival, but they can replicate themselves along with any DNA inserted in their molecule, and that has made them very important cloning vectors for recombinant DNA molecules. An *Hfr (high frequency of recombination)* is a cell with an F plasmid incorporated into the chromosome.

PROBLEM

In transduction,

a)　a male chromosome is injected into a female cell.

b)　a female chromosome is injected into a male cell.

c)　a bacteriophage transfers genetic material between bacteria.

d)　a DNA fragment from a donor strain is directly taken up by a recipient strain of bacteria.

Solution

Choice a) refers to bacterial conjugation. Choice b) is an impossibility. Choice d) refers to transformation. Choice c) is correct.

Genetic Mapping

Mapping means identifying the chromosome in which a gene resides. Although classical mapping meant making crosses, doing pedigree analyses, and doing cytogenetics research (karyotyping), today it has flourished with many new molecular techniques that now permit mapping complementary DNA/RNA sequences at the base-sequence level. By using genetic *probes* (a known sequence of DNA), often from distant species, complementary sequences (genes) can be located on specific chromosomes from many candidates. This section will review the basics that are still necessary to the understanding of mapping genes.

Linkage is the inheritance of genes on the same chromosome as a group; genes on the same chromosome are also said to be *syntenic*. Keep in mind that, unless techniques have been used to identify the specific chromosome on which a gene(s) reside, or such information is available from previous studies, only the relationships between two or more gene loci is determined, *not* the chromosome on which they are located. The distance between two genes on the same chromosome allows for crossing over between homologous chromosomes, and, therefore, recombination of the genes. The frequency of recombination can be used to map genes on the chromosome.

Understanding the basic rules of Mendelian inheritance permits us to distinguish between linkage and independent assortment by examining the ratio of offspring in a cross between heterozygous (*AaBb*) and homozygous recessive (*aabb*) individuals. When independent assortment occurs, the heterozygote produces four types of gametes in equal proportions (*AB, Ab, aB, ab*) and, therefore, four types of offspring in equal proportions (*AaBb, Aabb, aaBb, aabb*).

For linkage, most offspring will be similar to the parents (parental types), with the recombinant class occurring in lesser numbers. The frequency of the various classes tells us whether the genes are linked or located on different chromosomes. If the linked genes are very far apart, the recombination frequency may be so great that the genes may appear to inde pendently assort. In that case, linkage has to be tested with other genes located closer to the two genes in question.

The recombination frequency is defined by the formula:

$$\text{Recombination frequency} = \frac{\text{\# of recombinants}}{\text{total number of progeny}}$$

The distance between genes in map units can be determined by multiplying the recombination frequency by 100.

Three point crosses (involving three genes) were devised to collect as much data as possible in one experiment. However, mapping problems are most easily handled by analyzing two genes at a time. Autosomal linkage problems require that each gene be heterozygous in one individual and homozygous recessive in the other. For X-linkage, the heterozygous females are crossed with hemizygous males for the recessive gene. In either autosomal or X-linked inheritance, eight phenotypic classes are produced.

When preparing the key, the eight phenotypic classes should be listed with their respective symbols, rather than listing alleles separately. This will make reading of the key easier when translating symbols at a later time.

Organize the results of the cross by listing the symbols for the phenotypic classes with their respective numbers. Identify each class type:

Parental classes: The two largest classes can be identified as the parental types.

Double crossovers: The two rarest classes are due to double crossovers. In cases where double crossovers are inhibited, these two classes may not be present, resulting in six instead of the expected eight different classes. For ease of solution, you may want to list the double crossover phenotypes anyway, and indicate that the frequency is zero.

Single crossovers: The last four classes can be recognized as two different types of crossovers. Two classes are due to a single crossover between the first two genes and are recognized by the reciprocal nature of the phenotype and similarity in numbers. The other two classes are due to a single crossover between the second two genes and are, likewise, recognized by their reciprocal nature as well as by their number similarity.

The order of the genes can be determined by comparing the parental classes with double crossover classes. How would two crossovers produce the genotypes listed for double crossovers? Try the three orders possible (for example, *abc, bac,* or *acb*) and see which order gives the genotypes observed for the double crossover classes.

Now let's work with the first two genes and ignore the third gene. Add the four classes that show recombination between these genes, divide by the total offspring, and multiply by 100 to obtain the map units between the first two genes. Repeat this process two more times: for the second and third genes, then for the first and third genes.

If the gene order was not previously determined, it can be now. Two distances should approximately equal the third. The longest distance reflects recombination between the outermost genes. The third gene, therefore, is in between.

If the two distances added together give a longer distance than that observed, double crossovers were inhibited. The expected double crossover frequency can be calculated by multiplying the recombination frequencies for the two single crossovers.

A measure of the inhibition of one crossover by another is called interference. This is expressed as follows:

$$\text{Interference} = 1 - \text{coincidence}$$

where,

$$\text{Coincidence} = \frac{\text{observed double crossover frequency}}{\text{expected double crossover frequency}}$$

When there is no inhibition of crossover, observed and expected double crossover frequencies should be equal; in this case, coincidence equals 1 and interference equals 0. As inhibition increases, interference also increases.

It is possible to use the expected double crossovers, the coefficient of coincidence, and map distances to predict the number and types of offspring that should be observed for a particular cross. The expected double crossovers equal the distance between the first and second genes multiplied by the distance between the second and third genes multiplied by the total number of offspring. The number of observed double crossovers equals their frequency (as determined from the coefficient of coincidence) multiplied by the total offspring expected. To find the number of organisms due to a single crossover between the first two genes, perform the following:

$$\text{Double Crossover Frequency} = \frac{\text{map distance between first two genes}}{\text{total number offspring}}$$

Repeat for single crossovers between the second two genes. Subtract all crossovers from the total to find the number of parental types.

In the case of *Neurospora* (e.g., bread mold), the ascospore order can help us map genes on a chromosome. Since meiotic products remain in order in the ascospore, crossing-over will be reflected in a disruption of that order. If recombination occurs between the centromere and the gene being mapped, there will be a 2:2:2:2 order. The frequency of recombinants divided by the total gives the map distance to the centromere.

For human gene mapping, pedigree analysis can be used when the matings are between heterozygotes and recessive homozygotes (or hemizygous males). Other mapping techniques depend upon knowledge of modern molecular techniques. Somatic cell hybridization—the fusion of human and mouse cells—results in the loss of most human chromosomes during their slower mitotic division. By correlating the remaining human chromosomes or fragments (as determined by karyotyping) with human proteins, the responsible genes can be mapped. Recombinant DNA hybridization probes are created from cloned genes. Hybridization with fragments of human chromosomes indicates location.

As mentioned before, the recombination frequency can be used as a way of determining map distance. To determine whether frequency correlates with actual distance, we have to look at the experimental results. Genes have been mapped on *Drosophila* chromosomes using the difference in appearance of the genes in different allelic form. These cytologically determined distances are not the same as the distances determined by recombination frequencies. The differences have been attributed to variations in the crossover frequency along the chromosome length.

Practice what you have learned about mapping and Mendelian inheritance by working through these problems.

PROBLEM

In a given organism, two pairs of contrasting genes are under investigation: A vs. a and B vs. b. An F_1 individual resulting from a cross between two homozygous strains was testcrossed, and the following testcross progeny were recovered:

Phenotype	Number
$A\ B$	621
$A\ b$	87
$a\ B$	92
$a\ b$	610

(a) Are these two genes linked or independent?

(b) If linked, what is the amount of recombination that has occurred between them?

(c) What are the genotypes of the original homozygous strains?

Solution

(a) The determination of linkage or independence can be made very readily. The expected distribution of testcross progeny for independent gene pairs is a 1:1:1:1 ratio for each of the phenotypic classes. A glance at the data reveals that they are not in a 1:1:1:1: ratio; therefore, the two genes must be linked together in the same chromosomal unit.

(b) Having determined that the genes are linked, we can find the amount of recombination that has occurred by using the formula

$$\text{\% recombination} = \frac{\text{total number of recombinant progeny} \times 100}{\text{total number of textcross progeny}}$$

The recombinant progeny are represented by the phenotypic classes that have the lesser numbers. Looking at our data, we see that the classes *Ab* and *aB* have the lowest number of individuals, and are therefore the recombinant classes.

Thus, % recombination = 179/1410 $\times$ 100 = 12.7%.

(c) The genotypes of the original homozygous strains can be determined by finding the parental progeny that are represented by the phenotypic classes that have the larger numbers. These are the classes *AB* and *ab*, since these represent the original combinations of the two genes, the genotypes of the parent stocks were <u>A B</u> <u>a b</u> and A B a b.

PROBLEM

In fruit flies, black body color (*b*) is recessive to the normal wild-type body color (*b*$^+$). Cinnabar eye (*cn*) is recessive to the normal wild-type eye color (*cn*$^+$). A homozygous wild-type fly was mated to a fly with black body and cinnabar eyes. The resulting heterozygous F_1 fly was mated to a fly with black body and cinnabar eyes. These were the results in the offspring:

90 wild-type

92 black body and cinnabar eyes

9 black body and wild-type eyes

9 wild-type body and cinnabar eyes

What is the map distance between the gene for black body and the gene for cinnabar eyes?

Solution

If genes are on the same chromosome, they are said to be linked because they will tend to be inherited together. If linkage is involved, genes will tend to remain in the original parental combinations. Combinations unlike either of the original parents will tend to be much less frequent than would be expected from independent assortment. In this example, the categories containing only nine individuals represent individuals unlike either of the original parents. These types resulted from crossing over or exchange of chromatid segments between the two gene loci. The amount of crossing over is proportional to the distance between genes. The map distance is determined by dividing the number of crossover types by the total number of offspring.

This is converted to percent by multiplying by 100; moving the decimal point two places to the right. One percent of crossing over equals one map unit. These two gene loci are, therefore, nine map units apart.

PROBLEM

In *Drosophila,* the genes black (*b*) and vestigial (*vg*) are 20 chromosome map units apart. In an original cross between black, normal-winged females and normal-bodied, vestigial males, F_1 flies were recovered. If the F_1 flies are intercrossed, predict the phenotypic classes and the number of flies in each, if 1,500 F_2 progeny were classified.

Solution

Since we know the distance between two loci, we can predict the frequency of both the recombinant and the parental gametes produced. From this, we can predict the progeny genotypes and phenotypes, including their respective frequencies in the following manner:

The F_1 females will produce the parental gametes *b*$^+$ and +*vg* at a frequency of 0.4 each, and will produce the recombinant gametes *b vg* and + + at a frequency of 0.1 each. Since there is little or no recombination in the male of *Drosophila*, the F_1 males will only produce the parental gametes at a frequency of 0.5 each.

We now use the Punnett square to complete our prediction.

♀ \\ ♂	$b+$	$+vg$
$0.4\ b+$	$\dfrac{b+}{b+}$ 0.20	$\dfrac{b+}{+vg}$ 0.20
$0.4\ +vg$	$\dfrac{b+}{+vg}$ 0.20	$\dfrac{+vg}{+vg}$ 0.20
$0.1\ ++$	$\dfrac{b+}{++}$ 0.05	$\dfrac{+vg}{++}$ 0.05
$0.1\ b\ vg$	$\dfrac{b+}{b\ vg}$	$\dfrac{b\ vg}{+vg}$ 0.05

Table 3.9.

Summarizing the phenotypes that will result from this cross, we have the following:

$++$	0.50
$b+$	0.25
vg	0.25
$b\ vg$	0

Thus, we see that phenotypic distribution will be 2:1:1:0 because of the failure to observe recombination in the male. We can now convert these frequencies into numbers simply by multiplying them by the total number of flies recovered.

$++$	$(0.50) \infty 1{,}500$	$=$	750
$b+$	$(0.25) \infty 1{,}500$	$=$	375
$+vg$	$(0.25) \infty 1{,}500$	$=$	375
$b\ vg$		$=$	0
Total			1,500

PROBLEM

The gene *r*, for rosy eyes, is 12 map units away from the gene *k*, for kidney-shaped eyes. Both of these genes are recessive to their wild-type alleles. If a heterozygous wild-type fly, resulting from a cross between a homozygous wild-type fly and a fly with rosy, kidney-shaped eyes, is crossed to a fly with rosy, kidney-shaped eyes, what will be the types of gametes and the frequencies of each?

Solution

The cross is shown in Figure 3.4.

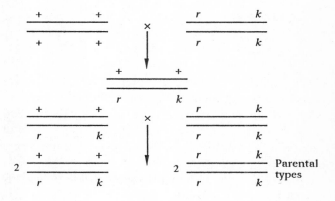

Figure 3.4.

Since the distance between the two loci is 12 map units, 12 percent of the progeny will undergo recombination to produce additional types of gametes.

Since crossing-over is a reciprocal exchange there should be two crossover types of equal frequency for both parental and recombinant classes.

The remaining 88 percent are parental types.

Thus, our gametes will be as follows:

$$\left.\begin{array}{l} 0.44 \; + + \\ 0.44 \; r \; k \end{array}\right\} \text{Parental types}$$

$$\left.\begin{array}{l} 0.06 \; r \; + \\ 0.06 \; + \; k \end{array}\right\} \text{Recombinant types}$$

PROBLEM

The actual physical distances between linked genes bear no direct relationship to the map distances calculated on the basis of crossover percentages. Explain.

Solution

In certain organisms, such as *Drosophila*, the actual physical locations of genes can be observed. The chromosomes of the salivary gland cells in these insects have been found to duplicate themselves repeatedly without separating, giving rise to giant bundled chromosomes, called *polytene* chromosomes (look in section 3.2). Such chromosomes show extreme magnification of any differences in density along their length, producing light and dark regions known as banding patterns. Each band on the chromosome has been shown by experiment to correspond to a single gene on the same chromosome. The physical location of genes determined by banding patterns gives rise to a physical map, giving absolute distances between genes on a chromosome.

Since crossover percentage is theoretically directly proportional to the physical distance separating linked genes, we would expect a direct correspondence between physical distance and map distance. This, however, is not necessarily so. An important reason for this is the fact that the frequency of crossing-over is not the same for all regions of the chromosome. Chromosome sections near the centromere regions and elsewhere have been found to cross over with less frequency than other parts near the free end of the chromosome.

In addition, mapping units determined from crossover percentages can be deceiving. Due to double crossing-over (which results in a parental type), the actual amount of crossover may be greater than that indicated by recombinant type percentages. However, crossover percentages are nevertheless invaluable because the linear order of the genes obtained is identical to that determined by physical mapping.

2. Chromosomes and Karyotypes

Structure and Composition

A **chromosome** is a single DNA molecule, composed of one long double helix (double stranded) and containing all (prokaryotes with one chromosome) or a part of the genome (eukaryotes, more than one chromosome, and visible during cell division). A DNA molecule can be several centimeters long yet is packaged into the nucleus or into a very small prokaryotic cell. It is believed that by complexing with histone proteins and forming "bead-like structures" called *nucleosomes*, the potential entanglement during cell division and maintenance is diminished. A eukaryotic chromosome is composed of about 30% DNA, a third is histone proteins, another 30–33% nonhistone-binding proteins, and the rest RNA. There are five types of histones: $H1$, $H2A$, $H2B$, $H3$ and $H4$. Each of these types of histones can be modified by methylation, ADP-ribosylation or phosphorylation. These *posttranslational modifications* (after synthesis of polypeptide) may be important in regulating the availability of DNA for replication and transcription by changing the charge or hydrogen-bonding capabilities.

Putting all these constituents of a chromosome together yields what is called *chromatin* (early cytologists noted the dark staining material when observing tissues with a microscope). Eukaryotic chromosomes may be further distinguished by small knobs at the tips called "satellites" and differential banding patterns which can act as markers for karyotyping (see previous sections, and below), mapping and identifying abnormalities.

Karyotype

Eukaryote species have a characteristic number of chromosomes. For example, each chromosome pair in a diploid species has distinguishing traits, including overall size, location of centromere, satellites (knobs on the end of a chromosome arm) and banding patterns due to differential staining of heterochromatin and euchromatin. Techniques are used that permit photos to be made, or using other imaging methods, so that each chromosome can be identified. An ordered, pictorial array is called the *karyotype* of a species. The human karyotype consists of the 22 pairs of autosomes and the pair of sex chromosomes, displayed from largest to smallest and centromere position. The location of the centromere in a chromosome provides further descriptors: *metacentric* means the centromere is near the center of the chromosome and the two arms are approximately equal, *acrocentric* has the centromere more toward one end or the other, and when *telocentric,* the centromere is near the tip of one arm.

Chromosomes are prepared with special stains (dyes that complex with the nucleic acids and proteins making them visible) so that their morphology can be identified. Because of the components, named above, and their varying proportions, differential staining occurs.

The general features of human chromosomes are arranged from large, metacentric to small acrocentric. There are no telocentric human chromosomes. Recall that acrocentric chromosomes have one arm that is longer than the other: the long arm is called the *q arm*, the shorter one the *p arm*. The regions called *heterochromatin* are considered to be more involved with maintaining the chromosome structure and shape while the *euchromatin* regions are associated with protein coding sequences. That is why breaks or other abnormalities are seen to have more severe effects when they occur in the euchromatic regions of a chromosome.

Chromosomal Aberrations

Since chromosomes contain genetic material, it is essential that the correct number and types be present for normal growth and development. During cell division, single DNA molecules coil around major chromosomal components, basic proteins called **histones**. These condensed DNA-histone complexes are visible in the cell as replicated chromosomes, each containing two chromatids.

When studying chromosome organization, remember that one chromosome contains one duplex DNA molecule. The linear arrangement of its genes is reflected in the characteristic banding pattern evident when the cell is stained. Changes in chromosome organization, such as *duplications, deletions, translocations, and inversions* are detectable by examining the banding pattern.

The basis for understanding chromosomal rearrangements is in knowing that chromosomes regularly break, producing sticky ends that are likely to reattach, either restoring the original arrangement or creating any number of rearrangements.

We can analyze the causes and consequences of chromosomal rearrangements by assigning letters in sequence to designated consecutive regions of a chromosome. Uneven crossing over of homologous chromosomes during prophase I of meiosis can result in duplication and deficiency. **Inversions** occur when a broken segment flips around and reinserts itself in reverse order. **Translocations** involve the movement of one chromosome or fragment to a nonhomologous chromosome. When chromosome breaks occur, it is possible to predict how reattachments will occur by recombining "sticky" ends in all possible combinations.

If crossing over produces one chromosome with no centromere and another with two, the "centromere-less" fragment cannot migrate in the cell and will be lost. In addition, the chromosome containing two centromeres will break as it is pulled to opposite poles during cell division.

The consequences of abnormal chromosome arrangement are determined empirically. Losses of *essential* genetic material result in an abnormal phenotype or failure of a fertilized egg (**zygote**) to develop. Duplications in genetic material may disrupt development (as in Down's syndrome) or create the raw material for evolution, and re-organization of genetic material may affect development if the position of the genes influences their functions.

Abnormal chromosome number may be due to the presence of multiple sets of chromosomes (**euploid**), or an alteration in a portion of a chromosome set (**aneuploid**). Two explanations for euploid—multiple fertilization of an ovum or mitosis without cytokinesis—are more easily understood if diagrammed.

Aneuploids, which are due to nondisjunction, are best visualized by diagramming chromosomes during metaphase of mitosis or meiosis. Attach spindle fibers from centromere to pole and draw the chromosomes as they move poleward. Be sure that one set of chromatids in mitosis or meiosis II, or one pair of homologous chromosomes in meiosis I, do not separate during anaphase, but move to one pole or the other.

The consequences of abnormal chromosome number are also determined empirically. To follow chromosome number through reproduction and development, diagram each stage, showing the chromosome count. Indicate chromosome sets as one set = n (haploid), two sets = $2n$ (diploid), three sets = $3n$, and so on. For aneuploids, designate the addition of a single chromosome as +1, of a pair of homologous chromosomes as +2, and the addition of two different chromosomes as +1+1. The loss of chromosomes is indicated by a negative number.

Changes in chromosome number and kind may alter development and viability, as well as provide the raw material for evolution. These changes can be correlated with particular phenotypes.

Aneuploidy in humans, for example, can be due to a translocation between two human autosomes. It is the genetic basis of Down's syndrome (there are other forms of aneuploidy in humans). About 10% of the cases of Down's syndrome are caused by a translocation. The translocation chromosome arises as a fusion between chromosomes 14 and 21. A woman who has such a translocation has a normal chromosome 14 and a normal chromosome 21. She also has the translocat-

ed chromosome which is designated $t(14;21)$. She will be phenotypically normal since she has a full diploid set of chromosomes (i.e., genetic information for normal function). However, in meiosis, synaptic pairing is asymmetric and hence the orientation of the spindle can become abnormal. Such a woman may produce eggs that have chromosome 14 but not chromosome 21, 21 but not 14, 21 and $t(14;21)$, or 14 and $t(14;21)$. If an egg with the constitution 21 and $t(14;21)$ is fertilized by a normal sperm containing both chromosomes 14 and 21, Down's syndrome will be the result. The child will have the three copies of chromosome 21 necessary to display Down's syndrome; the karyotype is trisomic for chromosome 21.

The chromosomes in a translocation heterozygote consist of two original chromosomes, *ABCDEF* and *GHIJKL*, and two translocated chromosomes, *ABCJKL* and *GHIDEF*. We can diagram and explain the viable and lethal chromosome combinations, as follows.

The homologous parts of a chromosome will associate with each other. As shown in Figure 3.5, such an association gives the gene sequence a definite order in a ring-shaped structure. At anaphase I of meiosis the ring is broken and two chromosomes move to each pole of the spindle. If the chromosomes were distributed at random, six different types of gametes would form. The only sets of gametes that would be viable would be the ones that had

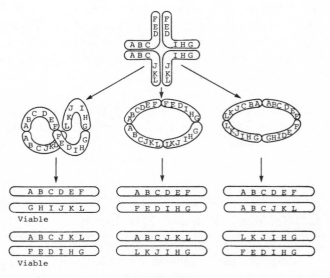

Figure 3.5.

the complete genetic sequences: *ABCDEF, GHIJKL* and *ABCJKL, FEDIHG.* The other types of gametes would be unbalanced and hence nonviable since they do not have a complete set of the genetic information. For instance, in the gamete set *ABCDEF, ABCJKL,* the sequence *ABC* occurs twice while the sequence *GHI* is missing entirely. The other gamete types are lethal for similar reasons.

Aneuploidy (trisomy) can also be demonstrated using the plant *Datura stramonium* (jimson weed). Consider the F_1 phenotypic ratio in the jimson weed, when a purple female (*PPp*) is crossed with a purple male (*PPp*). This problem involves a cross between two individuals that have three alleles for the color of their flowers. Since they have three alleles, they probably have the rest of the chromosomes to go with them. They are, therefore, trisomic. Jimson weed plants that are trisomic in their flower color allele have an extra copy of chromosome 9. In this gene purple is dominant over white. But since these are trisomics, the alleles will not segregate in the usual 3:1 Mendelian ratio, as will be seen. The female plant produces *P, Pp, PP,* and *p* megaspores (female gametes). The male, on the other hand, only produces functional *P* and *p* pollen; pollen with more or less than the normal 12 chromosomes will be nonfunctional.

The cross is shown in Figure 3.6 below.

PPp ♀ x PPp ♂

gametes: P, P, Pp, Pp, PP, p   P, P, p

Male gametes		Female gametes			
		2P	2Pp	PP	P
	2P	PP 4	PPp 4	PPP 2	Pp 2
	p	Pp 2	PPp 2	PPp 1	pp 1

Figure 3.6.

By counting the numbers of similar progeny, a genotypic ratio of 4*PP*:4*Pp*:5*PPp*:2*Ppp*:2*PPP*:1*pp* is obtained. The presence of one or more *P* alleles is expressed phenotypically as purple flowers. White flowers are only expressed in plants homozygous for the recessive *p* allele. Thus, the phenotypic ratio is 17 purple:1 white.

Polytene and "Lampbrush" Chromosomes

Polytene chromosomes in the salivary glands of *Drosophila* spp. (fruit flies) can be easily observed using light microscopy; besides cells in the salivary glands, other cells having polyteny include ovary nurse cells, follicle cells (near oocytes), abdominal histoblasts, fat body cells, and gut cells.

The four pairs of *Drosophila* chromosomes synapse during interphase prior to cell division and replicate without completing the cell (called endoreduplication) division cycle producing between 1000–2000 copies all oriented parallel to one another. When stained and karyotyped, distinct and reproducible banding patterns are plainly visible and they appear as if they are very large haploid cells, the homologues being in synapsis. Because fruit fly species produce large numbers of progeny from single pair matings and their generation times are measured in hours, they are ideally suited for genetic analyses and particularly the relationships between chromosomal abnormalities and gene mutations. The knowledge gained from these species has been ported over to many other organisms. There are thousands of recognizable bands in the Polytene chromosomes. *In situ* hybridization using cloned DNA of identified genes to banded polytene chromosomes help locate specific gene sequences within the chromosome and these are then correlated with banding patterns. However, nearly a third of the *Drosophila* genome is heterochromatin, and these regions are underrepresented in polytene chromosomes because these regions do not undergo endoreduplication.

"Puffing" of the polytene chromosomes describes structural changes that appear to be correlated with transcription and translation of proteins during development of the organism (e.g., appears to be induced by hormone ecdysone).

It is worth also noting that *Drosophila*, while having the most famous and researched polytene chromosomes, is not the only genus having them. Among other genera are *Anopheles* (mosquitos), and even plants (e.g., *Phaseolus* [bean]).

Another example of regulation of genetic activity at the chromosome level is exemplified by *lampbrush* chromosomes (LBC). Lampbrush chromosomes are transitory structures that exist during the first meiotic division in female gamete cells of most animals, *except* mammals (for more on this, please see *http://www.ex.ac.uk/ lampbrush/intro.htm*). LBCs were first seen in sections of salamander (*Ambystoma mexicanum*) oocytes by Flemming in 1882. The chromosomes go from a compact telophase form at the end of the last oogonial mitosis, become "lampbrushy" (feather-like tendrils projecting from the axis of the chromosome) and then contract again to form normal first meiotic metaphase bivalents (tetrads). Their most conspicuous feature is widespread RNA transcription from thousands of transcription units that are arranged at short intervals along the lengths of all the chromosomes. In these senses it has been possible to exploit LBCs in the study of chromosome organization and gene expression during meiotic prophase, and in studies of molecular RNA transcription.

3. Genome Sequence Organization

Introns and Exons

Genetic information is organized and acted upon differently in prokaryotes and eukaryotes. The molecular definition of a gene includes both the nucleotide sequences (DNA [coding region] and RNA) but also the sequences needed in order to get an mRNA transcript made that will produce a functional polypeptide.

While exceptions have been found, a majority of the eubacteria that have been analyzed contain only uninterrupted coding regions for a protein. Some archaebacteria have been found with introns in genes that encode information for synthesizing transfer RNA. Eubacteria also generally have genes located next to one another (contiguous) that are required for the transcription/regulation and translation of a protein. These transcription units are referred to as *operons* and the multi-gene sequences as *polycistronic units*.

Eukaryotes, on the other hand, commonly contain introns and the transcript message is processed prior to leaving the nucleus and the resulting functional message to be translated is made up of exons (i.e., coding regions that result in a functional protein, or sequences that are both transcribed and translated; commonly these are monocistronic having only a transcript for one protein). However, eukaryotes can have complex transcription units with multiple splice sites associated with overlapping exons. Transcription control sequences and genes a far distance from the protein sequence, and even in different compartments, can do the regulation of eukaryotic genetic transcription and translation (e.g., interactions between nuclear and organelle genes).

Single-Copy and Repetitive DNA

The majority of eukaryotic DNA in the genome of multicellular organisms is not in unique sequences (solitary genes) but occurs in some sort of multi-copy form. Estimates run between 25–50% of the protein coding DNA is in unique sequences but the remaining protein coding is in families of two or more similar genes (e.g., actin genes, important to the cytoskeleton, may contain between 5 and 30 proteins in a family). It is believed that gene families arise from duplications that can occur from unequal crossing over during meiosis I. So-called pseudogenes are duplications that have, through evolution, become nonfunctional if transcribed. Histone proteins, rRNA and tRNA genes are found in tandem repeated arrays of identical genes; it is believed that many copies of these are needed for very rapid and simultaneous, transcription and translation of proteins at many disparate genes.

Early DNA research uncovered some interesting properties that were applied to characterize similar sequenced DNA. After denaturing double stranded DNA by breaking it up into smaller, single stranded fragments then estimating the time and amounts of reassociation, scientists were able to classify it and make inferences about the degrees of sequence similarities. The reassociation fractions were called: (1) rapid reassociation or simple sequence DNA, composing ~10–15 % of DNA, in this case, in mammalian cells; sequences were 5–10 base pairs in tandem, arrays; (2) intermediate reassociation or intermediate repeat DNA, 25–40%, many copies of a few sequences of gene families; and (3) slow reassociation or single-copy DNA and from this it was estimated that only about 5–10 % of total human DNA codes for the unique functional proteins.

Transposable Elements

Intermediate-repeat DNA is scattered throughout the genome (contains two subclasses: short and long interspersed elements) and may be copied thousands of times in the genome. Many of these repeats can move or transpose to new sites within the genome. They are called transposable elements or mobile DNA elements. These repeated sequence appear to have little if any function and were called "selfish DNA" by Francis Crick because apparently they only exist to perpetuate themselves; it has been suggested that they are a repository of genetic information that may evolve into new useful genes. In humans, these mobile elements may account for as much as 30% of the genome. The first mobile elements were discovered by Barbara McClintock, who was studying genetics of corn (maize) in the 1940s and later their existence was clearly genetically demonstrated in *E. coli*. It is accepted now that transposable elements move about by one of two mechanisms: (1) as DNA alone or (2) as a double-stranded DNA sequence arising from an RNA message that has been acted upon by reverse transcriptase. When by the first method, these elements are called *transposons* (some have called them "jumping genes"), the second type are called *retrotransposons* (more about retroviruses and reverse transcriptase in a later section).

4. Genome Maintenance

DNA Replication

The structure of DNA is best appreciated by looking at the functions that it performs:

1. It must be able to replicate itself, which requires strand separation and the synthesis, via DNA polymerase, of a complementary daughter strand from each parent strand.

2. It must contain the information necessary for proteins to be synthesized that are able to direct the growth, development and maintenance of an organism.

Recall the simplicity of the DNA molecule: it is a duplex (double stranded) helical, polymer molecule composed of a phosphate: 5-carbon sugar (deoxyribose) backbone containing four nitrogenous bases, adenine (A), thymine (T), guanine (G), and cytosine (C). Each sugar-phosphate-base combination is called a nucleotide. Its integrity is largely dependent upon the cross-bonding between the two complementary strands of bases: A bonds with T, and G with C; more energy is required to separate the G-C pairs than the A-T pairs because the G-C have three hydrogen bonds while the A-T have only two.

The base composition of the genome is characteristic for each species. Also, because of complementarity, if we know the percentage of one base, we can calculate that of all others. For example, if a molecule of DNA has 20% adenine, we know also it has 20% thymine and thus, $(100 - 40\%) = 60\%$, remains as G-C (30% of each).

The process of replication is also based upon complementarity because when the strands uncoil to synthesize new strands, only A and T bond, and G and C bond in the new molecules.

Replication, as with other chemical reactions, is under enzymatic control. Some of the more important enzymes along with their functions are listed in Table 3.10 and diagrammed in Figure 3.7.

For DNA replication to occur, the two strands of the double helix must separate (acted upon by helicase). Each separated strand serves as a template for DNA polymerase in the synthesis of a new strand. Because of the rules of nitrogenous base pairing (i.e., A pairs only with T, and G pairs only with C), the sequence along the old strand dictates precisely the sequence along the new strand. The point where they are separated is called the replication fork in which both new strands are synthesized in a 5' to 3' direction. This results in continuous synthesis for the leading strand but discontinuous synthesis for the lagging strand (see Figure 3.7). Discontinuous synthesis results in Okazaki fragments, which are later joined to form a continuous strand. Each new double-stranded DNA molecule contains one old strand and one new strand; thus, the process of DNA replication is said to be semiconservative.

DNA Mutation and Repair

Chromosomal aberrations, like translocations discussed above, are obvious forms of *mutation or genetic change* that causes deviation from the normal or "wild type" phenotype; it is important to remember that not all mutations are "bad." More subtle, but sometimes no less damaging, are changes caused by simple point mutation, or a few base changes that can cause the incorrect amino acid to be inserted into a protein causing a change or loss of function. These are obviously not all deleterious because the genetic code is redundant with respect to more than one codon for each of the 20 amino acids. When you consider the many times that DNA is replicated, RNA transcribed, translated and tailored, it is quite amazing that we do not see more mutations than we do. Actually, there are more than we see because evolution has built in some very neat ways of repairing damage (luckily for us and many other plants and animals). Finally, many mutations that do not get repaired are never observed because they cause early death or abortion of the embryo. In this section you will also be reminded that there are physical and chemical factors (mutagens) that cause genetic change.

Mutations are changes in the sequence of A, T, G, and C along a DNA strand. Mutations are changes in the genotype; they may or may not change the phenotype (the observable characteristics). Only those mutations that occur in the germ-line cells, which may become gametes, can contribute to the next generation. Somatic mutations, while not uncommon, are not inherited in the progeny.

A *silent* mutation does not alter the phenotype. Mutations may be neutral, harmful, or beneficial in their

**Table 3.10.
Sequence of Enzymes and Functions
During DNA Replication.**

Enzyme	Function
Helicase	Unwinds and holds apart DNA section to be replicated
Primase	Begins new DNA molecule with RNA primer
DNA polymerase	Brings in new DNA nucleotides
Exonuclease	Removes RNA primer
Ligase	Seals sugar-phosphate backbone

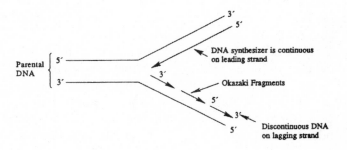

Figure 3.7.

effect. Spontaneous mutations that occur naturally appear to be due to random errors in base pairing during DNA replication. Induced mutations occur as a result of exposure to a mutagen that is a chemical substance or physical factor that increases the mutation rate; it causes permanent damage to the DNA.

Mutation rate is the probability that a mutation will occur in a gene when the cell divides. Different genes have different mutation rates. Point mutation (base substitution) is a change in a single nucleotide; this may result in no change in amino acid (silent mutation), substitution of a different amino acid (missense mutation), or the creation of a "stop" codon (nonsense mutation) that causes premature termination of a functional protein. A frameshift mutation is when a base pair is inserted or deleted; this generally affects the mRNA from that point on (i.e., it can change all triplets for the rest of the sequence). The changes may include silent, missense, and nonsense mutations. Frameshift mutations are more likely to result in a deficient protein product than are point mutations.

In controlled settings, mutants can be identified by selecting or testing for an altered phenotype. Positive selection is when mutant cells are selected and nonmutant cells are inhibited. Negative selection, using the replica plating technique, is used to identify mutants that cannot grow under the selective conditions (i.e., some specific growth factor is omitted). For example, a histidine auxotroph is a bacterium that cannot grow unless histidine is added to the medium.

The following are examples of chemical *mutagens*: base analogs such as 2-aminopurine, which incorporates in place of thymine; base-pair mutagens such as nitrous acid, which converts adenine into hypoxanthine. Others include alkylating agents, deaminating agents, and acridine derivatives. Ionizing and ultraviolet radiation are also mutagens. Ionizing radiation can cause base substitutions, disrupt the sugar-phosphate backbone, or create reactive free radicals. Ultraviolet radiation causes thymine dimers (i.e., bonding between adjacent thymines).

DNA repair is facilitated by specific repair enzymes. There are two kinds of repair: light repair, in which a light-activated enzyme breaks thymine-thymine bonds; and dark repair, in which several different enzymes are involved in excising defective DNA and resynthesizing the DNA strand based on the nonmutated strand.

Carcinogens (chemicals that cause cancerous cell growth) can be identified by specially design protocols and using certain strains of bacteria. One of these is the Ames test for the identification of carcinogens. Many mutagens are also carcinogens. The test involves exposure of histidine auxotrophs of *Salmonella* to a suspected mutagen, followed by selection for nonmutant cells. The presence of nonmutant *Salmonella* indicates a positive test for mutagenicity. The Ames test is quick and relatively inexpensive. Recall that an auxotroph is

an organism/clone unable to synthesize specific chemicals necessary for growth (i.e., histidine).

Regulation of Gene Expression in Bacteria

Constitutive enzymes are those that are always present (i.e., being transcribed and translated). **Constitutive genes** are those that continue to produce proteins regardless of other factors, including the concentrations of substrate and end product. For example, most glycolytic enzymes (and the genes that produce them) are constitutive.

Other enzymes are regulated at the genetic level and are described as **inducible**. Regulatory mechanisms such as **induction**, **repression**, and **attenuation** are ways to control the activity of genes to determine which mRNAs are synthesized and, therefore, which proteins will be made.

The Operon Model. An operon consists of three segments: a *promoter*, an *operator*, and *structural genes*. When the repressor, which is the protein that binds to the operator, is removed, transcription can take place. Gene expression occurs once transcription of the structural genes is set in motion.

Enzyme Induction. In the *operon model* for an inducible system, the presence of an inducer activates an operon and the cell synthesizes more enzymes. In some cases, the inducer binds to the repressor so that it cannot bind to the operator and prevent transcription. Beta-galactosidase and the *lac operon* is an example of an inducible system (see Figure 3.8). Beta-galactosidase is an enzyme involved in the metabolism of lactose. Lactose is the inducer for the operon controlling production of beta-galactosidase. When lactose is absent, a repressor is produced that binds to the operator and inactivates the operon—beta-galactosidase is not produced. When lactose is present, it inactivates the repressor, thus allowing transcription to occur and beta-galactosidase is produced. The *lac operon* can be under both negative and positive control. Negative control comes from substances that inhibit gene activity, positive control arises from those that promote activity (see regulator genes below).

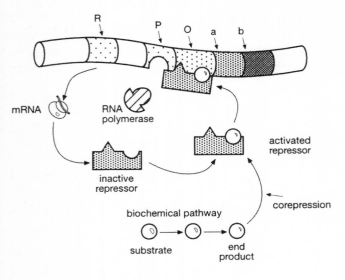

Figure 3.8.

Enzyme Repression. In the operon model for a repressible system, a repressor binds to the operator and prevents transcription; in some cases the repressor cannot bind to the operator site without a corepressor, and it is the presence or absence of the corepressor that controls synthesis of a message, then protein. The repressor is often a synthetic product that inhibits further production of the enzyme responsible for its synthesis. Tryptophan and the trp operon illustrate enzyme repression. When tryptophan is present, it attaches to and activates a regulator protein that represses the trp operon. When tryptophan is not present, the repressor is not activated and transcription of the trp operon can occur. Catabolite repression—the presence of glucose (or some other preferred nutrient) represses synthesis of the enzymes necessary to metabolize an alternative substance.

Attenuation. Attenuation is a form of regulation wherein mRNA synthesis is terminated at a point called the attenuator site.

In addition to structural genes that code for proteins, or transfer and ribosomal RNA, ***regulator genes*** also exist. These regulator genes either enhance the

transcription of structural genes (*positive control*) or inhibit transcription (*negative control*).

When looking at regulation in prokaryotes, keep in mind that the structural genes for a particular biochemical pathway are located in sequence next to each other. You should also be familiar with the regulatory genes:

1. The repressor gene produces a repressor protein that binds on the operator to inhibit transcription. When the substrate to be catabolized (think back to Chapter 2) by the gene product is present, it binds with the repressor protein and prevents repressor inhibition of transcription. For anabolic reactions, the repressor is inactive until it combines with a metabolic end-product; the active repressor then binds to the operator preventing further synthesis.

2. The operator site is located next to the structural genes. When bound to the repressor protein, RNA polymerase cannot bind. Thus, transcription cannot occur. The operator has its effect on the structural genes distal to it on the same chromosome.

3. The promoter region is located next to the operator site, and is the attachment site for RNA polymerase. Positive control occurs when components enhance gene activity; enhancers are DNA sequences that bind transcription factors, also called enhancer-binding proteins. The enhancer complement of a gene determines both its level of expression, and in eukaryotes, its tissue specificity. Hormones, and special proteins like catabolite activator protein (CAP) and cyclic AMP (cyclic adenosine monophosphate) enhance gene transcription. (see Figure 3.9). The promoter region has its effect on its neighboring operator and structural genes.

Evidence also exists for regulation in viruses. Regulation in lambda phage determines whether it will undergo lysis or lysogeny. Lysis depends upon the production of new protein coats and replication of the chromosome, while lysogeny involves the incorporation of viral DNA into the host genome. In this viral genome, DNA strands function as sense strands, one reading to the right and the other to the left.

When solving problems involving prokaryote regulation, draw a map of the structural and regulatory genes, and list all symbols for mutants with their description. The following symbols are useful to know:

+ = wild type,

− = mutant,

c = constitutive (always functioning).

Genetic Regulation in Eukaryotes

Understanding how genes are transcribed and proteins made in multicellular, multitissue, multiorgan eukaryotes is not so straightforward. Think for a moment about the "construction" of us humans, from our beginning as a two-celled zygote to a multibillion-celled organism (estimates are ~10^{12} cells in a human). And, in order to function satisfactorily, we must start and stop growing at specific times, must continuously have genes turned on and turned off, and all of these events are based upon highly organized signal processes that originate both internally and externally.

In addition to the previously described chromosome indicators of regulation of transcription (puffing and lampbrush), the "heat shock response" is another example of the modes of regulation of gene transcription in eukaryotes. Heat shock is a response found in both prokaryotic and eukaryotic cells and many of the genes are highly conserved across species (e.g., a gene called Hsp70 in *E. coli* has about half the same sequence as humans, and the gene in *Drosophila* has ~85% identity with humans). When the temperature is raised, the cells stop transcribing the genes that were previously active and instead transcribe a new set of genes called heat shock genes. Translation of the original genes also stops and the heat shock genes are preferentially translated. This event occurs in *Drosophila*, yeast, corn, and cultured mammalian cells. *Drosophila*'s giant (recall polytene chromosomes) salivary gland chromosomes show nine new chromosomal puffs as a result of high temperatures.

Again, referring to gene enhancers and transcription factors, many enhancer binding proteins are activated (i.e., transcribed into mRNA) or inactivated (no transcription) during cell signaling, which makes the gene responsive to a hormone. Different genes in a cell can be turned on by different hormonal stimulation by virtue of different enhancers for different genes. The same hormone can induce the expression of different genes in different target cells as a result of differences in the enhancer-binding-protein population within various cell types.

RNA and Protein Synthesis

The linear sequence of every protein a cell produces is encoded in the DNA of a specific gene. But DNA does not make proteins directly—it can only encode the synthesis of RNA or copies of itself.

The fundamental characteristic of the genetic code (see Table 3.11), is that it is a triplet code with three adjacent nucleotide bases, termed a codon, specifying each amino acid. The three nucleotides specify one of the 20 common amino acids.

The mRNA codons make up the **genetic code,** which is essentially identical among all living things. The code comprises 64 codons—61 code for amino acids (these are the **sense codons**), while 3 do not code for amino acids but function as "stop" signals for the translation process (these are the **nonsense codons**). The 61 sense codons code for 21 amino acids; thus, some amino acids are coded for by more than one codon. Hence, the code is said to be **degenerate**. The **start codon** is **AUG** and codes for the amino acid methionine.

It was once believed that the genetic code that was first deciphered for prokaryotes held for all species. Work in recent years has pointed to important differences in the code among species and most particularly, in the mitochondrion genome; recall that the chloroplast also has a small genome, but it appears to use the "standard" code, and some proteins are synthesized in the stroma, some in the cytoplasm. For example: the "standard" nuclear/prokaryotic codon for terminating mRNA transcription, the "stop" codon, is UGA. However, in the mitochondrion of mammals, *Drosophila*, Neurospora and yeast, UGA encodes the amino acid trp (tryptophan). In mitochondria of plants, the "standard" nuclear codon for "stop" is used.

Transcription is the synthesis of single-stranded RNA (ribonucleic acid) molecules based on a DNA sequence (the template). The two strands of the DNA must pull apart temporarily, allowing **RNA polymerase** to access the DNA for use as a template for RNA production. RNA is synthesized from adenine (A), guanine (G), cytosine (C), and uracil (U). Thymine is not present in RNA.

Three kinds of RNA may be produced: (1) **ribosomal RNA (rRNA)**, which combines with proteins to form ribosomes (where new proteins are synthesized); (2) **transfer RNA (tRNA)**; which transports amino acids to the ribosome for assembly into proteins; and (3) **messenger RNA (mRNA)**, which dictates the sequence of amino acid assembly.

Transcription occurs in the cytoplasm of prokaryotes (no nucleus), while it occurs in the nucleus of eukaryotes.

Eukaryotic mRNA contains regions that are not used for protein synthesis (the **introns**), as well as those regions that are (the **exons**). Prior to mRNA transport out of the nucleus, enzymes remove the introns, and the **spliceosome** connects the exons into a functional mRNA that is then exported to the cytoplasm.

Translation is the process wherein information in the form of nitrogenous bases along an mRNA is translated into the amino acid sequence of a protein. The sequence of nucleotides along the mRNA is "read" in groups of three (triplets or codons). Codons in mRNA pair with anticodons found in tRNA molecules. The triplet **anticodon** is located at one point on the tRNA molecule while the corresponding amino acid is attached to the tRNA at another point. The mRNA and tRNA are brought together at the ribosome. The ribosome moves along the mRNA strand during the synthesis of the **polypeptide**. A polypeptide is a chain of

Table 3.11.

		Second Position of Codon					
		T	C	A	G		
FIRST POSITION	**T**	TTT Phe [F]	TCT Ser [S]	TAT Tyr [Y]	TGT Cys [C]	**T**	**THIRD POSITION**
		TTC Phe [F]	TCC Ser [S]	TAC Tyr [Y]	TGC Cys [C]	**C**	
		TTA Leu [L]	TCA Ser [S]	TAA Ter [end]	TGA Ter [end]	**A**	
		TTG Leu [L]	TCG Ser [S]	TAG Ter [end]	TGG Trp [W]	**G**	
	C	CTT Leu [L]	CCT Pro [P]	CAT His [H]	CGT Arg [R]	**T**	
		CTC Leu [L]	CCC Pro [P]	CAC His [H]	CGC Arg [R]	**C**	
		CTA Leu [L]	CCA Pro [P]	CAA Gln [Q]	CGA Arg [R]	**A**	
		CTG Leu [L]	CCG Pro [P]	CAG Gln [Q]	CGG Arg [R]	**G**	
	A	ATT Ile [I]	ACT Thr [T]	AAT Asn [N]	AGT Ser [S]	**T**	
		ATC Ile [I]	ACC Thr [T]	AAC Asn [N]	AGC Ser [S]	**C**	
		ATA Ile [I]	ACA Thr [T]	AAA Lys [K]	AGA Arg [R]	**A**	
		ATG Met [M]	ACG Thr [T]	AAG Lys [K]	AGG Arg [R]	**G**	
	G	GTT Val [V]	GCT Ala [A]	GAT Asp [D]	GGT Gly [G]	**T**	
		GTC Val [V]	GCC Ala [A]	GAC Asp [D]	GGC Gly [G]	**C**	
		GTA Val [V]	GCA Ala [A]	GAA Glu [E]	GGA Gly [G]	**A**	
		GTG Val [V]	GCG Ala [A]	GAG Glu [E]	GGG Gly [G]	**G**	

amino acids. A functional protein can be one or more polypeptides.

This process is not the same in eukaryotic cells, in which transcription occurs in the nucleus and translation occurs in the cytoplasm.

Processing and Modifications of Both RNA and Protein

Bacteria simultaneously synthesize protein (translate the mRNA codons from the triplets in the DNA template) while transcription of mRNA is occurring. A **polyribosome** is an mRNA with many ribosomes attached. Due to the absence of a nuclear membrane, prokaryotic translation can begin even before transcription is completed. These processes are more complex in eukaryotes because they are compartmentalized (e.g., transcription occurs in the nucleus and translation [protein synthesis] is in the cytoplasm). These steps are summarized in Table 3.12 and in Figure 3.10 (the steps are similar for prokaryotes and eukaryotes, but the ribosome sizes differ), below:

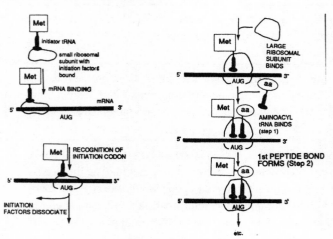

Figure 3.9.

Protein modification. Just as the single stranded mRNA polynucleotide undergoes posttranscriptional modification (above), proteins may go through what is called "posttranslational modification." This covers many different kinds of modifications. Modifications that may happen include cystein bridges (disulfide bond between

two cysteines, modifies 3-D protein structure), leader peptide cleavage, and additions to certain amino acids. Leader peptides appear involved in regulating gene expression and specifying in which cells that occurs.

As soon as a polypeptide is completed and released from the ribosomal complex it begins assuming different shapes depending upon its active form. It may also combine with other polypeptides to make a functional protein (e.g., haemoglobin is composed of four subunits each a polypeptide). "Chaperone proteins" can help stabilize the newly synthesized polypeptide/protein by guiding the folding of the molecule. Folding, like the primary sequence of amino acids, is critical to the normal function of a protein. For example, an incorrectly folded protein (alpha-1-antitrypsin) is involved in persons with a predisposition to one form of emphysema. Similar posttranslation difficulties can cause a form of cystic fibrosis.

5. Gene Expression and Regulation: Effects

Control of Normal Development

We, as humans, are constantly being reminded of the effects of genetic expression. At the personal level, we are confronted every time we look at ourselves in a mirror (i.e., our phenotype is the culmination of our genotype and our environment). Deviations from "normal development" are more than simply looking different from friends, family and the rest of our species; they are the result of genetic and environmental variation. From the time of conception, whether plant, animal or bacterium, the phenotype is under some level of genetic regulation. Almost 70% of all human abnormalities arise during the embryonic phase. Many, those not aborted, are not observed until infancy, maybe early childhood or even middle age (the latter are referred to as late onset). Not all disruptions of normal development in an organism are due to errors caused by gene mutations.

To understand biochemical genetics we must be familiar with how chemical reactions occur in the cell.

Table 3.12.
Steps in Transcription, Translation, and Protein Synthesis in Eukaryotes

Location	Process	Activity
Nucleus	Transcription, synthesizing a mRNA molecule from the DNA template.	Initiation—RNA polymerase attaches to DNA strand at the promoter region.
Nucleus	Transcription, continues.	When transcription is under way ("start" codon is AUG which codes for the amino acid methionine) a "cap" sequence of methyl-guanosyl triphosphate (GTP) is added to the 5888 ends.
Nucleus	Transcription, continues.	Elongation—RNA polymerase adds nucleotides ($3'$ to $5'$ direction of DNA template) to the growing mRNA transcript. On the $3'$ end, a few hundred adenosines are added, so-called "poly A tail" (i.e., polyadenylation); Poly A tail prevents the mRNA being degraded before reaching the cytoplasm.
Nucleus	Transcription, continues.	Termination—"stop" codon signifies end of transcription, the mRNA molecule is cleaved.
Nucleus	Transcription, continues.	Native mRNA (called pre-mRNA) is *not* directly transported from the nucleus to the cytoplasm, it is first processed. Nucleotide sequences that are transcribed but that will not be translated are spliced out of the pre-mRNA molecule, these are called *introns*. They vary greatly in size (<100 to >100,000 bases). Introns are excised by ribozymes that are associated with other proteins to form "small nuclear ribonucleoproteins", a.k.a. "snurps". These enzymes excise introns and also help put together the remaining sequences, called *exons*, which are later translated. Introns are not common to prokaryotes.
Cytoplasm (ribosomal complex)	Translation, synthesis of protein.	The mature, post-processed mRNA transcript moves through the nuclear pores into the cytoplasm where protein synthesis occurs at the ribosome complex.
Cytoplasm (ribosomal complex)	Translation, continues.	Initiation complex is formed: mRNA, small ribosome subunit (30S[1] in prokaryotes, 40S in eukaryotes;) and tRNA[met], bearing met amino acid (methionine, "start" codon, always the first amino acid in polypeptide). In eukaryotes, three initiation factors are required: IF1, -2 and -3.
Cytoplasm (ribosomal complex)	Translation, continues.	Large ribosomal subunit (50S in prokaryotes, 60S in eukaryotes) binds to the initiation complex; the small-large ribosomal complex has two sites to accommodate the tRNA bearing each amino acid: P = peptidyl for linking the peptide bonds and A = aminoacyl for aligning each amino acid with the complex. The total small-large ribosomal complex is 70S in prokaryotes and 80S in eukaryotes[2]. Once the tRNA[met] is in place, the second tRNA arrives with the second aminio acid. The first and second amonioacids are attached by peptide bonds. The first tRNA detaches and leaves the complex.
Cytoplasm (ribosomal complex)	Translation, continues.	The small-large ribosomal complex, with the mRNA molecule between "marches" down the transcript, one codon at a time, adding amino acids to the growing polypeptide until the terminator codon is reached.

[1] S denotes the sedimentation constant expressed in svedbergs; it is a measure of size based upon the shape and density of a molecule.
[2] The "S" numbers are not strictly additive because recall that sedimentation also accounts for shape differences.

These reactions depend upon the acquisition of nutrient precursors that are converted to various intermediates and, finally, to an end-product. Specific enzymes mediate the reactions. Any step can be altered if the gene for that enzyme is defective or missing. Modification of any step carries through to all subsequent steps.

We can determine the sequence of intermediates in a biochemical pathway by using mutant strains of an organism that affect specific steps. Let's assume there are three enzymatic reactions in the pathway. If the mutant strain cannot perform the first reaction, then none of the intermediates or end-products will form. For the mutant strain that cannot perform the second reaction, there will be a buildup of only the first intermediate. For the mutant strain that cannot perform the third reaction, the first two intermediates will accumulate, but no end-product. We can test for the presence of these chemicals in microorganisms by plating them on a minimal medium instead of on a medium that contains nutritional supplements. Then we can determine the step in the pathway that a mutant affects by the supplement(s) required for growth.

Cancer and Oncogenes

Cancerous growths, whether in animals or the undifferentiated growth of a gall on the branch of a tree, are the result of mitosis, which when properly functioning is a normal process that results in growth, maintenance and repair (e.g., healing of a wound). The differences lie in how the process may be altered due to changes in the genes that regulate that process, both in location within the organism and in time with respect to its stage of development.

Oncogenes (onco—meaning cancer) arise from changes in proto-oncogenes which are normal genes involved with controlling the mitotic cell cycle. The changes may involve only a single base that alters the protein. One form of bladder cancer results from a base change in a proto-oncogene.

Tumor suppressor genes help control cell division but when a mutation occurs in the suppressor, cell division goes unchecked leading to tumor growth.

When DNA repair mechanisms fail, the result can lead to creation of oncogenes from proto-oncogenes and loss of suppression function by tumor suppression genes. One of the most studied of these types are the familial breast cancer causing genes, BRCA1 and BRCA2 (BRCA for, breast cancer predisposition gene). It is also thought that variations on these may be combinations of factors that are both tumor suppressor and/or faulty DNA repair mutations.

A viral infection is implicated in acute T-cell leukemia. This results when the virus inserts next to a proto-oncogene and the unregulated transcription of the viral particles also causes the adjacent proto-oncogene to act as an oncogene.

There are a number of diseases that result in a high incidence of cancer. Many of these diseases are also associated with an increased frequency of chromosomal instability. This instability causes gaps, breaks, exchanges in, and rearrangement of chromosomal structure. Chromosomal instability may not "cause" cancer but it may create an environment that is highly susceptible to certain forms of cancer. For example, Burkitt's Lymphoma in humans (common in parts of Africa) is caused by a translocation of a proto-oncogene on chromosome 8 next to an antibody gene on chromosome 14 where it is overly expressed along with the antibody gene.

Xeroderma pigmentosum is a disease in which the mechanism that repairs ultraviolet-induced damage in DNA is defective. In cell culture, clones of affected cells with chromosomal rearrangements have been found. Various skin cancers have been reported in patients with this disorder.

Bloom's syndrome is characterized by dwarfism. Individuals who are homozygous for this recessive trait show a high frequency of spontaneous chromosome breaks in their cells. Individuals with Bloom's syndrome have a high incidence of leukemia and malignant neoplasms.

Individuals who are homozygous or heterozygous for Fanconi's anemia are also cancer-prone. This disorder is characterized by chromosomal aberrations, anatomical defects, and mental retardation.

Victims of these diseases show an instability in their chromosomal structures. They also show a heightened susceptibility to cancer. Since the chromosomes of individuals with these afflictions are easily and effectively disrupted, they must have various defects in the repair pathways that normally protect chromosomes. Although chromosomal instability is not necessarily a cause of cancer, cancer can be one result of chromosomal instability.

Signaling Mechanisms in Cells

Communication between cells and the regulation of cell division are under genetic control (i.e., protein-laden pathways). One of the methods for cell regulation is signal transduction. First, a cell surface receptor binds with an incoming molecule called the *first messenger*. This induces nearby regulator proteins that in turn activate enzymes that result in what is called the *second messenger*. This is the important step: it determines how the cell will respond to outside stimuli. Mutations that affect the molecules in signaling are implicated in several human illnesses (e.g., neurofibromatosis type I).

Practice and review: Here are some "case studies" in biochemical genetics; try them and see how you are doing.

Case 1: What is the genetic basis for albinism?

Albinism is characterized by a failure to form the black pigment melanin. Normally, melanin is formed from the product of the hydroxylation of tyrosine. This product, L-dihydroxyphenylalanine, or dopa, is converted to melanin in a series of reactions by the enzyme complex tyrosinase. These reactions occur in melanocytes. Albinos have the normal number of melanocytes in their skin, but tyrosinase activity is not evident.

Albinism can result from two different mutant recessive genes. Gene *a* fails to code for the enzyme necessary to convert tyrosine products into melanin; gene *i* prevents normal absorption of tyrosine. When hair, including roots, is cultured in a tyrosine-rich medium, the homozygous *a/a* still fails to produce pigment but homozygous *i/i* does produce pigment.

There are a number of genes that are involved in determining the amount of pigment deposited in the skin, hair, and eye cells. This accounts for the great variability in human coloring. However, only one gene locus is involved in the production of melanin. A person who is homozygous for the recessive allele of this gene, which codes for the enzyme that converts tyrosine to dopa, cannot produce melanin.

Albinism can be caused by another genetic factor. Melanin production is dependent on the presence of tyrosine in a cell. A person who is homozygous for an allele at a different locus, will not have membranes permeable to tyrosine. People with this type of albinism have a small amount of melanin due to the creation of phenylalanine from tyrosine that occurs inside the cell.

Case 2: How does the measurement of enzymatic activity help to detect galactosemia?

Galactosemia is an inherited disorder in which dietary galactose accumulates because its metabolism is blocked. The absence of the enzyme galactose 1-phosphate uridyl transferase blocks the metabolism of galactose. This enzyme is usually present in such large amounts that a 50 percent reduction does not produce clinical symptoms. Heterozygotes have between 60 and 70 percent of the enzymatic activity and thus show no symptoms. Homozygotes, however, produce no enzyme and usually fail to thrive. When milk is consumed, the affected infants vomit or have diarrhea. Many are mentally retarded. The absence of galactose 1-phosphate uridyl transferase in the red blood cells is a definitive test for the disease.

This disease can be treated with a diet that excludes galactose. Almost all of the clinical symptoms, except for the mental retardation, regress with a galactose-free diet.

Case 3: Tay-Sachs disease is an inborn error of metabolism. How can heterozygotes be detected?

Tay-Sachs disease is a recessive disease in which the nervous system degenerates. Most affected individuals do not live past age three. Homozygous recessives do not produce the enzyme b-N-acetylhexosaminidase. This enzyme is responsible for the removal of the terminal sugar of one type of lipid. This ganglioside, as it is called, accumulates in the cerebral cortex causing striking pathological changes.

The absence of the enzyme can be detected prenatally by amniocentesis. Heterozygous parents can be detected since the enzyme is only present in 50 percent of the normal amount.

Case 4: When a person eats food containing phenylalanine or tyrosine, his urine turns black when exposed to air. What is the genetic basis?

This condition is due to an inherited metabolic disorder called alcaptonuria. This disorder is another inborn error of metabolism; it is caused by an inherited defect in a single enzyme. The enzyme that is defective in alcaptonuria is homogentisate oxidase.

A person who is homozygous for the recessive allele at the alcaptonuria gene produces urine that turns black because he lacks the enzyme that converts homogentisate to acetoacetate in the phenylalanine and tyrosine catabolic pathway. An affected individual excretes homogentisate into his urine. Homo-gentisate is oxidized and polymerized to a melanin-like compound when it remains exposed to air. This gives the urine its black color.

Alcaptonuria is relatively benign. It has no very harmful effects in young people. However, later in life it usually results in degenerative arthritis. This may be caused by the crystallization of homogentisate in the cartilages of the body.

6. Immunobiology

The primary function of the immune system is to provide resistance from attack by infectious agents such as bacteria and viruses. Most immune responses in higher organisms involve the production of antibodies (i.e., acquired immunity), but other innate mechanisms for killing infectious agents also exist. The acidic digestive juice of the stomach is also effective in killing infectious agents introduced by swallowing.

Antigens, Antibodies, and Antigen-Antibody Reactions

Acquired immunity depends upon the production of recognition molecules that can distinguish "self" from "nonself." Acquired immunity does not develop until after contact is made with an invading agent. The two forms of acquired immunity are as follows:

(1) humoral immunity and

(2) cellular immunity.

Cellular immunity refers to the production of lymphocytes that bind to the invading agent and cause the destruction of the invading agent. Humoral immunity refers to the production of protein molecules (antibodies) that bind to the invading agent and mark it for destruction.

A typical antibody consists of four polypeptide chains. There are two identical "light chains" and two identical "heavy chains," and the four chains are held together by disulfide bonds as shown in Figure 3.10(a) to form a Y-shaped molecule. As shown in Figure 3.10(b), both the heavy and light chains of an antibody are built up from a structurally similar "domain" or polypeptide subunit of about 220 amino acids. Each light chain has two such domains: a "constant" domain and "variable" domain. Similarly, each heavy chain has three (sometimes two) constant domains and one variable domain. The variable domains are at the amino-terminal ends (see Figure 3.10) of both heavy and light chains and the amino acid sequence in this region is very variable. The variable regions provide the specificity that enables an antibody to bind to a very specific region of another molecule (i.e., the antigen). The constant regions of antibodies provide a mechanism for the binding of the antibody to other cells (such as macrophages) or binding to elements of the complement system.

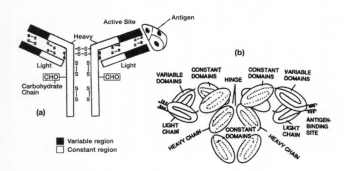

Figure 3.10.
The Structure of Typical IgG Antibody Showing the Constant and Variable Polypeptide Segments:
(a) shows position of disulfide bonds and
(b) shows arrangement of domains.

When antibodies bind to an antigen on an invading organism they mark it for destruction by either the complement system or by macrophages. The proteins of the complement system destroy an invading organism by perforating its cell membrane; they can also inactivate an invading organism by agglutination (multiple antigenic sites are bound together to form a clump), precipitation (the water-soluble antigen complexes with the antibody and the complex is insoluble), or by neutralization (the antibody binds to and covers a toxic site). Macrophages destroy invading organisms by ingestion.

Tissues, Cells, and Cellular Basis for Immunity

Lymphoid tissues form the "organ system" responsible for immunity. Lymphoid tissue contains a high concentration of lymphocytes. Lymphoid tissue is widely distributed throughout the body for example in lymph nodes, the thymus, the bone marrow, and the spleen. Lymphocytes can move between lymphoid tissue and the blood.

T-lymphocytes and B-lymphocytes. Lymphoid tissues contain two types of lymphocytes called T-lymphocytes and B-lymphocytes. The T-lymphocytes form the sensitized cells of the cellular immune system and the B-cells play a key role in the production of antibodies that provide for humoral immunity. See Figure 3.11. T-cells have another class of recognition molecules called T-cell receptors, which will only recognize cells that bear both "self" and "nonself" markers. Both T- and B-lymphocytes arise from embryonic stem cells, but before becoming part of lymphoid tissues they require a maturation process. For T-cells this maturation occurs in the thymus gland and for B-cells the exact site is not known but is thought to be in the bone marrow.

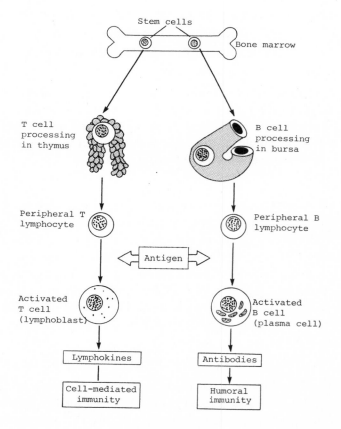

Figure 3.11.

After the maturation process the T- and B-cells migrate to and become imbedded in the lymphoid tissues. The immunological events following an infection by an infectious agent such as a virus are as follows:

(1) Macrophages ingest a number of viruses and display some specific viral "markers" or antigens on their surface. Some helper T-cells in circulation have the proper T-cell receptors to recognize the processed viral antigens on the macrophage surface and these T-cells become activated.

(2) The activated helper T-cells multiply and also stimulate the multiplication of killer T-cells and activated B-cells that can also recognize the same processed viral antigens. The activated B-cells multiply and differentiate into plasma cells that produce antibodies to the viral antigen. Some of the activated B-cells become memory cells that permit a rapid response to any future infection by the virus.

(3) The killer T-cells will destroy host cells that have become infected with the virus and thereby inhibit viral replication. The antibodies produced by the B-cells will also bind to the virus and prevent them from infecting additional host cells.

(4) When the infection is contained, suppressor T-cells halt the immune responses, and memory T-cells and memory B-cells remain in the blood and lymphatic system.

The AIDS virus is particularly damaging to the immune system because it invades and kills helper T-cells.

PROBLEM

The functional difference between B-cells and T-cells is that

a) B-cells differentiate from stem cells and T-cells differentiate from lymphocytes.

b) T-cells differentiate from stem cells and B-cells differentiate from lymphocytes.

c) T-cells secrete antibodies in response to introduced antigens and B-cells direct the cell-mediated response.

d) B-cells secrete antibodies in response to introduced antigens and T-cells direct the cell-mediated response.

Solution

d) T-cells and B cells are the cells involved in the immune responses of the body. Both cell types differentiate from stem cells of the bone marrow. The stem cells that migrate to the thymus become T-cells and are responsible for cell-mediated immunity. Those stem cells that migrate to the bursa (in birds) or lymphoid tissues (in mammals) become B-cells and are the cells of the humoral immune system.

Genetics of antibody production. To understand the genetics of antibody production, we must first be familiar with an antibody's structure. Refer back to Figure 3.11 and then practice drawing a diagram of the antibody indicating light and heavy chains.

We need to explain the tremendous variability of antibodies. In understanding the *clonal selection theory*, we see that exposure to a foreign antigen stimulates cell division of preexisting B-cells that are programmed to make the corresponding antibody. However, we must still explain how the cell genetically controls production of the original B-cells.

Hybridization experiments provide evidence supporting one of these theories of antibody production. An understanding of this technique is important. Remember that mRNA complements the DNA from which it was transcribed. When the two strands of DNA are experimentally separated and mixed with mRNA, the latter forms hydrogen bonds with its complement. We can detect this hybridization by using radioactively labelled mRNA. By using a messenger that codes for a particular antibody, we can locate the corresponding portion(s) of the genome by hybridization.

The variable regions of the light and heavy chains have been found to form by splicing a certain number of DNA sequences together. Each of these DNA sequences exists in variable forms in the genome. Multiply the number of sequences by the number of forms it has to obtain the number of variable regions that can be made for a particular chain.

The genetics of T-cell production are based upon understanding the major histocompatibility complex (MHC). One portion of the MHC codes for the T-cell receptor proteins that recognize foreign antigens. These T-cells can cause skin graft rejection, kill tumor cells, and kill virus-infected cells.

Evidence for a genetic basis for transplant incompatibility, such as skin graft rejection, can be obtained by using inbred strains of animals. Such inbred strains have many homozygous gene loci as the result of breeding siblings for many generations. Different inbred strains can be developed that are homozygous for different alleles. A cross between two different inbred strains will produce offspring heterozygous for many gene loci. We test the effects of different genetic backgrounds for the skin graft by choosing either parent or heterozygous offspring as graft recipients.

In predicting the immune response to a virus-infected cell, you should realize that such a cell will be recognized as foreign by the body.

As part of T-cell production, the body is able to distinguish between self and non-self. Self is defined by a set of proteins found on the surface of most cells in the body. These proteins are coded for by another portion of the MHC. Normally T-cells only form against foreign proteins, but when they form against self, autoimmune disease can result.

PROBLEM

What is the difference between T-cells and B-cells?

Solution

T-cells and B-cells are activated forms of lymphocytes that result in response to antigens. Morphologically, T- and B-cells are indistinguishable from each other.

They arise from the same type of undifferentiated stem cell. Their difference lies in the location in which they differentiate and in their function.

Antibody Diversity and Synthesis

PROBLEM

Any protein can elicit an antibody response. How can an organism make so many specific antibodies?

Solution

The clonal selection theory was proposed in the 1950's by Sir Macfarlane Burnet to explain immunity. His theory states that millions of specific lymphocytes exist in the body. Each secretes an antibody that is specific for a single antigen. When the antigen is introduced to the body, its specific antibody becomes activated. It divides mitotically to form a clone of identical cells (see Figure 3.12). Each of these cells is specific for that antigen. When these lymphocytes are no longer needed, they circulate in a relatively unreactive state throughout the body

The next time the same antigen is encountered, the cell's immune response is much quicker since the supply of the specific antibody-producing lymphocytes has been greatly increased by the mitotic process. The persistence of these long-lived lymphocytes is the basis for immunological memory.

PROBLEM

Supposing that the clonal selection theory is correct, how can an organism genetically make such diverse proteins?

Solution

There are three theories that account for this diversity. They are the germline hypothesis, the somatic mutation theory, and the rearrangement theory.

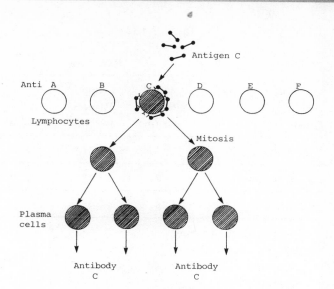

Figure 3.12.

The *germline hypothesis* states that all of the possible sequences are carried in the DNA. This means that each mammalian genome would need to carry the information to make more than 10^8 antibodies.

The *somatic mutation* theory says that the actual number of antibody genes was initially small. The diversity arose through mutation. This implies that the immunoglobulin system is very random.

The *rearrangement theory* suggests that separated germline sequences come together in all possible combinations via recombination. The variable gene segments can come together in this way to create immense diversity.

To some extent, all of these theories are based on actual mechanisms. These mechanisms, among others, diversify the antibody sequences so that the millions of antigens can be recognized by a single class of protein.

PROBLEM

How are the variable portions of heavy chains combined to create diversity in the antibody sequences?

Solution

The variable portions of heavy chains are organized like the variable portions of light chains. The major difference between them is that three separate DNA regions must be combined in heavy chains. These regions are called V, D (for diversity), and J_H. In heavy chains two joining events, V–D and VD–J, must occur.

DNA hybridization studies have detected 15 D regions. The current estimates for V and J_H regions are 250 and 4 respectively. The possible V–D–J_H joining reactions can make $250 \times 15 \times 4 = 15,000$ different variable genes for heavy chains.

Since the sequences for the heavy chains and light chains are on separate chromosomes, recombination between them is rare. This almost ensures that a light chain's variable sequence does not combine with a heavy chain's J sequence.

PROBLEM

How can the genes for an IgM molecule encode for an IgG molecule with the same specificity later in the immune response?

Solution

The five immunoglobulin classes, IgM, IgG, IgD, IgA, and IgE, differ only in the constant regions of their heavy chains. These constant regions are encoded by a cluster of genes on the same chromosome. Class switching is the term used to describe the change from one immunoglobulin class to another.

There are two recombination events that are needed in order for IgG to be produced, see Figure 3.13. The first event attaches a V_H gene to a C_μ gene at one of the J segments. This sequence is used to synthesize IgM. The second recombination event attaches $V_H J_H$ to a C_γ gene. The IgG molecule which is synthesized from this sequence has the same variable sequences as the previous immunoglobulin and, hence, the same specificity.

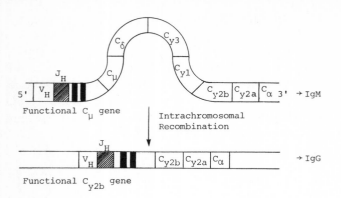

Figure 3.13.

Table 3.13.

Donor	Recipient		
	Male Parent Strain	**Female Parent Strain**	**Progeny**
Male parent strain	+(1)	−(2)	+(3)
Female parent strain	−(2)	+(1)	+(3)
Progeny	−(4)	−(4)	+(3)

(1) grafts within a strain
(2) grafts between strains
(3) grafts from parents or siblings
(4) grafts from progeny

Note: + indicates a successful graft, − indicates a rejected graft.

PROBLEM

How can crosses involving strains of highly inbred mice be used to determine a genetic basis for transplant incompatibility?

Solution

Mouse strains that are homozygous at nearly all loci can be obtained by mating brothers and sisters for many generations. Different inbred strains will have different alleles that are homozygous. We can test two such inbred strains for histocompatibility.

Any skin transplant made within a strain is readily and permanently accepted. A transplant between two individuals of different strains does not take. When these strains are crossed and the F_1 generation is tested for hybrid graft rejection, the results show that a graft from either parent or from the hybrid will be accepted when the recipient is a hybrid. (See Table 3.13.) The results also show that a graft from the F_1 hybrid will not take on either of the parents, but it will on siblings.

This experiment suggests that transplant rejection (or acceptance) has a genetic basis. It also suggests that codominant alleles are involved because the progeny exhibit antigenic properties of both of the parents.

7. Bacteriophages, Animal Viruses, and Plant Viruses

Viral Genomes, Replication, and Assembly

Knowledge of the properties and characteristics of viruses, bacteria, and fungi is essential to understanding the causes of many diseases as well as the therapeutic approaches used to alleviate these diseases. Modern molecular approaches are proving to be particularly useful in this regard.

Viruses are important because they:

(1) provide insight into evolution,

(2) are important tools for understanding the molecular biology of normal cells,

(3) are important in many diseases such as AIDS, and

(4) may provide powerful molecular tools for combating certain diseases.

Some viruses have been shown to cause cancer in animal models. Viruses have been obtained in homogeneous states and some viruses have been crystallized and their three-dimensional structure fully determined by X-ray crystallography. Recently, much emphasis has been placed on the possibility of using viruses to transmit selected genetic information into eukaryotic cells to correct defective genes (i.e., gene therapy).

DNA, RNA, and Protein Components

Viruses are the simplest supramolecular complex capable of initiating replication. They contain nucleic acids (either DNA or RNA but not both) with a surrounding protein coat called the capsid, which protects the encapsulated nucleic acid from damage. Some animal viruses also have an envelope of lipid and glycoprotein surrounding the capsid. An extracellular viral particle (or virion) cannot independently reproduce itself and requires a host cell for this function. It accomplishes this task by diverting the biosynthetic machinery of the host cell to synthesize its own components. In some RNA viruses, the viral mRNA preferentially binds to the host ribosomes. Hence, synthesis of viral proteins is favored over synthesis of host proteins.

There are four classes of RNA eukaryotic viruses that are distinguishable by the relationship of their viral RNA to their mRNA (see Table 3.14). mRNA is designated as (+) RNA and its complementary RNA as (−) RNA. Class I viruses contain (+) RNA which, in turn, is the template (+) mRNA. The parental RNA also functions as mRNA since it is capable of polymerizing ribonucleotides from an RNA template. For class I, as well as class II and class III viruses, this is accomplished by a viral RNA-directed-RNA polymerase (or RNA replicase).

Class II viruses contain (−) RNA which is transcribed into monocistronic mRNAs by a viral RNA transcriptase contained in the virion. One of these mRNAs

codes for an RNA replicase, which generates double-stranded RNA from the parental (−) RNA. The RNA replicase also synthesizes progeny (−) RNA strands from the double-stranded RNA.

Class III viruses contain double-stranded RNA, and the (−) strand provides the template for (+) mRNA. Class IV viruses are particularly important because the flow of genetic information is from (+) RNA to DNA and then back to RNA (see Table 3.14). Class IV viruses are called retroviruses and they code for a RNA-directed DNA polymerase (or reverse transcriptase). The HIV virus, which causes AIDS, is a retrovirus.

An important property of some RNA retroviruses (class IV RNA viruses) is their ability to induce tumors in animal models. Some DNA viruses (i.e., Simian virus 40 and polyoma virus) can also cause tumors. Cancer-causing viruses (i.e., oncogenic viruses) transform their host cells by inserting their viral specific genes into the host chromosome. Normal cells stop multiplying when in close contact with one another (i.e., contact inhibition). Transformed cells no longer exhibit contact inhibition and, therefore, grow continuously.

Table 3.14.
Classes of RNA Viruses

Class	Viral RNA	Flow of Genetic Information
I	(+) RNA →	(−) RNA → (+) mRNA
II	(−) RNA →	(+) mRNA
	(±) RNA →	RNA
III	(±) RNA →	(+) mRNA
IV	(+) RNA →	(−) DNA → (±) DNA → (+) mRNA

In the DNA or RNA viruses, the DNA provides the template for the synthesis of mRNA molecules, which preferentially use the host ribosomes to synthesize viral proteins and the enzymes necessary for viral DNA synthesis.

Viruses contain very few genes (between 3 and 240) and, therefore, construct much of their molecular machinery from identical protein subunits. For example, the protein coat of the TMV (Tobacco mosaic virus), which contains only 6 genes, is made up of 2,130 identical protein subunits. Coat-protein subunits usually arrange themselves into either rods or spheres or a combination of these shapes.

PROBLEM

Viruses differ from living organisms because

a) viruses possess no bounding membrane.

b) viruses lack all metabolic machinery.

c) viruses lack all reproductive machinery.

d) All of the above.

Solution

d) Viruses differ from living things in many ways. They do not have any membranes because they have no need to take in or expel material. Viruses lack all metabolic machinery and do not produce ATP because they do not perform energy-requiring processes. Viruses do possess either DNA or RNA, but cannot independently reproduce. They must rely on host cells for reproductive machinery and components.

Bacteriophage

Bacteriophages (or phages) are bacterial viruses that have either RNA or DNA genomes. Figure 3.14 illustrates the structure of a typical bacteriophage, which has a head, tail, and tail fibers. Infection of a bacterium (1 to 10 μm in length) begins when a phage (100–300 nm) attaches its tail fibers to a surface receptor on the bacterium. The DNA, which is tightly packed in the phage head, is subsequently injected through the cell wall and the cell membrane into the bacterium (see Figure 3.15). In only a few minutes all the metabolism of the infected bacterium is directed towards the synthesis of new phage particles. About 30 minutes after infection the bacterium undergoes

lysis and hundreds of completed bacteriophages are released.

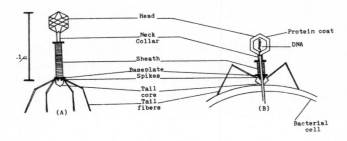

Figure 3.14.
The Structure of a Typical Bacteriophage

The complex coordination of the phage life cycle is a result of different phage genes being expressed at different times. The early phage genes are expressed before phage DNA synthesis begins. For many phages some of these gene products shut down the biosynthetic capacity of the bacterium. One of the early phage gene products that helps to shut down the metabolism of the host cell is a nuclease specific for bacterial DNA but not the phage DNA.

The late gene products are associated with the synthesis of viral DNA, capsid formation, packaging of the viral DNA into preformed heads, and the synthesis of lysozyme to degrade the bacterial cell wall (thus causing lysis). Not all phages cause immediate lysis of the infected bacterium. In some cases the phage DNA incorporates itself into the bacterial chromosome and is only replicated when the host chromosome is replicated. This process is called lysogeny. Viruses that exhibit this state are called temperate or moderate viruses.

The viral DNA incorporated into the host chromosome is called a provirus, or prophage. In the case of bacteriophages this prophage can be induced to become virulent and lyse its host bacterium. The resulting infectious phages may carry small amounts of bacterial chromosome, which can be transferred to newly infected bacteria. The process whereby DNA is

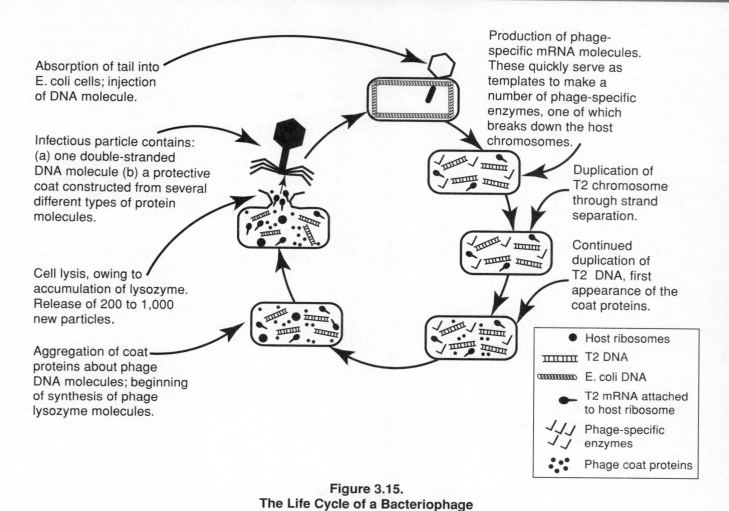

Figure 3.15.
The Life Cycle of a Bacteriophage

Absorption of tail into E. coli cells; injection of DNA molecule.

Infectious particle contains: (a) one double-stranded DNA molecule (b) a protective coat constructed from several different types of protein molecules.

Cell lysis, owing to accumulation of lysozyme. Release of 200 to 1,000 new particles.

Aggregation of coat proteins about phage DNA molecules; beginning of synthesis of phage lysozyme molecules.

Production of phage-specific mRNA molecules. These quickly serve as templates to make a number of phage-specific enzymes, one of which breaks down the host chromosomes.

Duplication of T2 chromosome through strand separation.

Continued duplication of T2 DNA, first appearance of the coat proteins.

● Host ribosomes
▥ T2 DNA
∽ E. coli DNA
•— T2 mRNA attached to host ribosome
ノノノ Phage-specific enzymes
⁞∴ Phage coat proteins

transferred from one bacterium to another by a phage is called transduction.

PROBLEM

Moderate viruses

a) replicate DNA only when the host replicates.

b) induce tumors.

c) cause immediate lysis of infected bacteria.

d) contain DNA and RNA.

Solution

a) In moderate viruses the phage DNA is incorporated directly into the host chromosome, and thus replicates only when the host does. RNA retroviruses may induce tumors. Most viruses, with the exception of moderate ones, cause immediate lysis of infected bacteria. Viruses contain either DNA or RNA, not both.

Retroviruses

Retrovirus research over the past eight decades has greatly contributed to our knowledge of cancer, the cell cycle, signal transduction, gene expression, genetics, and immunology. However, it was only the relatively recent implication of the human immunodeficiency virus (HIV) as the causative agent of the Acquired Immunodeficiency Syndrome (AIDS) that hurled them into the spotlight of public attention. Unfortunately, despite our increasing understanding of how the virus operates, no cure or vaccine is yet available for HIV.

Retroviruses are divided into three official taxonomic groups, assignments into which are purely functional. The three taxons into which retroviruses are grouped are: *Oncoviridae*, whose members induce tumors; *Lentiviridae,* whose members cause slow disease; and *Spumiviridae*, a largely unknown class whose members cause foamy vacuoles in infected cells. The known human retroviruses are the human T-cell leukemia viruses (HTLV-1, HTLV-2) and the human immunodeficiency viruses (HIV-1, HIV-2). All belong to the lentiviruses.

The structure of a retroviral virion is composed of the RNA genome, a surrounding protein capsid, and an outer lipid bilayer envelope. The RNA genome consists of two identical mRNA-like RNA molecules joined near their 5' ends. A tRNA molecule specific for each type of RNA is hybridized at this same site to each RNA molecule. Tightly associated with the RNA are about two thousand copies of a low molecular weight, basic protein. Another protein forms a structure called a capsid around the coated RNAs. This whole structure is called a nucleocapsid. The nucleocapsid can be of either helical or icosahedral symmetry, depending on the virus in question. Surrounding the nucleocapsid is a host-cell, membrane-derived lipid bilayer known as the envelope. The region between the nucleocapsid and the envelope is the matrix. The viral enzymes reverse transcriptase, protease, and integrase are found here. Two viral proteins are found associated with the envelope. One is a transmembrane protein (*TM*) and the other is a surface, outer-envelope glycoprotein linked to the virion via disulfide bonds with *TM*. This surface protein (*SU*) predictably determines the virus host range. The enveloped particle is of icosahedral symmetry.

Retroviruses were found to violate that rule until the mid-1960s. In 1964, Howard M. Temin discovered

that retroviruses, which are RNA viruses, replicate through a DNA intermediate. This meant that the DNA intermediate (i.e., the provirus) had to be copied from an RNA molecule. This was in direct violation of the **central dogma** of molecular biology, which states that genetic information invariably passes from DNA to RNA to protein. Dr. Temin's discovery was initially met with disbelief and led to a fervent controversy. In the end Temin was proven correct when he and David Baltimore each independently isolated **reverse transcriptase**. Temin and Baltimore shared the 1975 Nobel Prize in medicine for their work in this area.

Ecotropic retroviruses infect only cells of the species producing the virus. Amphitropic retroviruses can infect several species including the one currently producing the virus. Xenotropic viruses infect species other than the one producing the virus. The latter is a rather confusing concept and is largely a laboratory phenomenon. A researcher can artificially introduce the retrovirus' genetic material into a cell type that it normally does not infect. This host cell produces progeny virus which cannot reinfect it or other cells from the same species.

Researchers categorize retroviruses by their host range. All retroviruses that enter their host cell via the same receptor belong to the same interference group. The name "interference group" comes from an early observation that cells infected with one type of retrovirus resist infection by certain other types of retroviruses. Retrovirologists later learned that these viruses all used the same receptor on the host cell. The viral envelope protein produced by the infected cell mediates this interference. Some of the envelope protein escapes from the membrane and binds to the receptor. The occupied receptor is unavailable for binding of free virus. It should be clear that members of an interference group also belong to the same host range group. However, the reverse is not true. There are typically more than one interference group per host range category.

Some HIV isolates differ in their ability to infect various alternative hosts. Strains may differ in that respect and genetically at one or more gene clusters in the genome. One of these regions of variability is far more prominent than others. The major class of these mutations lies in the *env* gene. Specifically, you would expect them to fall in that part of the gene that encodes the *SU* protein, gp120. The observed mutations do fall in this region, but somewhat surprisingly, they do not seem to fall in the exact same region believed to interact with CD4. Rather, they occur in a region next to the CD4-binding domain. This finding gives credence to the notion that HIV recognizes more than one receptor.

8. Recombinant DNA Methodology

Terminology

Genetic recombination—rearrangement of genes from separate groups of genes. Genes from two chromosomes (or two DNA molecules) are recombined into one chromosome containing some genes from each of the original chromosomes.

Genetic engineering—manipulated gene transfer in the laboratory.

Recombinant DNA—DNA that has been artificially altered to combine genes from different sources.

Biotechnology—application of genetic engineering and use of recombinant DNA in research, medicine, industry, and agriculture.

Vector—piece of DNA used to transfer DNA between organisms. The gene of interest is inserted into the vector (usually a plasmid or a viral genome), which is then transferred into a new cell. The new cell is used to grow a **clone** from which large amounts of the gene or its product can be harvested. Some vectors contain antibiotic-resistance genes, or **markers**, that can be used to identify cells containing the vector.

Restriction enzyme—an endonuclease enzyme that recognizes and cuts a specific DNA sequence (note: exonuclease digests DNA from the end of the molecule); *restriction* alludes to the natural function of these enzymes which evolved to destroy the function of bacterial invaders (e.g., phages). They are generally named for the microorganism in which they were first discovered (e.g., the restriction enzyme *EcoRI* was isolated from *Escherichia coli;* it recognizes and cuts the sequence GAATTC).

Palindrome—short array of bases that read the same in both directions in a DNA or an RNA molecule. These are often the sites where restriction endonuclease will cleave a polynucleotide.

Gene library—an entire genome cut with restriction enzymes and inserted into vector molecules.

cDNA—DNA synthesized from a strand of mRNA by reverse transcription; the required enzyme is **reverse transcriptase**, which was originally isolated from a retrovirus.

Clone identification—can be accomplished through replica plating.

Restriction Endonucleases

Restriction endonucleases (restriction enzymes), ligases, and reverse transcriptase are three of the most important enzymatic, molecular tools available to modern molecular genetics. Endonucleases provide the means to cut DNA into small fragments at highly specific sites (these may be palindromes) and a ligase acts as the "glue" that is used to connect the inserted fragments into a new, recombinant molecule. The recognition sites may vary from four to six bases, may be present in some organisms and not others. The cuts may be blunt or staggered. The latter produces sticky ends that are very useful when inserting DNA fragments into plasmids in preparation for molecular cloning. Reverse transcriptase enables us to synthesize DNA from RNA.

In *Situ* Hybridization

Once a DNA sequence is cloned it can be used in many ways. One that provides a means to locate a gene to a specific location on a chromosome(s) is called *in situ* hybridization. Cloned fragments are labelled with biotin conjugates or fluorescent dyes and then placed upon microscope slides bearing chromosome preparations. Following special treatment, the complementary regions between the cloned fragment and the chromosome can be visualized; thus the cloned gene(s) is mapped to a specific location and to a specific chromosome.

Restriction Fragment Length Polymorphisms

When DNA is cut by one or more restriction enzymes, the distance between the sites where the cuts are made will vary and, of course, the sequence of DNA is identified because of the enzyme's specificity. When the pieces of DNA are radioactively labelling at the 3′end of the DNA and an electrical charge (electrophoresis) applied, they will migrate in varying distances in a gel (agarose or acrylamide gels) depending upon size; with appropriate chemicals and UV light, the labelled fragments can be seen and photos taken. These different segments can then be "mapped" according to the length/number of bases between restriction sites where the cut was made. These polymorphisms can be visualized and mapped as if they were alleles and correlated with other gene markers of interest. Such maps are powerful tools when selecting and cloning specific genes. They can be compared among families, populations, species and so forth for studying mutation rates and evolution of the DNA under investigation. And, very importantly, inserting the fragments into a plasmid or other type of vector molecule can yield "libraries" of cloned fragments that can be sequenced base-by-base and/or otherwise utilized for the particular genes that they contain.

Similar protocols apply to making cDNA (complementary DNA) libraries that originate from mRNA molecules.

DNA Cloning, Sequencing, and Analysis

Plasmids are extrachromosomal (outside the bacteria's chromosome), circular pieces of bacterial DNA. They are self-replicating accessory chromosomes. They carry genes for antibiotic resistance, metabolism of natural products, and production of toxins. Plasmids have become important components in the manipulation of genes in cloning experiments.

When the restriction enzyme *EcoRI* is used to cut a piece of DNA, it leaves "sticky" ends: see Figure 3.16. Such a fragment can be inserted into a circular plasmid DNA that has been similarly treated. As shown in Figure 3.17, these hybrid plasmids can be used to infect bacteria.

A plasmid can be used as a means of introducing a foreign piece of DNA into a bacterial cell. Since the plasmid is self-replicating, the foreign DNA will remain a part of the plasmid and a part of the bacterial strain as long as the plasmid stays in a cell. Thus, clones of cells containing a piece of foreign DNA can be produced. This is the basis of recombinant DNA technology.

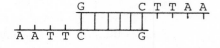

**Figure 3.16.
Sticky Ends of a Piece of DNA
Cut with the Restriction Enzyme *EcoRI***

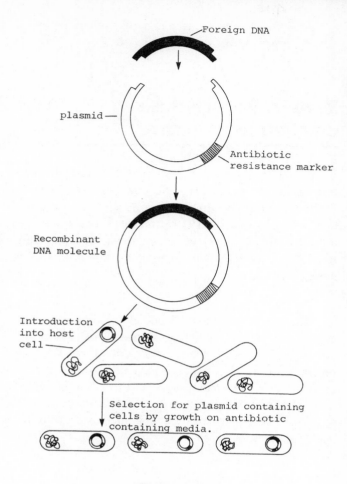

**Figure 3.17.
Inserting and Cloning
Foreign DNA in a Plasmid Vector**

Cloned fragments can be sequenced so that individual bases are mapped within a fragment. The popular press often refers to base sequence analyses as "DNA fingerprinting." The following is a brief outline of the procedure, but today the methods have been highly automated.

1. The 5′ end of DNA can be identified if radioactively labelled with ^{32}P.

2. The weak hydrogen bonds holding DNA together can be broken, giving two single strands. The strand containing more purines is the heavy strand. (Remember that purines are double-ring structures and pyrimidines are single-ring structures.) Analysis

of only one strand is enough, since the strands are complementary.

3. Different chemicals can be used to cause breaks at different sites in the DNA: guanine, adenine, both cytosine and thymine, and cytosine. Varying fragment lengths result, since the chemicals do not cleave all possible sites in each sample.

4. Varying fragment lengths can be separated by gel electrophoresis, the smallest, and, therefore, the lightest, moving the farthest.

5. The position in the gel tells us the length of the fragment. Therefore, we know the position of the particular base as measured from the 5′ end. By reading the bands produced in each of the four gels (one for each chemical), we can determine the base sequence. To determine the position of thymine we must look at the gel that shows cleavage of both cytosine and thymine and compare it with the one for cytosine (there is no chemical that cleaves only thymine).

RNA nucleases aid in RNA sequencing by cleaving specific sites. For example, pancreatic RNase breaks RNA after a pyrimidine. Thus, we know that each fragment produced must end in cytosine or thymine. Further treatment of the fragments cuts the sugar-phosphate backbone after the 3′ phosphate when travelling in a 5′ → 3′ direction. The first base (at the 5′ end) can be recognized because it will have both 5′ and 3′ phosphates. The last base (at the 3′ end) can be recognized because it will have neither. Using this information, we can put the bases within each fragment, and the fragments themselves, in order.

Polymerase Chain Reaction (PCR)

Polymerase chain reaction (PCR) technology uses a cell's DNA replication processes to make millions of copies of a specific DNA fragment that is targeted by the scientist (i.e., gene amplification). This is now one of the most powerful tools because a single cell's DNA can rapidly be increased to usable amounts for analyses. The steps of PCR are briefly described here:

1. Denature. DNA is heated to make single stranded DNA.

2. Temperature is lowered, two short sequenced DNA primers are added to the mix, plus Taq1 DNA polymerase. The primers bind to the complementary single stranded target DNA.

3. More bases (A,T,G,C) are added and the DNA polymerase adds bases to the primers and a sequence is constructed that is complementary to the target DNA molecule.

4. These newly constructed strands act as templates in the next round of replication. This is induced by increasing the temperature.

5. Through replication, DNA fragments increase exponentially and the number of DNA pieces becomes approximately 2^n, where n = the number of temperature cycles. Thus, after 20 cycles, there are ~1 million copies of the piece of target DNA.

6. Keep in mind that most all this procedure is now automated. Also, because of its high sensitivity, precautions must be used to avoid contamination of the target sample of DNA.

Reverse Transcriptase and Retroviral Vectors

In 1970, Howard Temin and David Baltimore independently discovered the enzyme *reverse transcriptase*. Since its discovery, this viral enzyme has been used in many ways by research biologists. Its unique ability to catalyze the synthesis of DNA from RNA has established it as an important tool in genetic research.

Reverse transcriptase is coded for by the nucleic acid of retroviruses. When retroviruses infect a bacterial host, their RNA genomes are transcribed into DNA by reverse transcriptase, which is present in small amounts in the viral protein coat. Once in the DNA form, the viral genome integrates into the host chromosome as a provirus. The provirus is then replicated as a part of the bacterial genome.

Figure 3.18.
Components of a Retroviral Vector

LTR = Long Terminal Repeat
U3 = Unique 39 region (promoter and enhancer)
R = Repeat
U5 = Unique 59 region (poly A signal)
PBS = Primary Binding Site
PPT = Polypurine Tract
ψ = encapsidation sequence

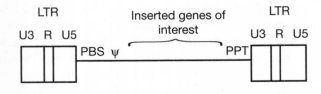

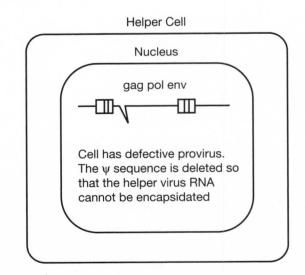

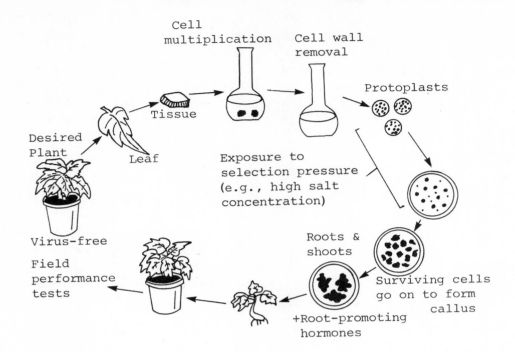

Figure 3.19.
The process of plant propagation from single cells in culture can produce plants with selected characteristics. These selections must be tested in the field to evaluate their performance.

The mechanism of this enzyme is similar to that of other DNA polymerases. The DNA is synthesized in the 5′ to 3′ direction. It needs a primer to begin this synthesis. The primer is a noncovalently bound tRNA that was picked up from the host during the previous round of infection. The DNA chain can be added to the 3′ end of this tRNA molecule. Somehow, double-stranded circular DNA molecules are made from this single-stranded DNA copy. These circular proviruses are now ready for insertion into the host genome.

This enzyme has proven to be a useful tool to the molecular biologist. It can be used to make DNA copies of any RNA that can be purified. This function can be useful in many types of experiments. Purified mRNA can be obtained from cells that are specialized to make specific proteins (e.g., hemoglobin from red blood cells and insulin from mammalian pancreas cells). The DNA copy of mRNAs from these cells can be used to find the sequence in the cells' own DNA. Thus, a specific area of the mass of DNA can be focused on. In similar ways, split genes and differences between hnRNA and mRNA can be studied. But perhaps most importantly, reverse transcriptase is a necessary reagent for genetic engineering with recombinant DNA. The complementary DNA (cDNA) that is transcribed from an mRNA is what is inserted into plasmids to create recombinant DNA molecules. Thus, this enzyme is a very significant part of genetic technology.

What is a *retroviral vector?*

Retroviral vectors are essentially retroviruses engineered to carry genes chosen by a researcher "in place" of their own genes (see Figure 3.18). The discovery that certain highly oncogenic, replication-defective retroviruses transduce cellularly derived genes (oncogenes) gave the first inclinations that retroviruses could be commandeered as gene transfer vehicles. The finding that the transducing virus could be packaged if its deleted proteins were supplied in trans by a coinfecting nontransducing wild-type virus (a helper virus) led to the idea that cell lines could be established to express all the virus proteins for the purpose of passaging the vector. Such cells, termed helper or packaging cells, supply all virus proteins but produce no replication-competent virus. The vector must possess the encapsidation sequences, PBS, PPT, U3 attachment site, U5, and R. It must also have either U3 or a promoter to generate full-length genomic transcripts. However, U3 itself is not necessary; any promoter will do.

Applications of Recombinant Molecules

How can genetic engineering be used to improve agricultural crops?

Agriculture has been using the principles of Mendelian genetics for thousands of years. The yields of wheat and corn have steadily increased over the past 50 years by the use of close inbreeding and rigorous artificial selection. This produces distinct and genetically uniform varieties with both desired and undesired traits. These varieties have improved crop yields and introduced hybrid vigor. For instance, genes were introduced, by sexual crosses, from a semi-dwarf variety of Japanese wheat to a Mexican wheat strain. The hybrid had greater wind resistance and needed less growth before flowering because the straw was shorter and stiffer; this meant greater adaptability to the region and greater disease and insect resistance.

The recent discovery of the crown gall plasmid system has created the possibility for recombinant DNA techniques to be applied to plants. This would be much faster than conventional breeding and would enable the introduction of genes from one sexually incompatible species to another. However, the isolation of specific genes that affect crop yield is extremely difficult.

Thus, plant breeders have been using genetics for a long time. The techniques of recombinant DNA may eventually be used to improve agricultural crops with greater specificity and speed than is available in the traditional methods. The plant propagation process is shown in Figure 3.19.

Animal Structure, Function, and Organization

Overview and Terminology

Structure (form) and function are interdependent and become organized in a species through natural selection over time. Variability among living organisms in how they adapt to their environments is demonstrated by the biological diversity (Chapter 8) within and among evolutionary lines (i.e., phylogenies). Therefore, structure and function are the consequences of long-term processes, both biological (genetics) and physical (climates and habitats). However, the way in which organisms cope with short-term, everyday variation in their local environments is a function of the balancing between genotypes and physiological buffering capacities.

This chapter will review animal structure, function and how cells become organized to perpetuate and maintain the organism. Recall the repetitive hierarchy in biology: atoms comprise molecules, molecules form cells, cells comprise tissues and tissues form organs and organs become intertwined in systems to yield a viable organism. Tables 4.1 and 4.2 provide some basic terminology and comments to aid your review during Chapters 4 and 5.

Most animals (excepting sponges and few other invertebrates) contain organs that have specialized functions. These include connective tissues and muscles that support and maintain the integrity of organs within the organism (e.g., the diaphragm separates the thoracic [heart and lungs] from the abdominal cavity). A broader level of organization are the organ systems, an interdependent system of organs described in Table 4.2.

1. Exchange with Environment

Survival, from the cellular to the organismal levels, is dependent upon maintaining an internal steady state called *homeostasis*. This internal balance, which is dynamic, is largely maintained by negative feedback control coordinated by the central nervous system, it is a continuous balancing between an organism's internal and external environments. For example, if the external temperature becomes colder, as human mammals, we can adjust via consumption of energy-laden foods, conserve heat by constricting blood vessels, thus maintain a constant body temperature; obviously, being less hirsute than some of our fellow mammals, we also resort to adding clothing to assist our physiology or we take it

Table 4.1.
Four Types of Tissues in Animals

Type of Tissue	Function and Description	Subtypes and Comments
Epithelial tissue	This tissue often provides protective cell layer(s) via commonly densely packed cells having tight junctions between cells. Epithelia are classified according to number of cell layers: simple (one layer) and stratified (multilayer).	<u>Pseudostratified columnar epithelial</u>—involved with secretion and/or absorption (e.g., nasal passages). <u>Simple cuboidal epithelial</u>—cells in kidney, thyroid, other glands. <u>Simple squamous epithelial</u>—thin cell membrane, easy diffusion of gas and liquids (e.g., blood and lungs). <u>Simple columnar epithelial</u>—functions in secretion and absorption (e.g., lining of intestines). <u>Stratified columnar epithelial</u>—function as excretion cells (e.g., inner lining of urethra). <u>Stratified squamous epithelial</u>—associated with surfaces of tissues that are subject to sloughing, abrasion (e.g., outer skin, esophagus, anus).
Connective tissue	These act to bind and support other tissues and organs; scattered among extracellular, web-like matrix. Connective tissues are of one or more of *three types of protein fibers*: collagen fiber, elastic fiber, and/or reticular fiber.	<u>Loose connective tissue</u>—holds organs in place, includes all three types of fibers and is the most abundant connective tissue in vertebrates. <u>Loose Adipose connective tissue</u>—special connective tissue that stores fat. <u>Fibrous connective tissue</u>—mainly collagen fibers, it is dense and inelastic, necessary for function of tendons and ligaments. Collagen fibers are extremely strong. Tropocollagen is the basic structural unit from which collagen is constructed. Tropocollagen has a unique triple helix structure and each polypeptide strand is called an alpha-chain. <u>Cartilage connective tissue</u>—this is made up of collagen and a matrix of chondroitin sulphate that is secreted by cells called chondrocytes. Cartilage must be strong but flexible (e.g., ears and nose), and it serves as cushion for knees and between vertebrae. There are three types of cartilage (i.e., hyaline, elastic, and fibrocartilage). <u>Bone connective tissue</u>—this is mineralized connective tissue that provides support and protection for the body (i.e., skeleton, rib cage). <u>Blood connective tissue</u>—there is a connective matrix in blood called plasma; contains two cell types, erythrocytes (red cells) and leucocytes (white cells).
Nervous tissue	These tissues sense and transmit signals (impulses) from one part of an organism to another (i.e., brain to muscle).	Neuron is the primary cell type and it transmits impulses. It contains a cell body and two or more processes (each comprised by an axon and dendrite). An axon transmits an impulse from one neuron to another, or to a structure such as muscle (the effector).
Muscle tissue	These are the most abundant type of tissue in most animals. They are composed of long cells called muscle fibers that can contract when signalled by a neural impulse. Within cells are parallel fibers called myofibrils made of proteins called actin and myosin.	Vertebrates have three types of muscle tissues: 1. Striated-Skeletal: adults have a fixed number and these facilitate voluntary movement. Thus, exercise increases their size, not the number of cells, which is fixed. 2. Cardiac: these form the wall of the heart, and are both striated and branched. Intercalated discs that insure cell-to-cell communication during a heartbeat join these cells. 3. Smooth: these lack striations and form the linings of structures, for example, that make up the digestive tract. Smooth muscle contracts in response to involuntary impulses (i.e., peristalsis).

Table 4.2.
Eleven Organ Systems in Mammals[1]

System	Components	General Functions	More Specific Functions
Respiratory	Lungs, trachea, other breathing tubes.	Exchanges/interactions with the environment.	Intake oxygen, exhale carbon dioxide.
Immune and lymphatic	Bone marrow, lymph nodes, thymus, spleen, lymph vessels, leucocytes (white blood cells).	Exchange/interactions with the environment.	Resistance and protection from infections and cancer.
Digestive	Mouth, pharynx, esophagus, stomach, intestines, liver, pancreas, anus.	Internal transport and exchange.	Food processing (ingestion, digestion, absorption, elimination).
Excretory	Kidneys, ureters, urinary bladder, urethra.	Internal transport and exchange.	Disposal metabolic wastes, maintain osmotic balance of blood.
Circulatory	Heart, blood, blood vessels.	Internal transport and exchange.	Internal distribution.
Integumentary	Skin and associated tissues (e.g., hair, claws, skin glands).	Support and movement (including flagellar and ciliary).	Protect against mechanical injury, infection and drying out.
Skeletal	Skeleton, including bones, tendons, ligaments and cartilage.	Support and movement (including flagellar and ciliary).	Body support and protection of internal organs.
Muscular	Skeletal muscle.	Support and movement (including flagellar and ciliary)	Movement and locomotion.
Nervous	Brain, spinal cord, nerves, sensory organs.	Integration, regulation and control.	Coordinate body activities; detect and respond to stimuli.
Endocrine	Pituitary, thyroid, pancreas, other hormone secreting glands.	Integration, regulation and control.	Regulate, coordinate functions such as digestion and metabolism.
Reproductive[2]	Ovaries, testes, associated organs.	Perpetuate the species.	Sexual reproduction.

[1] This table is based upon Table 40.1 (p. 840) in Campbell, N. and Reece, J. (2002) *Biology*, 6th ed., Benjamin Cummings Pub., San Francisco, 1247 pp.

[2] Refer to Chapter 5 to review this system.

off when we need to cool down. Components that must be kept in control by the *endocrine and nervous systems* to maintain homeostasis include:

— nutrition

— oxygen and carbon dioxide

— waste products

— hydrogen ion concentration [pH]

— salts and other ions

— temperature

— cell volume and pressure

Nutrient, Salt, and Water Exchange; Gas Exchange; Energy

Respiratory System

Oxygen and carbon dioxide are the important respiratory gases. While a small amount of oxygen is carried dissolved in the blood, most of the oxygen is carried by hemoglobin, an important protein found in red blood cells. Carbon dioxide (CO_2) is also dissolved in the blood and carried by hemoglobin, but most carbon dioxide is carried in the blood as bicarbonate ion (HCO_3).

There are chemoreceptors that are sensitive to changes in the chemical composition of the blood. Central chemoreceptors in the medulla oblongata are sensitive to levels of hydrogen ion (H^+) or CO_2. When H^+ or CO_2 increase, the receptors are excited and they in turn send signals to the breathing centers in the medulla oblongata, stimulating breathing. In addition, there are peripherally located chemoreceptors located in the aortic arch and carotid arteries. They, too, can stimulate the medullary breathing centers when there is an increase in H^+ (decrease in pH) or an increase in CO_2, as well as when there is a decrease in oxygen in the blood. The increase in breathing will function to increase the oxygen and decrease the carbon dioxide levels in the blood.

The lungs are the sites of gas exchange in the pulmonary circulation. The alveoli or air sacs of the lungs are thin-walled, as are the pulmonary capillaries that supply them. Hence, the respiratory gases can easily diffuse through these walls. Gases diffuse from a region of high partial pressure to one of lower partial pressure. When one inhales, there is an increase in oxygen in the alveoli; when one exhales, there is a decrease in carbon dioxide in the alveoli. Since the blood entering the pulmonary capillaries has a low level of oxygen and a high level of carbon dioxide, oxygen will diffuse from the alveoli into the pulmonary capillaries and carbon dioxide will diffuse from the pulmonary capillaries into the alveoli. Hence the blood returning to the heart from the lungs will be replenished with oxygen, and the body tissues will be ridded of carbon dioxide, the major waste product of cellular metabolism.

Breathing Structures and Mechanisms. The respiratory system in humans and other air-breathing vertebrates includes the lungs and the tubes by which air reaches them (Figure 4.1, a and b). Normally, air enters the human respiratory system by way of the external nares or nostrils, but it may also enter by way of the mouth. The nostrils, which contain small hairs to filter incoming air lead into the nasal cavities, which are separated from the mouth by the palate. The nasal cavities contain the sense organs of smell, and are lined with mucus-secreting epithelium that moistens the incoming air. Air passes from the nasal cavities via the internal nares into the pharynx, then through the glottis and into the lar-

ynx. The larynx is often called the "Adam's apple," and is more prominent in men than women.

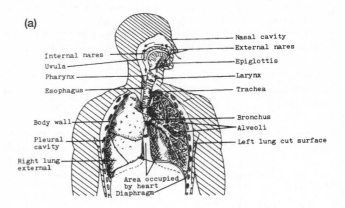

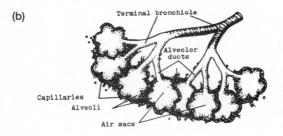

Figure 4.1.
Diagram of the Human Respiratory System

The opening to the larynx is always open except in swallowing, when a flap-like structure (the epiglottis) covers it. Within the larynx are the vocal cords. The glottis is the opening between the vocal cords. Muscles of the larynx move the vocal cords together, closing the glottis or move the vocal cords apart allowing air to pass through the glottis. Leading from the larynx to the chest region is a long cylindrical tube called the trachea, or windpipe. In a dissection, the trachea can be distinguished from the esophagus by its cartilaginous rings that serve to hold the tracheal tube open. In the middle of the chest, the trachea bifurcates into bronchi that lead to the lungs. In the lungs, each bronchus branches, forming smaller and smaller air tubes. Very small air tubes called bronchioles terminate in clusters of air sacs called alveoli. Each alveolus is a cup-shaped cavity. The wall of an alveolus is covered with a rich network of capillaries. The walls of the alveoli are thin and moist and molecules of oxygen and carbon dioxide diffuse readily between the air in the alveolus and the blood in the capillaries. The total alveolar surface area across which gases may diffuse has been estimated to be greater than 100 square meters.

Each lung, as well as the cavity of the chest in which the lung rests, is covered by a thin sheet of smooth epithelium, the pleura. The pleura are continuous with each other at the point at which the bronchus enters the lung. Thus, the pleura is more correctly a sac than a single sheet covering the lungs. The space within this sac is filled with a thin layer of fluid; this fluid enables the lungs to move without much friction during breathing.

The chest cavity is closed and has no communication with the outside. It is bounded by the chest wall, which contains the ribs on its top, sides, and back, and the sternum anteriorly. The bottom of the chest wall is covered by a strong, dome-shaped sheet of skeletal muscle, the diaphragm. The diaphragm separates the chest region (thorax) from the abdominal region, and plays a crucial role in breathing by contracting and relaxing, changing the intrathoracic pressure.

Respiration in Other Organisms

Amoeba—Simple diffusion of gases between the cell and water is sufficient to take care of the respiratory needs of the amoeba. Oxygen diffuses into the cell and carbon dioxide diffuses out of the cell.

Paramecium—The paramecium takes in dissolved oxygen and releases dissolved carbon dioxide directly through the plasma membrane.

Hydra—Dissolved oxygen and carbon dioxide diffuse in and out of two cell layers through the plasma membrane.

Grasshopper—The grasshopper carries on respiration by means of spiracles and tracheae (Figure 4.2). Blood plays no role in transporting oxygen and carbon dioxide. Muscles of the abdomen pump air into and out of the spiracles and the tracheae.

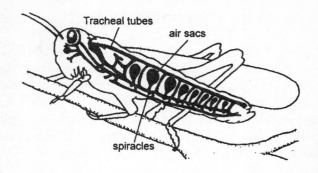

**Figure 4.2.
Respiration in the Grasshopper**

Earthworm—The skin of the earthworm is its respiratory surface. Oxygen from the air diffuses into the capillaries of the skin and joins with hemoglobin dissolved in the blood plasma. Carbon dioxide from the tissue cells diffuses into the blood. When the blood reaches the capillaries in the skin again, the carbon dioxide diffuses through the skin into the air.

Refer to Table 4.3 for some comparisons among different organisms.

Fish—The gills are specialized organs of respiration in fish. Fish pump water through the mouth and over the gills. As water passes over the gills, oxygen in the water is picked up by the capillaries within the gills and is distributed as blood flows throughout the body. At the same time, carbon dioxide in the blood diffuses out of the gill capillaries and into the water.

**Table 4.3.
Comparison of Respiratory Surfaces
of Different Organisms**

Organism	Respiratory Surface Present
Protozoan	Plasma membrane
Hydra	Plasma membrane of each cell
Grasshopper	Tracheae network
Earthworm	Moist skin
Human	Alveoli in lungs

Immune and Lymphatic System

The lymphatic system provides an important link with the cardiovascular system and the immune system. Fluid and proteins that are lost from the cells are returned to the blood via the lymphatic system. The fluid in the lymphatic system is call lymph. The lymph system begins with blind ended lymph capillaries (Figure 4.3) which join together and enter large vein-like lymph vessels. At certain locations, the lymph vessels are expanded into structures called lymph nodes. The lymph nodes are an important part of the body's defense system. There are also lymph capillaries called lacteals in the villi of the small intestine that absorb fats and transport them from the digestive tract to the circulatory system. The lymph from the upper and lower portion of the body flows into veins in the thorax.

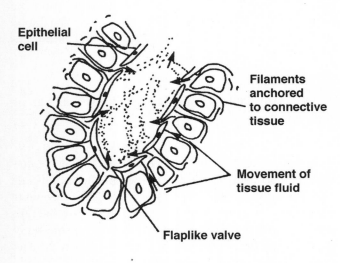

Epithelial cell

Filaments anchored to connective tissue

Movement of tissue fluid

Flaplike valve

**Figure 4.3.
A Lymphatic Capillary**

Lymph capillaries have many "flaps" or "valves" that permit the inflow of fluid and particulate matter but prevent their "back flow" into the interstitial spaces. Edema, which is the accumulation of fluid in tissues, can result if fluid flow through the lymphatic system is blocked.

Bacteria and viruses that have infected tissues cannot be directly absorbed via the blood capillaries. These pathogens enter lymph and are subsequently transported into lymph nodes that contain large numbers of lympho-cytes and macrophages that defend against these microorganisms.

The lymphatic system transports all the following EXCEPT

a) plasma.   b) interstitial fluid.

c) pathogens.   d) macrophages.

Solution

a) Plasma travels in the circulatory system. Interstitial fluid (lymph), pathogens, and macrophages may be transported in the lymphatic system.

2. Internal Transport and Exchange

Circulatory System

The circulatory system has a major role in maintaining the stability of the body's internal environment, (i.e., homeostasis). The fluid in the body can be divided into intracellular fluid and extracellular fluid, which have different compositions. The extracellular fluid can be further divided into interstitial fluid and the fluid of the circulatory system (i.e., plasma). The circulatory system is responsible for the movement and mixing of the extracellular fluid. Some major roles of the circulatory system are as follows:

(a) the delivery of oxygen and required nutrients

(b) the removal of metabolic waste products

(c) the transport of regulatory molecules such as hormones

(d) the transport of protective chemicals and enzymes. Vitamin E is an example of a protective chemical

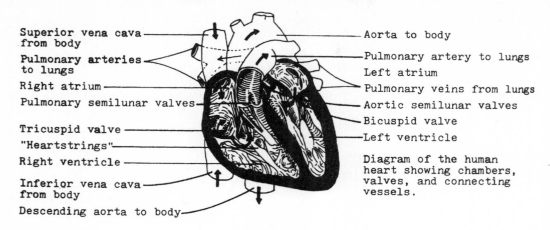

Superior vena cava from body
Pulmonary arteries to lungs
Right atrium
Pulmonary semilunar valves
Tricuspid valve
"Heartstrings"
Right ventricle
Inferior vena cava from body
Descending aorta to body

Aorta to body
Pulmonary artery to lungs
Left atrium
Pulmonary veins from lungs
Aortic semilunar valves
Bicuspid valve
Left ventricle

Diagram of the human heart showing chambers, valves, and connecting vessels.

Figure 4.4.
The Four-Chambered Heart

that inhibits free radical damage to cell membranes and macromolecules

(e) the transport of molecules and cells essential to the immune system

(f) thermoregulation

PROBLEM

The circulatory system does all the following EXCEPT

a) deliver oxygen.   b) regulate blood pressure.

c) transport enzymes.   d) remove waste products.

Solution

b) The circulatory system transports materials to and from tissues; it does *not* regulate blood pressure. Blood pressure is controlled, in part, chemically.

The four-chambered heart contains two pumps: one pumps blood to the lungs and the other pumps blood to the systemic circulation (see Figure 4.4). The four-chambered heart is composed of two atria and two ventricles. The atria are filling chambers and pump blood to the ventricles, which provide the main contractile force needed to move blood through the circulatory system. Between the atria and ventricles

are the atrioventricular valves. At the exits from the heart, between the right ventricle and pulmonary artery and left ventricle and aorta are the semilunar valves.

Deoxygenated venous blood from the inferior and superior vena cava flows into the right atrium. Atrial contraction moves the blood through the atrioventricular valve to the right ventricle. The contraction of the right ventricle pumps blood to the lungs where the blood is oxygenated and carbon dioxide is removed. The oxygenated blood is returned to the heart via the pulmonary veins and enters the left atrium. With the contraction of the left atrium the blood enters the left ventricle. The contraction of the left ventricle pumps blood into the systemic circulatory system.

The heart is supplied with oxygenated blood by two branches of the aorta (i.e., the right and left coronary arteries). A major cause of cardiovascular disease is the accumulation of atherosclerotic plaques in the coronary arteries. This severely narrows the lumen of these arteries and a small blood clot can block these arteries resulting in a cut-off of oxygenated blood to the heart. This results in a heart attack.

The cardiac cycle is the period from the end of one contraction to the end of the next contraction. The period of relaxation is called the diastole and the period of contraction is the systole.

PROBLEM

Which of the following is *not* a true statement?

a) Blood enters the heart through the superior (anterior) vena cava or through the inferior (posterior) vena cava.

b) The pulmonary artery carries oxygenated blood.

c) Oxygenated blood first enters the left atrium of the heart.

d) The systemic circulation contains oxygenated blood.

Solution

b) Blood enters the right atrium of the heart through the superior or inferior vena cava. When this chamber is filled, the blood is forced through the tricuspid valve and into the right ventricle. From there, this deoxygenated blood travels through the pulmonary artery to the lungs where it exchanges carbon dioxide for oxygen. Once oxygenated, the blood travels to the left atrium through the pulmonary veins. It travels through the bicuspid valve to the left ventricle, out through the aorta, and is then distributed throughout the body. [Statement d is also incorrect.]

Arterial and Venous Systems, Capillary Beds, and Systemic and Diastolic Pressure

Arteries carry blood away from the heart under high pressure. Arteries have strong walls, which pulse with heart pulsations. The maximum pressure reached during the arterial pulse is called the systolic pressure and the lowest pressure is called the diastolic pressure. The arteries end in arterioles, which control the flow of blood into the capillary beds. The capillaries have a very thin wall that permits the exchange of nutrients, hormones, electrolytes, and other substances between blood and the interstitial spaces between cells. The deoxygenated blood from the capillary beds collects in the

venous system and returns to the heart. The venous system is under low pressure and veins have thinner walls than arteries.

The arterial pulse pressure is influenced primarily by the stroke volume of the heart and by the compliance of the arterial vasculature. The stroke volume of the heart is the amount of blood pumped out of the heart with each heartbeat. The greater the stroke volume output, the greater the arterial pulse pressure. The compliance of the arterial vasculature refers to the distendability of the arteries to a pressure load. The greater the arterial compliance, the lower the arterial pulse pressure.

PROBLEM

The only artery in the human body that carries deoxygenated blood is the

a) pulmonary artery.   b) right coronary artery.

c) left coronary artery.   d) carotid artery.

Solution

a) The pulmonary artery carries blood to the lungs to be cleaned of its carbon dioxide. All other arteries carry oxygenated blood.

Composition of Blood

Whole blood can be separated by low speed centrifugation into a cell-free fluid called serum (or plasma if a blood anticoagulant is present) and a pellet containing cells and platelets. Plasma is about 92% water and contains electrolytes, lipoproteins, proteins, hormones, other nutrients, and vitamins. The lipoproteins are lipid-protein complexes. Lipoproteins are the primary transport molecules for lipids. Lipoproteins are divided into very low-density lipoprotein, low-density lipoprotein, and high-density lipoprotein. High plasma levels of low-density lipoprotein are associated with atherosclerosis and cardiovascular disease. In contrast, high plasma levels of high-density lipoprotein are thought to protect against atherosclerosis.

The primary proteins found in plasma are albumin, globulins, and fibrinogen. Albumin is the most abundant plasma protein (about 60%) and is a carrier molecule for nonesterified fatty acids. Albumin also plays a role in maintaining the osmotic pressure of blood. The globulins are divided into alpha-, beta-, and gamma-globulins. The gamma-globulin fraction contains molecules that function as antibodies in the humoral immune system. Fibrinogen functions in clot formation.

The red blood cell (or erythrocyte) is the primary cell found in blood. This unique cell has a plasma membrane but no other membranous organelles and does not have a nucleus. The primary function of red blood cells is oxygen transport to tissues and the removal of carbon dioxide. The oxygen-carrying molecule in the red blood cell is hemoglobin. The red blood cell has a biconcave shape and is extremely deformable and able to move through very small capillaries. In anemia the number of red blood cells in a given volume of blood is low resulting in a decreased ability to deliver oxygen to tissues. Nutritional and/or genetic factors can contribute to anemia.

Blood also contains white blood cells and platelets. White blood cells (or leukocytes) include monocytes, lymphocytes, neutrophils, eosinophils, and basophils. Neutrophils and eosinophils, as well as monocytes, are phagocytic cells. Lymphocytes also play a key role in the immune system. Platelets function in clot formation.

Role of Hemoglobin in Oxygen Transport

Hemoglobin is the primary molecule found in red blood cells and its primary function is in the transport of oxygen. The three-dimensional structure of hemoglobin is known in detail from X-ray crystallographic studies. Hemoglobin is a tetramer with two identical alpha subunits and two identical beta subunits (alpha$_2$beta$_2$). Both the alpha and beta subunits have a structure similar to myoglobin. Myoglobin is the monomeric oxygen binding protein of muscle. Each of the hemoglobin subunits has a heme group containing iron in the ferrous (Fe^{+2}) state. Each heme group can bind a single oxygen molecule. Oxygen binding does not change the oxidation state of the heme iron.

Hemoglobin is an allosteric protein. The binding of oxygen to hemoglobin is regulated by other molecules such as protons (H^+), carbon dioxide (CO_2), and 2,3-diphosphoglycerate (DPG). These molecules exert their influence on oxygen binding by binding to sites that are distinct from the oxygen binding sites. A key feature of oxygen binding to hemoglobin is the cooperative nature of this binding (see Figure 4.5). Cooperative binding occurs when the binding of each oxygen molecule facilitates the binding of the next oxygen molecule. This cooperative binding results in a characteristic sigmoidal dissociation curve as shown in Figure 4.5. In contrast, the binding of oxygen to myoglobin is not cooperative (and the dissociation curve is a hyperbola), but myoglobin does have a stronger affinity for oxygen than does hemoglobin.

The cooperative binding of oxygen to hemoglobin plays an important physiological role in the delivery of oxygen to tissues. Hemoglobin is almost fully saturated with oxygen at the partial pressure of oxygen found in the lung (pO_2 = 100 mm Hg). Oxygen is readily dissociated from hemoglobin and delivered to myoglobin, which has a stronger affinity for the oxygen. As more oxygen is dissociated from hemoglobin, the affinity of the remaining oxygen is less. This follows since dissociation is just the reverse of binding. Thus, hemoglobin is able to deliver oxygen to tissues even at the low pO_2 levels found in capillaries (pO_2 = 20–26 mm Hg).

CO_2, DPG, and H^+ shift the oxygen dissociation curve to the right (i.e., the affinity of hemoglobin for oxygen is decreased). Tissues with a high metabolic activity, such as contracting muscle, generate large amounts of H^+ and CO_2. High H^+ and CO_2 lower the affinity of hemoglobin for oxygen and thereby increase the delivery of oxygen to these metabolically active tissues. This is called the Bohr effect.

PROBLEM

In muscles, oxygen leaves hemoglobin to bind with myoglobin because

a) the presence of H+ and CO_2 in muscles increases the affinity of hemoglobin for oxygen.

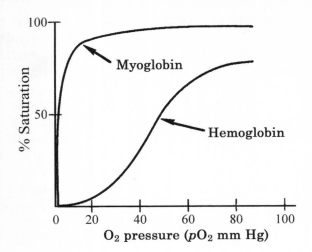

Figure 4.5.
Oxygen Binding to Hemoglobin and Myoglobin

b) the removal of oxygen from hemoglobin increases hemoglobin's affinity for the remaining oxygen.

c) myoglobin has a stronger oxygen affinity than hemoglobin.

d) the bonding of oxygen to myoglobin is cooperative.

Solution

c) Myoglobin has a stronger oxygen affinity than hemoglobin. Myoglobin does not have the cooperative oxygen binding. The presence of H+ and CO_2, as well as the removal of oxygen from hemoglobin, decreases the affinity of hemoglobin for oxygen.

Circulation Mechanisms in Other Organisms

Not all vertebrate animals have the same construction; see Figure 4.6 and note the progression from fish to mammals.

Fish—Fish have two-chambered hearts, with one atrium and one ventricle. Blood pumped from the ventricle travels to the gills, where oxygen from the water is taken up and carbon dioxide is removed. The oxygenated blood is carried to other parts of the body, and deoxygenated blood is returned to the atrium by veins. Thus, blood passes through two capillary beds, one in the gills and one in the other body organ.

Amphibians—Frogs and other amphibians have three-chambered hearts, with two atria and one ventricle. Blood that is pumped out of the ventricle enters either the pulmonary (to the lungs and skin) or the systemic (to all other organs except the lungs) circuit. The pulmonary circuit picks up oxygen in the lungs, and the oxygenated blood returns to the heart and is pumped to the rest of the body systems. This is known as double circulation.

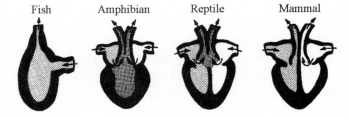

Figure 4.6.
The Hearts of Four Classes of Vertebrates (Note from fish to mammal that there is increasing separation between the two sides of the organ, and therefore a decreased amount of mixing between oxygenated and deoxygenated blood.)

Examples of Some Nonvertebrate Circulatory Systems

Protozoans—Most protozoans are continually bathed by food and oxygen because they live in water or another type of fluid. With the process of cyclosis or diffusion, digested materials and oxygen are distributed within the cell, and water and carbon dioxide are removed. Proteins are transported by the endoplasmic reticulum.

Hydra—Like the protozoans, materials in the hydra are distributed to the necessary organelles by diffusion, cyclosis, and by the endoplasmic reticulum.

Earthworm—The circulatory system of the earthworm (Figure 4.7) is known as a "closed" system because the blood is confined to the blood vessels at all times. A pump that forces blood to the capillaries consists of five pairs of aortic loops. Contraction of these loops forces blood into the ventral blood vessel. This ventral blood vessel transports blood toward the rear of the worm. The dorsal blood vessel forces blood back to the aortic loops at the anterior end of the worm.

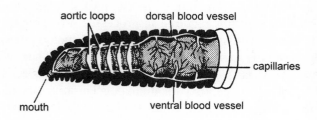

Figure 4.7.
"Closed" Circulatory System of the Earthworm

Grasshopper—The grasshopper possesses an "open" circulatory system in which the blood is confined to vessels during only a small portion of its circuit through the body (Figure 4.8). The blood is pumped by the contractions of a tubular heart and a short aorta with an open end. Blood from the heart flows into the aorta and then into sinuses. The blood then returns to the heart.

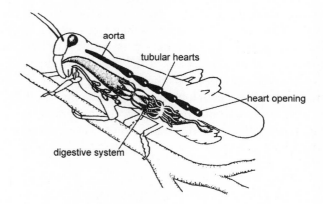

Figure 4.8.
"Open" Circulatory System of the Grasshopper

Gastrovascular and Digestive Systems

Foods consumed by humans and other animals are a combination of carbohydrates, fats, proteins, water,

minerals, salts and vitamins. All these are essential to life but only carbohydrates, fats and proteins can provide energy. Vitamins and minerals act as coenzymes and cofactors in metabolism.

The standard unit of energy is the calorie. One calorie is the amount of energy required to raise the temperature of one gram of water by one degree Celsius. Biology literature often uses kilocalories (kcal), the energy needed to raise the temperature of a liter of water by one degree Celsius. During catabolism, all carbohydrates and fast and most proteins can be degraded into carbon dioxide and water. The energy released during this metabolism is called the *heat of combustion*. Typical heat of combustion values for the major sources of energy are as follows:

— fat = 9.5 kcal/gram

— protein = 4.5 kcal/gram

— carbohydrates = 4.0 kcal/gram.

Obviously, fats contain almost twice the energy per gram as protein and carbohydrates.

A six-carbon monosaccharide, glucose, is the primary source of energy for all cells in the human body (and most higher animals). Recall in Chapter 2 that aerobic oxidation of glucose yields 36 ATP molecules (energy sources) while anaerobic glycolysis releases 2 ATP molecules, thus an 18:1 conversion ratio of a molecule of glucose. A mole of fat yields 463 ATP, a ratio of ~13:1 fat to carbohydrate energy content. Physiologically, more than half of the energy of combustion is lost as heat while about 38% is converted into cellular energy via ATP.

In animal cells, glucose is stored as glycogen, in plants as starch. When glucose (energy) is required, starch and glycogen are broken down into glucose and smaller molecules. Brain cells are some of the first to suffer from glucose deficiency because it is not stored there, whereas muscle tissues are less affected because they are able to store it as glycogen.

Both liver and muscle cells have the capacity to store excess glucose as glycogen. The main difference is the way it is utilized in the presence in the liver cells of an enzyme, glucose-6-phosphatase (see Figure 4.9).

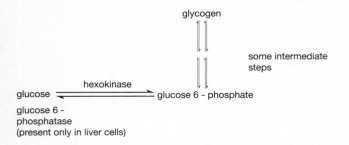

Figure 4.9.
Interconversion of Glucose
and Glycogen in Muscle and Liver Cells

Proteins yield almost as much energy as carbohydrates but are rarely fully metabolized because their amino acids are used for cellular protein synthesis. Therefore, it requires less energy for the cell to utilize existing amino acids than to use energy to completely degrade them and then synthesize new ones.

Nutrient exchange is primarily via blood vessels called capillaries. These have very thin walls (one epithelial cell thick) and thus permit diffusion of nutrients, hormones and oxygen from the blood to tissues; likewise, carbon dioxide and nitrogenous wastes are carried away from tissues.

About 9,000 mL of fluid normally enter the intestinal tract of an adult human from the stomach. This fluid (called chyme) is composed of digested material from the stomach, gastric secretions, and ingested fluids, either digested or not. The material is nearly isotonic, i.e., having the same solute concentration as plasma. However, because of the rapid absorption of monosaccharides, sodium and amino acids in the small intestine, the total concentration of solutes in the lumenal fluid drops, thus becoming relatively hypotonic to the plasma. This results in the water rapidly moving by diffusion across the highly permeable intestine wall from the lumen into the plasma. Additionally, other ions are absorbed, some by active transport. Other ions include: Cl^-, K^+, Mg^{2+} and Ca^{2+}. Additional water is absorbed in the large intestine where active transport of Na^+ from the lumen to the blood causes the osmotic absorption of water. The longer fecal material remains in the colon, the more water is reabsorbed.

Sodium (Na^+) is essential to the body for proper functioning. It is required for changes in membrane potential and thus required in processes involving excitation of cells. Recall that the plasma membrane is differentially permeable to potassium (K^+) and sodium ions. It also is able to actively pump sodium out of cells. The net effect is a higher sodium concentration outside the cell and a lower one inside. Potassium is also distributed unequally on either side of the membranes but not to the same degree as sodium. Sodium is also important in nerve impulse and muscle contraction (section 4.4). It is also important in helping to maintain solute concentration (osmolarity) of the blood plasma and other body fluids and this is vital to cell integrity (e.g., an increase in sodium concentration in blood plasma can cause water to move out of red blood cells resulting in their crenation).

Plasma sodium level is closely related to the plasma volume because of the principle of osmosis (i.e., if the Na^+ concentration in the blood increases, water enters the plasma from the surrounding tissues thus raising the plasma volume; conversely, if the plasma Na^+ level decreases, the osmolarity of the blood decreases). Water then leaves the plasma by osmosis, reducing the blood volume. thus sodium concentration in the body is regulated in the same way that blood volume is regulated via glomeruler filtration.

The primary role of the digestive system is to break down food into substances that can be absorbed through the lining of the digestive tract and thereby enter the body tissues. Any substances that cannot be absorbed from the digestive tract are eliminated in the feces. The digestive system consists of the alimentary canal and the exocrine organs that secrete digestive juices into the alimentary canal. The alimentary canal starts at the mouth and includes the pharynx, esophagus, stomach, small intestine, large intestine, and anus

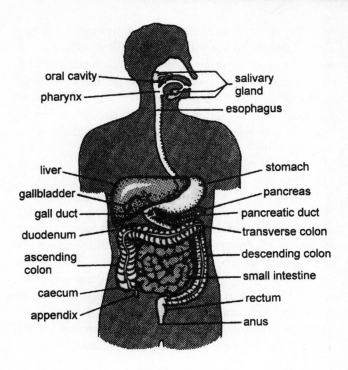

**Figure 4.10.
The Digestive System**

food particles and also initiates the digestion of carbohydrate. Amylase is the digestive enzyme in saliva; it splits starch and glycogen into disaccharides. In plants, glucose is stored as starch granules, and in animals, glucose is stored as glycogen granules. The two principal components of starch are the polysaccharides amylose and amylopectin.

PROBLEM

Which of the following organs is *not* a part of the human digestive system?

a) Esophagus   b) Thymus

c) Gall bladder   d) Pancreas

Solution

b) All the choices are a part of the digestive system except for the thymus. The thymus is part of the immune system.

Stomach

The food mass from the mouth passes through the pharynx and the esophagus by a peristaltic wave and enters the stomach. In the stomach the food mass is mixed with gastric secretions and the digestion of protein is initiated. The gastric glands secrete pepsinogen, mucus, hydrochloric acid, and intrinsic factor. Intrinsic factor is important in promoting the absorption of vitamin B_{12}.

Pepsinogen is the inactive form of pepsin. The inactive form of an enzyme is called a zymogen. Pepsinogen is activated to pepsin by cleavage of a 44-residue peptide (from the amino terminal end) and this activation occurs spontaneously at pH 2 and is also catalyzed by pepsin. Pepsin is an acid-protease and has maximal enzymatic activity at pH 2 to 3. The hydrochloric acid in the stomach functions to maintain an acidic pH (of 1) and also denatures dietary protein to make it more susceptible to protease attack. Pepsin primarily catalyzes the hydrolysis of peptide bonds with an aromatic amino

(see Figure 4.10). The salivary glands, pancreas, liver, and gallbladder are organs that secrete substances into the alimentary canal.

Ingestion: Structures and Their Functions

The first component of the alimentary canal is the mouth, which mechanically reduces the size of food materials and mixes the masticated particles with saliva from the salivary glands. The teeth are specialized structures for breaking up food particles and increasing the surface area of the food particles. The incisors cut large food sections, the canines serve to grasp and tear food, and the premolars and molars are effective in grinding food particles.

Saliva is secreted into the mouth by the salivary glands and it increases the moisture content of the acid residue such as phenylalanine, tryptophan, and tyrosine. The polypeptides produced by pepsin digestion are transported to the small intestine for further hydrolysis. The mixture of food mass and gastric juices in the stomach constitutes chyme. Chyme enters the small intestine, which also receives the secretions of the liver and pancreas.

PROBLEM

Which of the following is a zymogen?

a) Protease

b) Tyrosine

c) Chyme

d) Pepsinogen

Solution

d) Pepsinogen is a zymogen, an inactive form of an enzyme. Protease is an enzyme, tyrosine is an amino acid, and chyme is the mixture found in the stomach.

Digestive Glands

The pancreas, in addition to its role in the endocrine system (see the Endocrine System), also functions in the exocrine system. The acinar cells, which form most of the pancreatic mass, secrete pancreatic juice, which travels through the pancreatic duct to the duodenum. The bile duct from the liver and gallbladder enters the duodenum at the same site. Pancreatic juice aids the digestion of carbohydrate, fat, and protein. The pH of pancreatic juice is alkaline, and it neutralizes the acidic chyme from the stomach. Carbohydrate digestion is assisted by pancreatic amylase, which breaks down starch and glycogen into disaccharides. Triglycerides are hydrolyzed into glycerol and free fatty acids by the action of pancreatic lipase.

A mixture of zymogens are released into the small intestine. These zymogens include trypsinogen, chymotrypsinogen and procarboxypeptidase, and they are converted into their active forms (trypsin, chymotrypsin, and carboxypeptidase, respectively). Trypsin hydrolyzes the carboxyl side of peptide bonds with arginine and

lysine residues while chymotrypsin acts on the carboxyl side of peptide bonds with aromatic residues (phenylalanine, tryptophan, and tyrosine) as well as methionine. Carboxypeptidase A releases the carboxyl-terminal amino acids and carboxypeptidase B is restricted to peptides with an arginine or lysine carboxyl-terminal. The pancreas also secretes nucleases to breakdown nucleic acids into nucleotides.

The liver functions in the process of digestion by secreting bile salts that act like a detergent and emulsify fat. This emulsification increases the surface area of the fat droplets. Lipases can efficiently utilize triglycerides (fat droplets) that have been emulsified by bile salts. Bile salts are synthesized by hepatic cells from cholesterol and are secreted into the common bile duct. The common bile duct exits to the duodenum via a sphincter muscle (the sphincter of Oddi). Between meals, the sphincter of Oddi is closed and bile is stored in the gallbladder.

PROBLEM

Bile is secreted by the

a) stomach.

b) liver.

c) duodenum.

d) gallbladder.

Solution

b) Bile is an aqueous solution that contains various organic and inorganic solutes. Among the major organic solutes are bile salts, phospholipids, cholesterol, and bile pigments. The adult human liver produces about 15 mL of bile per kilogram body weight. The rate of synthesis and secretion is dependent mainly upon blood flow to the liver.

Small Intestine and Large Intestine

The small intestine receives chyme from the stomach as well as pancreatic juice and bile. The small intestine begins at the pyloric sphincter and ends where it joins the large intestine. The three segments of the small intestine are the duodenum, the jejunum, and the ileum. The small intestine has numerous villi that project into the intestinal lumen. These villi serve to mix chyme with intestinal juices and aid in the absorption of digested nutrients. The small intestine is the primary absorbing organ of the alimentary canal. The epithelial cells of the intestinal mucosa of the duodenum have a variety of digestive enzymes. These include peptidases (which break polypeptides into amino acids), and enzymes that convert disaccharides into monosaccharides. Also, the duodenum is where the pancreatic juices and bile enter the digestive tract.

The large intestine plays almost no role in the digestion of food but does reabsorb water and electrolytes. In addition, it stores feces until defecation of undigestible food components, such as fiber.

PROBLEM

Villi are finger-like protrusions of the

a) small intestine.   b) outer ear.

c) bronchioles.   d) capillaries.

Solution

a) Villi line the lumen of the small intestine and thereby increase the intestinal surface area. Most of the nutrient absorption during digestion occurs through the villi.

Muscular Control of Digestion

The smooth muscles of the alimentary canal promote both the mixing of food with gastric juices and the rhythmic wave-like movements (peristaltic contractions) that propel food through the lumen of the digestive tract. The regulation of this muscular contraction is primarily controlled by an "intrinsic nervous system." This intrinsic system also regulates much of the secretory functions required for digestion. In addition, nerve fibers from the parasympathetic and sympathetic branches of the autonomic system also interact with the intrinsic nervous system of the gut. In general, the para-

sympathetic system increases the activity of the gut and the sympathetic system decreases this activity.

The vagus nerve, which is part of the parasympathetic system, arises from the brain and innervates the esophagus, stomach, pancreas, and the proximal half of the large intestine. Parasympathetic nerve fibers also originate from the sacral segments of the spinal cord and innervate the distal segment of the large intestine. The sympathetic innervation also regulates the gastrointestinal tract. The preganglionic fibers originate in the spinal cord and the postganglionic fibers innervate all parts of the gut. The norepinephrine secreted by the ends of the sympathetic nerves inhibits the contractions of smooth muscles in the gut and also inhibits the intrinsic nervous system.

PROBLEM

Norepinephrine inhibits all the following EXCEPT

a) smooth muscle contraction.

b) peristalsis.

c) the intrinsic nervous system.

d) the parasympathetic nervous system.

Solution

d) Norepinephrine inhibits the intrinsic nervous system and smooth muscles in the gut (which produce peristalsis). Norepinephrine does not affect the parasympathetic nervous system.

Ingestion and Digestion in Other Organisms

Protozoans—Many protozoa of the Eukarya Domain ingest food using a specialized feeding structure called an oral groove, which leads to a mouth-like structure called the cytostome. Food is packaged and broken down within a vacuole that serves as a digestive compartment. Nutrients pass through the vacuole membrane into the cytoplasm, and waste is eliminated via exocytosis through the anal pore, a specialized region of the plasma membrane.

Hydra—The hydra possesses tentacles that have stinging cells (nematocysts) that shoot out filaments covered with a poison that paralyzes the prey. If successful in capturing an animal, the tentacles push it into the hydra's mouth. From there, the food enters the gastric cavity. The hydra uses both intracellular and extracellular digestion.

Earthworm—As the earthworm moves through soil, the suction action of the pharynx draws material into the mouth cavity. From the mouth, food goes into the pharynx, the esophagus, and then the crop, which is a temporary storage area (Figure 4.11). The food then passes into a muscular gizzard where it is ground and churned. The food mass finally passes into the intestine where absorption occurs; any undigested material is eliminated through the anus.

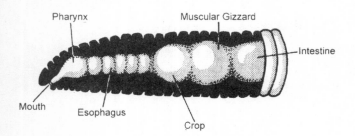

Figure 4.11.
The Digestive System of the Earthworm

Grasshopper—The grasshopper is capable of consuming large amounts of plant material. This plant material must first pass through the esophagus into the crop, a temporary storage organ (Figure 4.12). It then travels to the muscular gizzard where food is ground. Digestion takes place in the stomach. Enzymes secreted by six gastric glands are responsible for digestion. Absorption takes place mainly in the stomach. Undigested material passes into the intestine, collects in the rectum, and is eliminated through the anus.

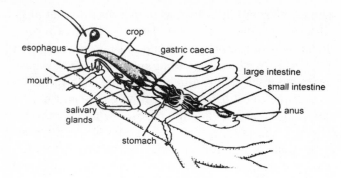

Figure 4.12.
The Digestive System of a Grasshopper

Excretory System

Metabolic waste products are often toxic and must be removed from the body to prevent tissue damage. Blood and lymph are the fluids that initially receive metabolic wastes. Gaseous waste products such a CO_2 are removed from blood by the respiratory system. Salts and nitrogenous wastes are removed from the circulatory system by the urinary system. The urinary system also plays a key role in body homeostasis by helping to regulate the volume and composition of extracellular fluid, blood pressure, and the production of red blood cells. The urinary system is composed of a pair of kidneys, a pair of ureters, a urinary bladder, and a urethra (Figure 4.13).

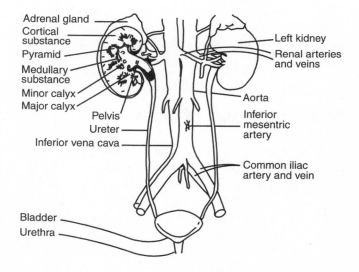

Figure 4.13.
The Kidneys

PROBLEM

Wastes are removed from the blood by

a) the respiratory system.   b) the urinary system.

c) the digestive system.   d) both a) and b).

Solution

d) The respiratory and urinary systems both remove wastes. The respiratory system removes gases from the blood and the urinary system removes liquid and dissolved materials from the blood. The digestive system breaks down food, but does not remove wastes.

Kidney: Structure and Function

The kidneys (see Figure 4.13) are located on both sides of the spinal column and are behind the parietal peritoneum. The surface of the kidney facing the spinal column (the medial surface) is concave and contains a deep sinus through which the renal artery (from the aorta) and the renal vein (from the inferior vena cava) enter. The ureter, which transports urine away from the kidney, also exits from the renal sinus.

The extrarenal ureter is expanded into a funnel-shaped sac (the renal pelvis) at its junction with the renal sinus. The intrarenal tubes of the ureter branch to form two major calyces, each of which further divides into minor calyces. Urine is delivered into the minor calyces by the renal papillae and then flows through the ureter into the bladder.

The interior of the kidney is divided into an inner renal medulla and an outer renal cortex. The renal medulla contains pyramidal structures whose apexes form the renal papillae. The renal cortex has a granular appearance due to the many small tubules of the nephrons (see Figure 4.14).

In addition to its role of removing metabolic wastes, the kidney also secretes erythropoietin, which stimulates the production of red blood cells. The kidney also secretes renin an enzyme involved in regulating blood pressure. The kidney converts the inactive form of vitamin D (25-hydroxyl-vitamin D) to the active form of vitamin D (1,25-dihydroxyvitamin D). The active form of vitamin D promotes Ca^{+2} absorption.

PROBLEM

The kidney can do all the following EXCEPT

a) remove metabolic wastes.

b) help activate vitamin C.

c) help regulate blood pressure.

d) help stimulate production of red blood cells.

Solution

b) The kidney does everything listed above but help in the activation of vitamin C. The kidney helps in the activation of vitamin D.

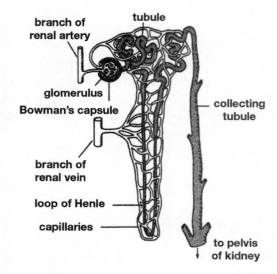

**Figure 4.14.
A Nephron**

Nephron: Structure and Function

Urine is formed by the nephrons, which are the functional units of the kidney. About a million nephrons are present in each kidney. The two major sections of a nephron (Figure 4.14) are the corpuscle (glomerulus enclosed within Bowman's capsule) and the tubule (proximal tubule, Loop of Henle, distal tubule and collecting tubule). The glomeruli are contained in the cortex of the kidney, and some of the tubules extend into the medulla (Figure 4.13).

PROBLEM

Which of the following is *not* part of the nephron in the human kidney?

a) Proximal convoluted tubule

b) Loop of Henle

c) Distal convoluted tubule

d) Major calyx

Solution

d) The nephron is the structural and functional unit of the kidney and consists of a renal corpuscle (a glomerulus enclosed within Bowman's capsule), and its attached tubule. The tubule consists of the proximal convoluted portion, the loop of Henle, and the distal convoluted portion. The nephrons empty into collecting tubules. There are approximately one million nephrons in each kidney. The major calyx is not part of the nephron, but is, rather, a part of the intrarenal collecting system.

Formation of Urine

Blood enters the glomerulus through the afferent arteriole and exits through the efferent arteriole. The blood in the glomerulus is under pressure (60 mm Hg) and this pressure forces some fluid out of the blood and into Bowman's capsule. The fluid in the Bowman's capsule flows into:

(1) the proximal renal tubule (in the cortex of the kidney),

(2) the loop of Henle,

(3) the distal tubule, and

(4) a collecting duct which, in turn, flows into the renal pelvis.

The end result of this fluid movement is the creation of urine. The solute composition of urine depends on how much water and each of various solutes needs to be excreted from the body to maintain proper concentrations of these substances in the body. The loops of Henle and the vasa recta provide mechanisms for regulating the osmolarity and volume of urine produced by the kidney.

The concentration of electrolytes and other substances in the fluid contained in Bowman's capsule (i.e., the glomerular filtrate) is very similar to that found in interstitial fluid. Most of the water and some of the solutes in the glomerular filtrate is reabsorbed by tubule cells, that is a tubule cell transports a substance from the lumen of the tubule into the interstitial fluid. Substances in the interstitial fluid then easily enter the peritubular capillaries. The solutes that are reabsorbed into the circulation are useful and their loss in urine would be wasteful. For example, glucose, amino acids, and many electrolytes are reabsorbed. Toxic, unwanted, or "excess" solutes are not reabsorbed and appear in urine. In addition to waste molecules that leave the blood when they are filtered into Bowman's capsule, some wastes are directly secreted by tubule cells, that is they are transported by tubule cells from the interstitial fluid into the tubule lumen.

Storage and Elimination of Wastes

Urea is a by product of amino acid metabolism and it is a main constituent of urine. Uric acid, which is formed from the catabolism of purines, is also eliminated in urine. Urine, from the renal pelvis flows through the ureter to the urinary bladder. The flow of urine through the ureter is promoted by peristaltic contraction of the muscular lining of the ureter. The process by which the urinary bladder empties is called micturition. Micturation occurs when the tension in the walls of the bladder (caused by an increasing volume of urine) reaches a threshold level. This can trigger a reflex that results in the emptying of the bladder. The micturation reflex can also be influenced by both inhibitory and stimulatory signals from the brain. In particular, the relaxation of the urethral sphincter is necessary before urination can proceed.

PROBLEM

Micturition is controlled by

a) reflex.

b) signals from the brain.

c) hormones.

d) both a) and b).

Solution

d) Micturition is controlled by reflex and signals from the brain, not hormones.

Excretion in Other Organisms

Protozoans—Elimination of metabolic wastes occurs by diffusion through the plasma membrane. The major waste products include ammonia, carbon dioxide, mineral salts, and water. Some protozoa have contractile vacuoles for the elimination of water and dissolved substances.

Hydra—The metabolic wastes of the hydra are excreted by simple diffusion. The major wastes of the hydra include ammonia, water, carbon dioxide, and salts.

Grasshopper—The excretory system of the grasshopper is made up of Malpighian tubules (Figure 4.15). Wastes such as water, salts, and dissolved nitrogenous compounds diffuse into the blood in the body cavity. The Malpighian tubules absorb these wastes. Water present in the Malpighian tubules is reabsorbed into the blood. Any remaining waste passes into the intestine

where it is eliminated. Salts and uric acid are the major metabolic wastes.

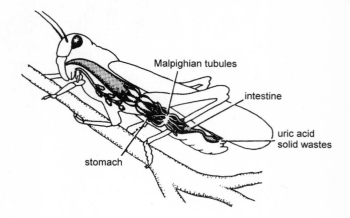

**Figure 4.15.
Excretory Organs of a Grasshopper**

Earthworm—The nephridia are the major excretory structures of the earthworm (Figure 4.16). They are needed for the elimination of water, urea, ammonia, and mineral salts. Each nephridium consists of a nephrostome, which lies within the body cavity. This body cavity is filled with fluid that enters the nephrostome and passes down a long tubule. As it travels down this tubule, useful materials are reclaimed by cells that line the tubule. Useless materials leave the earthworm by the nephridiopores, which are openings to the outside.

Refer to Table 4.4 for comparisons among organisms.

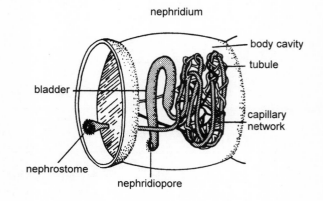

**Figure 4.16.
Excretory System of the Earthworm**

Table 4.4.
Summary of Excretion Among
Five Different Organisms

Organism	Major Organs for Excretion	Major Waste Products
Protozoans	Plasma membrane	water, ammonia, carbon dioxide, salts
Hydra	Plasma membrane	water, ammonia, carbon dioxide, salts
Grasshopper	Malpighian tubules and intestine	uric acid, salts,
Earthworm	Nephridia and intestine	urea, water, salts, ammonia
Human	Lungs, kidneys, skin, liver, large intestine	urea, salts, carbon dioxide, bile salts

Fish—Instead of conserving water like most terrestrial animals, freshwater fish need to release excess water because of their aqueous surroundings. To maintain ionic balance freshwater fish must reabsorb salts from the filtrate in the nephrons. Bony fishes that live in saltwater have the opposite problem and must excrete large amounts of ions, such as Na^+, Ca^{2+} and Mg^{2+}, and produce concentrated urine.

Frogs—In frogs, urine is produced by the kidneys and travels through the ureters to the cloaca, a common opening for the digestive, urinary, and reproductive tracts.

3. Support and Movement

Integumentary System

Human skin is composed of a comparatively thin, outer layer, the epidermis, which is free of blood vessels, and an inner, thick layer, the dermis, which is packed with blood vessels and nerve endings (Figure 4.17). The epidermis is a stratified epithelium whose thickness var-

ies in different parts of the body. It is thickest on the soles of the feet and the palms of the hands. The epidermis of the palms and fingers has numerous ridges, forming whorls and loops in very specific patterns. These unique fingerprints and palmprints are determined genetically, and result primarily from the orientation of the underlying fibers in the dermis. The outermost layers of the epidermis are composed of dead cells, which are constantly being sloughed off and replaced by cells from beneath. As each cell is pushed outward by active cell division in the deeper layers of the epidermis, it synthesizes large amounts of the fibrous protein keratin and becomes a flat (squamous), scalelike dead cell. Keratin makes the outer layer of the skin tough and durable.

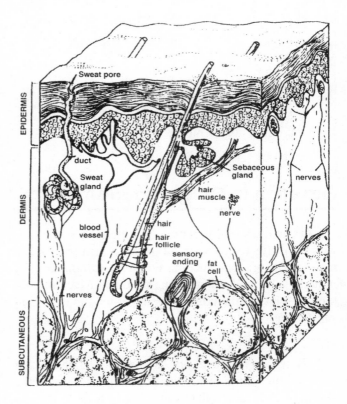

Figure 4.17.
Section of Human Skin

Scattered at the juncture between the deeper layers of the epidermis and the dermis are melanocytes, cells that produce the pigment melanin. Melanin serves as a protective device for the body by absorbing ultraviolet rays from the sun. Tanning results from an increase in melanin production as a result of exposure to ultraviolet radiation. All humans have about the same number of melanocytes in their skin. The difference between light and dark skin color is under genetic control and is due to the fact that melanocytes of a darker-skinned person produce more melanin.

The juncture of the dermis with the epidermis is uneven. The dermis throws projections called papillae into the epidermis. The dermis is much thicker than the epidermis, and is composed largely of connective tissue. The lower level of the dermis, called the subcutaneous layer, is connected with the underlying muscle and is composed of many fat cells and a more loosely woven network of fibers. This part of the dermis is one of the principal sites of body fat deposits, which help preserve body heat. The subcutaneous layer also determines the amount of possible skin movement.

The hair and nails are keratin (a protein) based derivatives of skin. Although the hair follicles are located in the dermis, they are derived from the epidermis. Similarly, nails are derived from the epidermis. Hair follicles are not found on soles and a few other regions. Individual hairs are formed in the hair follicles, which have their roots deep within the dermis (Figure 4.17). At the bottom of each follicle, a papilla of connective tissue projects into the follicle. The epithelial cells above this papilla constitute the hair root and, by cell division, form the shaft of the hair, which ultimately extends beyond the surface of the skin. The hair cells of the shaft synthesize and accumulate keratin, then die and form a compact mass that becomes the hair. Growth occurs at the bottom of the follicle only. Associated with each hair follicle is one or more sebaceous gland. The sebaceous glands secrete an oily secretion that makes the surface of the skin and hair more pliable. Like the sweat glands, the sebaceous glands are derived from the embryonic epidermis but are located in the dermis. To each hair follicle is attached smooth muscle called arrector pili, which pulls the hair erect upon contraction.

Skin functions (Protection and Thermoregulation)

Perhaps the most vital function of the skin is to protect the body against a variety of external agents and to help maintain a constant internal environment. The layers of the skin form a protective shield against blows, friction, and many injurious chemicals. These layers are essentially germproof and, as long as they are not broken, keep bacteria and other microorganisms from entering the body. The skin is water-repellent and therefore protects the body from excessive loss of moisture. In addition, the pigment in the outer layers protects the underlying layers from the ultraviolet rays of the sun.

In addition to its role in protection, the skin is involved in thermoregulation. Heat is constantly being produced by the metabolic processes of the body cells and distributed by the bloodstream. Heat may be lost from the body in expired breath, feces, and urine, but approximately 90 percent of the total heat loss occurs through the skin. This is accomplished by changes in the blood supply to the blood vessels of the skin. When the air temperature is high, the skin blood vessels dilate, and the increased flow of blood to the skin allows for heat loss from the blood to the environment. Due to the increased blood supply, the skin appears flushed. When the temperature is low, the blood vessels of the skin are constricted, thereby decreasing the flow of blood through the skin and decreasing the rate of heat loss. Temperature-sensitive nerve endings in the skin reflexively control the blood vessels diameters.

At high temperatures, the sweat glands are stimulated to secrete sweat. The evaporation of sweat from the surface of the skin lowers the body temperature by removing from the body the heat necessary to convert the liquid sweat into water vapor. In addition to their function in heat loss, the sweat glands also serve an excretory function. Five to ten percent of all metabolic wastes are excreted by the sweat glands. Sweat contains similar substances as urine but is much more dilute.

Epithelial cells form the inner and outer surfaces of the body. These cells have many specialized shapes and functions. Epithelial cells adhere to each other and to the basal lamina. Simple squamous epithelium (Figure 4.18a) forms a thin layer of cells that covers the inner lining of blood vessels. Simple cuboidal epithelium (Figure 4.18b) also consists of a single layer of tightly fitting cells but they have a cube-like shape. Cuboidal epithelium lines the ducts of many glands. A layer of elongated simple columnar epithelial cells forms the lining of the stomach, the cervix, and the small intestine (Figure 4.18c). Goblet cells are found in simple columnar epithelium and they secrete mucus. Stratified epithelia are several cell layers thick and form the surface of the mouth, the esophagus, and the vagina. The outer layer of skin consists of stratified squamous epithelial cells that have undergone a process of keratinization. Intestinal epithelial cells are very specialized and the cell surface facing the lumen of the small intestine has many microvilli that are important in the absorption of nutrients. The microvilli contain actin filaments, which help maintain their rigidity.

A major function of all epithelial cells is to provide a boundary between different cell types. The basal lamina provides a distinct boundary between the epithelial cells and the cells that underlay the basal lamina.

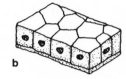
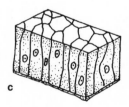

Figure 4.18.
Types of Epithelial Cells
(a) Simple Squamous Epithelium
(b) Simple Cuboidal Epithelium
(c) Simple Columnar Epithelium

Skeletal System

Bone Structure. Bone, like other connective tissues, consists of cells and fibers, but unlike the others its extracellular components are calcified, making it a hard, unyielding substance ideally suited for its supportive and protective function in the skeleton.

Upon inspection of a long bone with the naked eye, two forms of bone are distinguishable: cancellous (spongy) and compact (Figure 4.19). Spongy bone consists of a network of hardened bars having spaces be-

tween them filled with marrow. Compact bone appears as a solid, continuous mass, in which spaces can be seen only with the aid of a microscope. The two forms of bone grade into one another without a sharp boundary.

In typical long bones, such as the femur or humerus, the shaft (diaphysis) consists of compact bone surrounding a large central marrow cavity. In adults, the marrow in the long bones is primarily of the yellow, fatty variety, while the marrow in the flat bones of the ribs and at the ends of long bones is primarily of the red variety and is active in the production of red blood cells. Even this red marrow contains about 70 percent fat.

The ends (epiphyses) of long bones consist mainly of spongy bone covered by a thin layer of compact bone. In a long bone that is still growing the epiphysis contains a cartilaginous region known as an epiphyseal plate. The epiphyseal plate and the adjacent spongy bone constitute a growth zone, in which all growth in length of the bone occurs. The surfaces at the ends of long bones, where one bone articulates with another, are covered by a layer of cartilage, called the articular cartilage. It is this cartilage which allows for easy movement of the bones over each other at a joint.

Compact bone is composed of structural units called Haversian systems. Each system is irregularly cylindrical and is composed of concentrically arranged layers of hard, inorganic matrix surrounding a microscopic central Haversian canal. Blood vessels and nerves pass through this canal, supplying and controlling the metabolism of

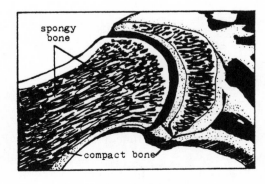

Figure 4.19.
Longitudinal Section of the End of a Long Bone

the bone cells. The bone matrix itself is laid down by bone cells called osteoblasts. Osteoblasts produce a substance, osteoid, which is hardened by calcium, causing calcification. Some osteoblasts are trapped in the hardening osteoid and are converted into osteocytes, which continue to live within the bone. These osteocytes lie in small cavities called lacunae, located along the interfaces between adjoining concentric layers of the hard matrix. Exchange of materials between the bone cells and the blood vessels in the Haversian canals is through tiny canals (canaliculi). Volkmann's canals penetrate and cross the layers of hard matrix, connecting the different Haversian canals to one another (see Figure 4.20).

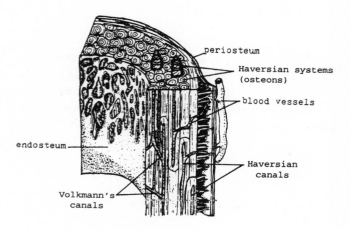

Figure 4.20.
Cross-Section of a Long Bone
Showing Internal Structures

With few exceptions, bones are covered by the periosteum, a layer of specialized connective tissue. The periosteum has the ability to form bone, and contributes to the healing of fractures. Periosteum is lacking on those ends of long bones surrounded by articular cartilage. The marrow cavity of the diaphysis and the cavities of spongy bone are lined by the endosteum, a thin cellular layer which also has the ability to form bone.

Haversian type systems are present in most compact bone. However, certain compact flat bones of the skull, such as the frontal, parietal, occipital, and temporal bones, and part of the mandible, do not have Haversian systems. These bones, termed membrane bones, have a different architecture and are formed differently than bones with Haversian systems.

Skeletal Structure. The axial skeleton (see Figure 4.21) consists of the skull, vertebral column, ribs, and the sternum. The primary function of the vertebrate skull is the protection of the brain. The part of the skull that serves this function is the cranium. The rest of the skull is made up of the bones of the face. In all, the human skull is composed of 28 bones, six of which are very small and located in the middle ear. At the time of birth, several of the bones of the cranium are not completely formed, leaving five membranous regions called fontanelles. These regions are somewhat flexible and the cranium can undergo slight changes in shape as necessary for the passage of the infant through the birth canal.

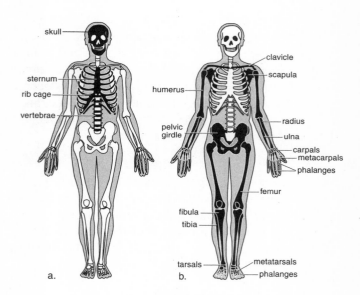

Figure 4.21.
Diagrams of Human Body Showing (a) the Bones
of the Axial Skeleton and (b) the Bones of the
Appendicular Skeleton

The human vertebral column, or spinal column, is formed from vertebrae, which differ in size and shape in different regions of the spine. In the neck region there are 7 cervical vertebrae; in the thorax there are 12 thoracic vertebrae; in the lower back region there are 5 lumbar vertebrae; in the sacral or hip region 5 fused vertebrae form the sacrum to which the pelvic girdle is attached; and at the end of the vertebral column is the coccyx or tailbone, which consists of four, or possibly five, small fused vertebrae. The vertebrae forming the sacrum and coccyx are separate in childhood, with fusion occurring by adulthood.

A typical vertebra consists of a basal portion, the centrum, and a dorsal ring of bone, the neural arch, which surrounds and protects the delicate spinal cord that runs through it. Each vertebra has projections for the attachment of ribs or muscles or both, and for articulating (joining) with neighboring vertebrae. The first vertebra, the atlas, has rounded depressions on its upper surface into which fit two projections from the base of the skull. This articulation allows for up and down movements of the head. The second vertebra, called the axis, has a pointed projection that fits into the atlas. This type of articulation allows for the rotation of the head.

In humans there are twelve pairs of ribs, one pair articulating with each of the thoracic vertebrae. These ribs support the chest wall and keep it from collapsing as the diaphragm contracts. Of the twelve pairs of ribs, the first seven are attached ventrally to the breastbone, the next three are attached indirectly by cartilage, and the last two, called "floating ribs," have no attachments to the breastbone.

The bones of the appendages and the girdles, which attach the appendages to the rest of the body, make up the appendicular skeleton. In the shoulder region the pectoral girdle serves for the attachment of the arms; in the hip region, the pelvic girdle serves for the attachment of the legs.

The pectoral girdle consists of two collarbones, or clavicles, and two shoulder blades, or scapulas. Articulating with the scapula is the single bone of the upper arm, the humerus. Articulating with the other end of the humerus are the two bones of the forearm: the radius and the ulna. The ulna has a process on its end called the olecranon that is often referred to as the "funny bone."

The wrist is composed of eight small bones called the carpals. The palm of the hand consists of five bones, known as the metacarpals, each of which articulates with a bone of the finger, called a phalanx. Each finger has three phalanges, with the exception of the thumb, which has two.

The pelvic girdle consists of the two hip bones. The hip bones are attached to the sacrum. Each hip bone is formed from three fused bones, the ilium, ischium, and pubis.

The pattern of bones in the leg and foot is similar to that in the arm and hand. The upper leg bone, the femur, articulates with the pelvic girdle. The two lower leg bones are the tibia (shinbone) and fibula. Ventral to the joint between the upper and lower leg bones is another bone, the patella or knee cap, which serves as a point of muscle attachment for upper and lower leg muscles and strengthens the joint (Figure 4.22). This bone has no counterpart in the arm.

The ankle contains seven irregularly shaped bones, the tarsals, corresponding to the carpals of the wrist. The foot proper contains five metatarsals, corresponding to the metacarpals of the hand, and the bones in the toes are the phalanges, two in the big toe and three in each of the others.

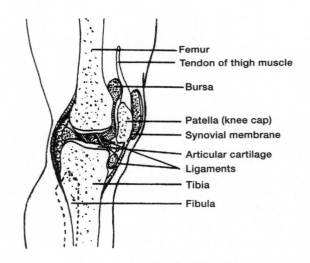

Figure 4.22.
The Structure of a Knee Joint

The point of junction between two bones is called a joint (Figure 4.22). Some joints, such as those between the bones of the skull, are immovable and extremely strong, owing to an intricate intermeshing of the edges of the bones. Some joints are freely movable. Movable joints are of several types. Some are ball and socket joints, such as the joint where the femur joins the pelvis, or where the humerus joins the pectoral girdle. These joints allow free movement in several directions. Both the pelvis and the pectoral girdle contain rounded, concave depressions to accommodate the rounded convex heads of the femur and humerus, respectively. Hinge joints, such as that of the human knee, permit movement in one plane only. The pivot joints at the wrists and ankles allow freedom of movement intermediate between that of the hinge and the ball and socket types. (Figure 4.23.)

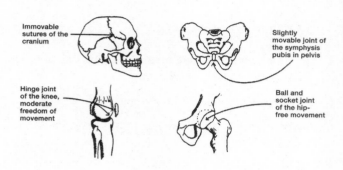

**Figure 4. 23.
Diagrams Illustrating the
Types of Joints Found in the Human Body**

Connective tissue strands called ligaments hold the different bones of a joint together. Skeletal muscles, attached to the bones by means of another type of connective tissue strand known as a tendon, produce their effects by bending the skeleton at the movable joints. The ends of each bone at a movable joint are covered with a layer of smooth cartilage. The joint cavity is filled with a liquid lubricant, called the synovial fluid, which is secreted by the membrane lining the cavity.

Movement Systems

Flagellar and Ciliary Movement

Cell motility in both plants and animals is accomplished by hair-like appendages called flagella and cilia. If there is one, longer structure it is a flagellum, if many, they are cilia. Movement is by whip-like strokes (power stroke, then recovery) as with flagella (e.g., sperm), or more vibrating or beating-type as with ciliary movement (e.g., paramecium, or the ciliated motion for clearing by the mucous cells of the windpipe in mammals). Both flagella and cilia are formed into microtubules, hollow tubes, constructed of protein called tubulin, a dimer molecule composed of two subunits: alpha and beta-tubulin. These structures are very similar to the microtubules that form the centrioles that move chromosomes during cell division because they have two in the center and nine sets of doublet-microtubules arranged in a cylinder (so-called 9+2 arrangement), but in centrioles they are triplets. Microtubules are also part of the cytoskeleton of cells and are involved in cytoplasmic streaming and organelle movement within the cell. They are one of three main types of fibers composing the cytoskeleton, the others being microfilaments and intermediate filaments.

Flagella and cilia have motor molecules that extend from each microtubule doublet to the next. These are made of a protein called dynein. Bending these motor molecules is what produces the movement by flagella and cilia. The motion of the dynein is referred to by some as "walking" and is functionally similar to the muscle movement produced by fibers during walking by terrestrial animals.

Prokaryotic (bacteria) microtubules are structurally different from those in eukaryotes. A prokaryote microtubule does not have a cytoplasmic matrix and has only a single microfibril instead of the 9+2 structure found in eukaryotes.

Muscular System

Muscle Tissues

Vertebrates have three types of muscle tissue: skeletal, smooth, and cardiac. The specialized muscle cells in muscle tissues all have the ability to contract using actin and myosin filaments. ATP hydrolysis provides the energy for muscle contraction.

Skeletal muscle connects the bones of the skeleton and is under voluntary control. Skeletal muscles are made of long muscle fibers (myofibers) and each fiber is considered a large single cell that is formed by the fusion of many separate cells. Each myofiber has many nuclei and bundles of myofibrils. As shown in Figure 4.24a, a myofiber (1–40 mm in length and 10–50 μm in width) has a striated (striped) appearance. The striated appearance is due to bundles of aligned myofibrils, which have dark bands (A bands) alternating with light bands (I bands). The narrow line that bisects each I band is called the Z-line.

Smooth muscle is under involuntary control. Smooth muscle is involved with movements of the small and large intestine and the bladder; smooth muscle also controls the diameter of blood vessels. Smooth muscle cells are not striated and have one nucleus per cell.

Cardiac muscle is found only in the heart. The cells are striated. Cardiac muscle is under involuntary control. Cardiac muscles produce the synchronous contraction of the heart (i.e., the heartbeat).

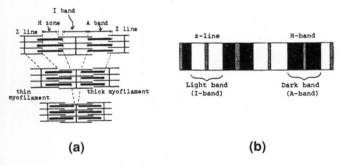

(a)   **(b)**

Figure 4.24.
The Structure of Two Adjacent Myofibrils
A sarcomere is the region between the two Z-lines.

The sarcomere is the basic unit of contraction in striated muscles, and it is the region between two Z-lines. Myofibrils are made up of repeating sarcomere units. The sarcomere has thick myosin filaments as well as thin actin filaments. The thin filaments are attached to the Z-line. Muscle contraction occurs as the result of the sliding of the thick and thin filaments past one another; note in Figure 4.24 that the individual filament lengths remain unchanged and the width of the A-band remains constant. Note also in Figure 4.24 that as contraction proceeds (top to bottom in the drawing) the I band narrows as the thin

filaments approach the center of the sarcomere. Thus, the shortening of the sarcomeres in the myofibril is the direct cause of the shortening of the whole muscle. The sliding of the filaments is facilitated by the myosin cross bridges and these are comprised of the myosin molecules of the thick filaments. Like strokes of an oar, these cross bridges must undergo repeated movements during muscle contraction (Figure 4.24).

PROBLEM

During muscular contraction,

a) the thick and thin filaments move past one another thus shortening the sarcomere.

b) myofibrils move between Z-lines.

c) myosin and actin filaments move between sarcomeres.

d) Z-lines are made of myofibrils.

Solution

a) Muscle contraction occurs as the result of the sliding of the thick and thin filaments past one another.

Types of Muscles

The primary function of all muscle tissue is contraction, during which some chemical energy is converted to mechanical energy and some is dissipated as heat. The contraction of muscle fibers causes tension on the body parts to which they are attached. Skeletal muscles function by applying tension to their attachment points on bones. Bones and muscles form lever systems, which control body movements and help maintain posture. Muscles also function to control the movement of fluids in the circulatory and excretory systems and help maintain body temperature.

Skeletal

Almost half the body is muscle mass, with the vast majority being skeletal muscle (90–95%). Smooth

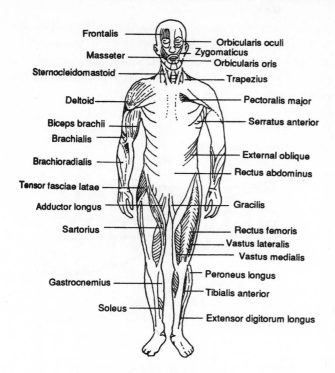

Figure 4.25a.
Anterior View of Superficial Skeletal Muscles

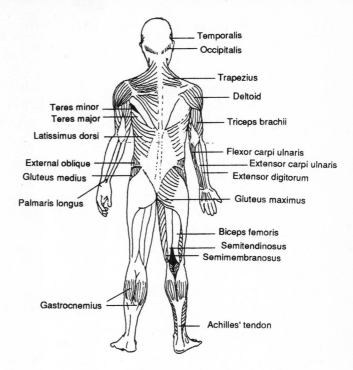

Figure 4.25b.
Posterior View of Superficial Skeletal Muscles

muscle and cardiac muscle account for about 5–10% of body mass. Skeletal muscle fibers are about 50–150 microns in diameter and extend the entire length of a muscle. The multinucleated cells of skeletal muscles are striated. Individual skeletal muscles are separated from each other by a surrounding fascia, which can also extend beyond the muscle to become part of a tendon. The tendon functions to connect the muscle to bone.

Control of Contraction. The contraction of skeletal muscles is under voluntary control and is initiated by neurotransmitter released from a neuron. Neurotransmitter triggers an action potential in the plasma membrane of the myofiber. The action potential rapidly spreads along the transverse (or T) tubules and causes changes in the sarcoplasmic reticulum that surrounds each myofibril. The sarcoplasmic reticulum releases Ca^{+2} ions, which causes all the myofibrils in the myofiber to simultaneously contract.

Some major skeletal muscles and their actions are summarized below. See also Figure 4.25 a and b.

Biceps brachii—this muscle in the upper arm has two heads (immovable origins) that originate on the scapula. The muscle follows the humerus and is connected to the radius by a tendon. Contraction causes the arm to bend at the elbow.

Pectoralis major—this large muscle of the chest connects the humerus (in the upper arm) to the bones of the thorax.

Deltoid—this triangular muscle is located on the shoulder; its contraction moves the upper arm at the shoulder joint.

Extensor digitorum—extensor muscles act to straighten body parts away from the main body.

Extensor digitorum muscles extend either the fingers or toes.

Sternocleidomastoid group—these muscles connect the sternum and mastoid and act primarily on the head, moving it to the side and flexing it.

Smooth Muscle

Smooth muscles contain fibers that are smaller (i.e., 2–5 microns in diameter and 50–200 microns in length) than skeletal fibers and they have only a single nucleus. Smooth muscles contract and relax more slowly than skeletal muscles.

Visceral and multiunit are the two major types of smooth muscles. Visceral smooth muscle cells are in contact with each other (at points called gap junctions). When one smooth muscle cell is stimulated the action potential can spread via gap junctions to other smooth muscle cells. Visceral smooth muscle is found in the intestines, the bile ducts, the ureters, and the uterus. Visceral smooth muscle is responsible for the peristaltic contractions of the intestinal tract.

In contrast, each multiunit smooth muscle cell acts independently and is usually innervated by a single nerve that controls its contraction. Multiunit smooth muscle is found in the walls of blood vessels and in the iris of the eye.

Cardiac Muscle

The primary function of cardiac muscles is the rhythmic pumping action of the heart. The ventricles provide the primary force for pumping blood through the blood vessels. Cardiac muscle exists only in the heart. Cardiac muscle cells are striated and have a single nucleus. Cardiac muscle cells contain intercalated discs (Figure 4.26), which are specialized cell junctions which anchor cardiac cells to each other and allow the action potential to spread from cell to cell. Cardiac muscle has two major types: (1) contractile cells that make up the bulk of the atria and ventricles and gener-

ate the force that pumps the blood, and (2) conducting cells that are specialized for developing and conducting action potentials. The conducting cells provide the mechanism for the rapid transmission of excitatory impulses throughout the heart.

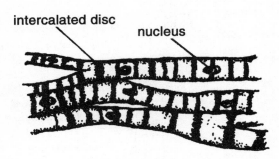

Figure 4.26.
The Intercalated Disc of Cardiac Muscle

The intercalated discs of cardiac muscle contain gap junctions and so allow the rapid transmission of action potentials. Action potentials rapidly propagate throughout the interconnected cells. A group of cells connected by gap junctions is called a syncytium. The heart is composed of the atrial syncytium and the ventricular syncytium. These syncytia are separated by fibrous tissue except by a band of conducting cells called the A-V bundle that allows the action potential to spread from the atria to the ventricles.

4. Integration Mechanisms and Control

Nervous System

The neuron is the fundamental cell type of the nervous system. Neurons can transmit information from inside and outside the body to processing centers in the brain and spinal cord. The processed signals can evoke responses, also transmitted by neurons, by muscles and glands. The coordination and integration of these events leads to behavioral adaptation to environmental changes and helps maintain a stable internal environment (i.e., homeostasis). The brain and spinal cord form the central nervous system. The nerves that connect the central nervous system to other body parts are called the peripheral nervous system.

PROBLEM

The central nervous system is composed of the

a) brain and spinal cord.

b) spinal cord and nerves.

c) neurons, synapses, and spinal cord.

d) sense organs, spinal cord, and brain.

Solution

a) The organs of the brain and spinal cord form the central nervous system. The nerves that connect the central nervous system to other body parts are called the peripheral nervous system.

The Human Nervous System

The nervous system is divided into the central nervous system (CNS) and the peripheral nervous system (PNS). The central nervous system consists of the brain and the spinal cord. The peripheral nervous system is all parts of the nervous system outside the brain and spinal cord, for example the nerves and ganglia.

The Central Nervous System (CNS)

The brain is divided into three regions: the forebrain, the midbrain, and the hindbrain.

Forebrain. The most prominent portions of the forebrain are the cerebral hemispheres (Figure 4.27).

All sensory and motor activity is interpreted here. It is also involved in memory, emotions, speech, and learning. The surface of the cerebrum is known as the cerebral cortex. The thalamus, hypothalamus, and pineal gland are part of the lower forebrain. The hypothalamus regulates the internal environment of the body. It works closely with the pituitary gland and controls the release of the various hormones produced by this gland. The hypothalamus is also involved in the regulation of thirst, hunger, water balance, behavior, and temperature.

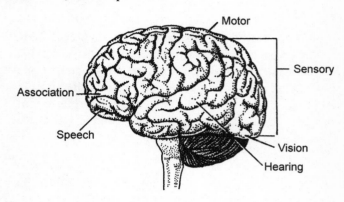

Figure 4.27.
The Surface of the Human Brain

The axons of CNS neurons are organized into bundles (tracts) that are white in appearance because of the presence of myelin. This tissue is known as white matter. The bodies of these cells are localized in the outer region known as gray matter. The cerebrum contains both sensory and motor areas and is highly convoluted in most mammals. The proportionate size of the cerebrum from smallest to largest is as follows: fish, amphibians, reptiles, birds, and mammals, with the primates having the largest cerebrum of all mammals.

Midbrain. The midbrain is one of the smallest regions of the brain. It acts as a relay station, transferring nerve impulses between the two other parts of the brain.

Hindbrain. The main regions of the hindbrain are the medulla oblongata, pons and the cerebellum.

1) **Medulla oblongata**—Its functions are respiratory and circulatory regulation, cough reflex, swallowing, sneezing, and the vomiting reflex.

2) **Pons**—It functions with the medulla oblongata to control ventilation (breathing).

3) **Cerebellum**—It is involved in the coordination of muscle activity, locomotion, and maintenance of balance. It is initiated by impulses originating in the forebrain.

Lateralization refers to the fact that each side of the brain controls different functions. For most people, the right brain controls artistic concepts and spatial perception, while the left brain is the speech, language, and calculation center. The corpus callosum is a thick band of fibers that connects the two halves of the brain.

Endorphins are peptide hormones produced in the brain and anterior pituitary that inhibit pain reception. The drug morphine binds to the same cellular receptors as endorphins.

The Spinal Cord

The spinal cord runs from the medulla oblongata down through the vertebral column. It is composed of an interior of gray matter and an exterior of white matter. The gray matter consists of neuron cell bodies and the white matter consists of myelinated axons. The white matter includes descending and ascending tracts that run from the spinal cord to the brain. Throughout its length, it is enclosed by three meninges and by the vertebrae. Running vertically in the spinal cord center is a narrow canal filled with cerebrospinal fluid.

The spinal cord controls the centers for reflex acts occurring below the neck, and it provides the major pathway for impulses between the peripheral nervous system and the brain. It is also a connecting center between sensory and motor neurons.

The Peripheral Nervous System

The peripheral nervous system includes nerves (bundles of axons), ganglia (groups of neurons) and sensory receptors. The nerves of the PNS connect the brain and the spinal cord to the sense organs, glands, and muscles. Sensory receptors and sensory neurons are located in the PNS and are sometimes referred to as the sensory division of the nervous system. The term motor division of the nervous system refers to the nerves that carry information out to muscles and glands. The motor division can be subdivided into the somatic nervous system and the autonomic nervous system.

Somatic Nervous System

The somatic nervous system consists of nerves that transmit impulses from the central nervous system to the skeletal muscles of the body.

Nerve Control of Muscles. *Motor and Sensory Control.* Motor (or efferent) neurons carry nerve impulses out from the brain or spinal cord. Each skeletal muscle fiber is connected to a myelinated motor neuron at a region called the neuromuscular junction. The specialized region of the muscle fiber membrane that forms a junction with the axon of the motor neuron is called a motor end-plate (see Figure 4.28).

A nerve impulse (also known as an action potential) reaching the neuromuscular junction will cause the release of acetylcholine, which, in turn, triggers the generation of action potential in the muscle fiber. The acetylcholine released into the synaptic cleft between an axon terminal and the plasma membrane of the muscle fiber is rapidly destroyed by acetylcholinesterase.

The Motor Unit. Each motor neuron branches to form contacts with many muscle fibers. The neuron and the muscle fibers attached to it form a motor unit. When an impulse is transmitted through the motor neuron it will cause the simultaneous contraction of all muscle fibers to which it is attached. Very fine movements require the number of connections a motor neuron makes with muscle fibers to be small (e.g., about 10).

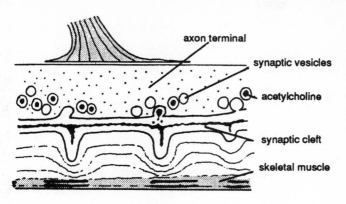

Figure 4.28.
The Motor End-Plate

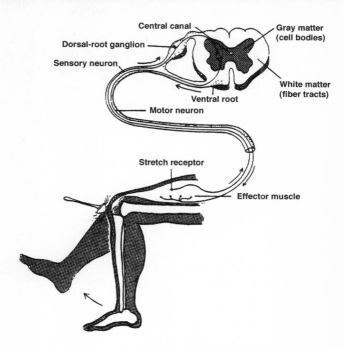

Figure 4.29.
A Simple Reflex Arc That Elicits
a Knee-Jerk Response

The Reflex Arc. Motor axons and sensory (or afferent) axons usually occur in the same "mixed" nerve. The simplest manner in which sensory and motor neurons are integrated to evoke behavior is in a reflex arc (see Figure 4.29). For example in the knee-jerk response, striking the patella stimulates a stretch receptor neuron, which sends an impulse to the spinal cord. Within the gray matter of the spinal cord the axon of the sensory neuron forms a synapse with a dendrite of a motor neuron. The impulse travels via the motor neuron axon to the quadriceps femoris muscle, which responds by contracting and causing extension of the leg.

Most of the sensory neurons that enter the spinal cord do not terminate on motor neurons but rather on interneurons. Interneurons are more numerous (by a factor of about 30) than motor neurons and they have numerous connections to each other and also to motor neurons. These interconnections provide the basis for complex reflex responses such as the withdrawal reflex.

Voluntary and Involuntary Control. Reflex behavior is unconscious and automatic but plays a critical role in homeostasis. Changes in vascular tone in response to hot or cold, sweating, and some motor functions of the gut are all examples of autonomic reflexes that occur in the spinal cord.

The brain stem, which connects the cerebrum to the spinal cord, has numerous nerve pathways that help regulate the involuntary functions involved with equilibrium, respiration, cardiovascular function, eye movements, and support of the body against gravity.

"Voluntary" control of motor functions is primarily under the control of the frontal lobes of the cerebral cortex and the cerebellum of the brain. The cerebral cortex contains a "pyramidal area" that contains very large pyramid-shaped cells. Motor signals from the brain originate in the pyramidal cells and travel through the brain stem and to the spinal cord via the pyramidal (corticospinal) nerve tract. Most of the pyramidal fibers terminate on interneurons in the cord gray matter. These interneurons, in turn, form synapses with motor neurons controlling various voluntary muscles.

In addition to the corticospinal tracts there are "extrapyramidal tracts" that also transmit motor signals from the brain. A specialized region of the frontal lobe that coordinates the muscular area involved with speech is called "Broca's area."

Autonomic Nervous System

The autonomic nervous system includes nerves that transmit impulses to smooth muscle, cardiac muscle and glands and thereby controls the responses of the internal organs. This system is further divided into the sympathetic and parasympathetic systems.

Sympathetic and Parasympathetic Nervous System.

The autonomic nervous system is not under voluntary control and functions independently. The contraction of smooth muscles, blood pressure regulation, temperature regulation, and the secretory function of most glands are under the control of the autonomic nervous system. The autonomic nervous system has been further divided into the sympathetic and parasympathetic nervous systems.

In general, the sympathetic subdivision serves to prepare an organism for energy expenditure while the parasympathetic system restores and maintains an organism in a resting state. Organs are innervated with nerve fibers from both the sympathetic and parasympathetic divisions. The sympathetic nervous system, for example, increases heart rate, and decreases intestinal secretions. These physiological adaptations are restored by the parasympathetic nervous system.

Most of the nerve fibers in the autonomic system are composed of motor neurons, and two neurons are used to connect the brain or spinal cord to the effector. The preganglionic axon comes from a neuron in the brain or spinal cord and forms a synapse with a ganglion outside the brain and spinal cord. The postganglionic axon comes from this second neuron and it goes to the effector.

Sympathetic nerves are adrenergic and secrete the neurotransmitter called norepinephrine at the end of their postganglionic fibers. Parasympathetic nerves are cholinergic and secrete acetylcholine at the ends of their postganglionic fibers.

PROBLEM

Which of the following is characteristic of stimulation of the sympathetic nervous system?

a) Elevated heartbeat

b) Increased saliva excretion

c) Elevated gastric secretion

d) All of the above.

Solution

a) In general, the sympathetic nervous system produces the effects that prepare an animal for emergency situations, such as quickening of the heart and breathing rates and dilation of pupils.

Some Problems of the Human Nervous System

Cerebral palsy—Impaired muscle control due to brain damage by infection or injury before or after birth.

Meningitis—Inflammation of the membranes (meninges) that surround the brain and spinal cord; may be caused by infection with a bacterium or virus.

Stroke—Caused by a blood clot in the brain or a decreased amount of arterial blood flowing to the brain; causes the portion of brain tissue supplied by the artery to die.

Poliomyelitis—A viral disease that attacks nerve cells and can result in paralysis. This disease is not seen very often due to vaccine.

Sensory Reception and Processing

Sensory and Effector Neurons. Neurons with a sensory function are called afferent or sensory neurons. The dendrites of these neurons either function directly as receptors for stimuli or their dendrites are in close association with specialized receptor cells that are affected by stimuli. Most sensory neurons have a unipolar structure (see Figure 4.30a).

The information from sensory neurons is transmitted, in the form of a nerve impulse, over peripheral nerves to the central nervous system. After integration of the sensory information a response can be transmitted by the peripheral nerves to effectors (i.e., muscles and/or glands).

Interneurons are neurons that get their input from other neurons and send their output to other neurons. Interneurons are found in the brain and spinal cord and are involved with processing and integration. Interneurons are multipolar (see Figure 4.30b).

Motor neurons, also called efferent neurons, transmit nerve impulses from the brain or spinal column to effectors. Motor neurons are usually multipolar.

The axons of neurons are bundled together to form nerves. Some nerves contain only motor axons, some only sensory axons. Most nerves however, contain both motor and sensory axons.

PROBLEM

Bundles of axons are known as

a) interneurons. b) association areas.

c) nerves. d) effectors.

Solution

c) A nerve is a bundle of axons. Interneurons are neurons that get their input from other neurons and send their output to other neurons. Effectors are muscles or glands that respond to nerve stimulation.

The special senses refer to smell, taste, hearing, equilibrium, and sight. The somatic senses refer to all other senses, for example touch, heat, pain.

Somatic Sensors

The somatic senses can be further divided into exteroreceptive, proprioceptive, visceral, and deep sen-

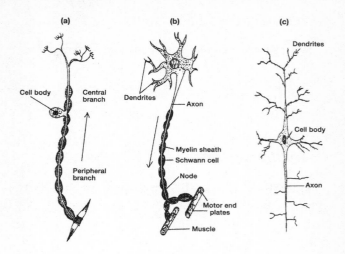

Figure 4.30

sations. The exteroreceptive sensations arise from the surface of the body. The proprioceptive sensations arise from muscles and tendons, the visceral sensations from the internal organs, the deep sensations from "deep" tissues (e.g., bones).

The somatic sensory receptors can be the following:

(1) mechanoreceptors, which respond to mechanical movement.

(2) thermoreceptors, which respond to hot and cold.

(3) pain (or nociceptors) receptors, which signal tissue damage.

(4) chemoreceptors which respond to changes in oxygen levels, carbon dioxide levels and pH.

A wide variety of mechanoreceptors exist. These include:

(1) free ends of sensory nerve fibers, which are found predominantly in epithelial cells that respond to touch and pressure.

(2) Meissner's corpuscles, which respond to light touch.

(3) Pacinian corpuscles, which respond to deep pressure and tissue vibrations.

PROBLEM

Somatic sensory receptors could *not* detect a

a) bright light.   b) hot stove.

c) stomach ache.   d) sunburn.

Solution

a) Light is detected by the specialized sensory receptors in the eye.

Olfaction and Taste

The neurons in the superior part of the nasal cavity that detect odors are called olfactory receptors. These bipolar (Figure 4.30c) neurons lie in a surrounding matrix of columnar epithelial cells. Bowman's glands, which secrete the mucous necessary for receptor functioning, are also embedded in the columnar epithelial cells. The mucosal ends of the olfactory neurons have many cilia, which are the primary receptor sites for gaseous molecules dissolved in the mucosal fluid.

The precise mechanism whereby different gaseous molecules are distinguished is not yet clearly known. When the cilia are stimulated, a receptor potential is generated that triggers a nerve impulse in the olfactory nerve fibers. This signal is transmitted to the central nervous system. The olfactory receptors undergo a sensory adaptation, which diminishes their response to a stimulus over time.

The sense of taste is generated by taste buds located in the tongue, and to a lesser extent, on the roof of the mouth. It is thought that taste consists of different combinations of four primary tastes, which in Western cultures are characterized as sour, salty, sweet, and bitter. The taste receptors are microvilli that protrude from taste cells that are specialized epithelial cells. The outer surface of the taste bud is covered with stratified squamous epithelial cells and the microvilli from the taste cells protrude from a pore on this surface. The taste cells

are replaced about every ten days. After stimulation the taste cells generate a receptor potential that, in turn, triggers a nerve impulse that is transmitted to the central nervous system.

PROBLEM

Following exposure to a strong odor over a long period of time, olfactory receptors exhibit a diminished response. This is due to

a) receptor stress.   b) sensory adaptation.

c) receptor death.   d) lack of mucosal fluid.

Solution

b) Sensory adaptation to strong stimuli reduces the response of taste and olfactory receptors.

Hearing and Equilibrium

Ear Structure. The ear, which functions in both hearing and balance, has external, middle, and internal components. The external ear consists of the auricle, which is funnel-shaped, and the auditory meatus, which is tube-shaped (Figure 4.31). These structures serve to funnel sound waves into the ear where they produce pressure oscillations on the eardrum. The middle ear is in the tympanic cavity of the temporal bone (Figure 4.32). The eardrum or tympanic membrane separates the external and middle ear.

Three small bones in the tympanic cavity transmit the vibration of the eardrum to the inner ear (Figure 4.33). The malleus or hammer is attached to the eardrum. The malleus causes the incus (or anvil) to vibrate and this movement is then transmitted to the stapes (or stirrup). It is the movement of the stapes that causes movement of fluid in the inner ear. The stapes is connected to an opening in the middle ear called the oval window.

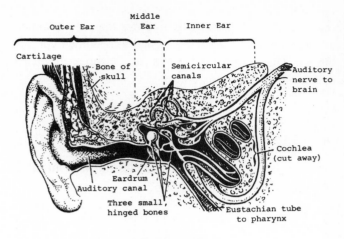

Figure 4.31.
The Structure of the Ear

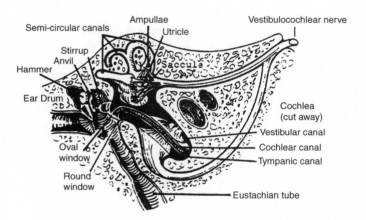

Figure 4.32.
Structure of the Middle Ear

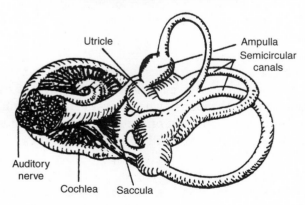

Figure 4.33.
The Labyrinth of the Inner Ear

Mechanism of Hearing. The organ of Corti generates a receptor potential when stimulated by the vibrations of the basilar membrane. The hearing receptor cells have cilia or hair-like structures that project into the endolymph of the cochlear canal. The sensitivity of the ear to different sound frequencies depends upon the differential sensitivity of the hair cells. The movement of the hairs causes a receptor potential to be generated that is transmitted to the cochlear nerve fibers. Some of the nerve impulses from each ear reach both sides of the brain.

PROBLEM

Which of the following items is not part of the human ear?

a) Malleus   b) Cochlea

c) Hyoid   d) Oval window

Solution

(c) Only the hyoid, which is a very small bone near the base of the tongue, is not a part of the human ear. The tectorial membrane is part of the cochlea, which is in the inner ear. The oval window is a membrane that separates the middle ear and the inner ear. The malleus is one of the small bones in the middle ear that conducts sound.

Vision

Eye Structure. The light receptors in the eye are extremely sensitive. A retinal rod cell can detect a single photon. The structure of the human eye is shown in Figure 4.34. The cornea is transparent and helps focus light, and provides mechanical protection to the other underlying tissues. The anterior chamber contains aqueous humor and lies between the cornea and the lens. The lens focuses light on the retina. The iris controls the diameter of the pupil and helps control the intensity of the light impinging on the retina. The vitreous (posterior chamber) humor in the eye cup helps control the internal pressure of the eye.

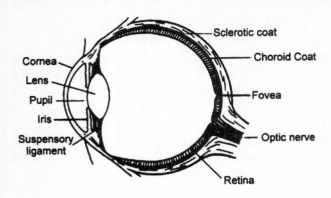

Figure 4.34.
Structure of the Human Eye

The retina contains the photoreceptor cells, which are either specialized for color vision (i.e., the cones) or for night vision (i.e., the rods). In the human retina there is a specialized region called the fovea that has a high density of cone cells. Before light reaches the photoreceptor cells it must pass through a number of other retinal layers. The rod photoreceptor (see Figure 4.35) are adjacent to the retinal pigment epithelium. The rod cells shed their tips each day and these tips are phagocytized by the retinal pigmented epithelium. Blood supply to the retina is by way of the choroid or by retinal blood vessels. The choroid is posterior to the retinal pigment epithelium.

The outer segment of the rod photoreceptor cell has numerous disc membranes that are not in direct contact with the plasma membrane. The photosensitive pigment, rhodopsin, is an intrinsic membrane protein found in the disc membranes.

Light Receptors. The rhodopsin molecules in the disc membranes are covalently linked with 11-*cis*-retinal. Retinal is an aldehyde form of vitamin A. Light causes an isomerization of the *cis*-retinal to the all-trans-retinal form. This isomerization triggers a change in the conformation of rhodopsin. The light-induced conformational change in rhodopsin causes Na^+-channels on the photoreceptor plasma membrane to close. In the dark, the photoreceptor cells are depolarized. This depolarization is due to open Na^+-channels that permit a constant influx of Na^+-ions. The result of a light stimulus is to close the Na^+-channels and thereby cause the receptor cell to become hyperpolarized. This action potential causes a decreased release of inhibitory neurotransmitter.

PROBLEM

The part of the eye that regulates the amount of incoming light is the

a) retina.   b) lens.

c) iris.   d) cornea.

Solution

c) Light entering the eye passes through the cornea and enters the lens via a small opening called the pupil. The size of the pupil can be changed by a diaphragm-like muscular structure, the iris, so that the amount of incoming light can be regulated. The iris may contain various colored pigments. The light then falls on a light-sensitive region, the retina, which is located at the rear of the eye.

Nervous Systems in Other Organisms

Protozoans—Protozoans have no nervous system; however, their protoplasm does receive and respond to certain stimuli.

Hydra—The hydra possesses a simple nervous system known as a nerve net (see Figure 4.36). The hydra lacks centralization of a nervous system. A stimulus applied to a specific part of the body will generate an impulse that will travel to all body parts.

Earthworm—The earthworm possesses a central nervous system that includes a brain, a nerve cord (which is a chain of ganglions), sense organs, nerve fibers, muscles, and glands (see Figure 4.37).

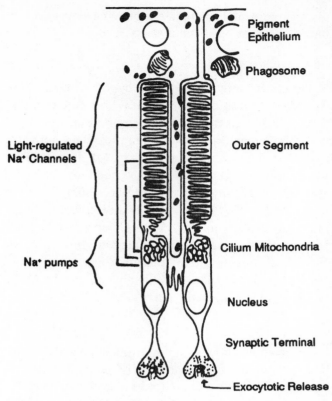

Figure 4.35.
Structure of a Rod Photoreceptor

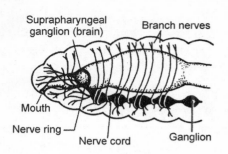

Figure 4.37.
Nervous System of the Earthworm

Grasshopper—The grasshopper's nervous system consists of ganglia bundled together to form the peripheral nervous system. The ganglia of the grasshopper are better developed than those of the earthworm (see Figure 4.38).

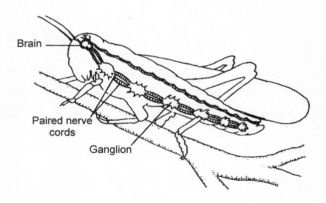

Figure 4.38.
Nervous System of a Grasshopper

Fish—The brains of fish are composed of a pair of olfactory lobes (smell), a cerebrum with two hemispheres, two optic lobes (sight), a cerebellum, and a medulla. Additionally, the lateral lines that run the length of both sides of a fish contain mechanoreceptors that are sensitive to movement, particularly to the water-current pressure.

Frogs—The brains of frogs are similar to those of fish in that they also are composed of olfactory and optic lobes, a cerebrum, a cerebellum, and a medulla, though the cerebrum of a frog is larger than that of a fish.

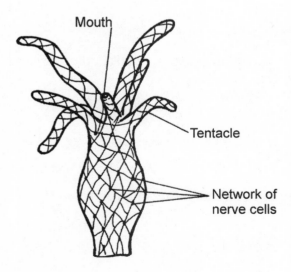

Figure 4.36.
Nerve Net of the Hydra

Birds—The optic lobes and cerebellum of birds are relatively large to accommodate their well-developed sense of sight and to enable the functions of coordination and balance in flight, respectively.

Endocrine System

Function of the Endocrine System

Bodily functions are controlled by the nervous system and the endocrine system as well as the interaction between these two systems. The nervous system, as detailed above, relies on electrical signals. The endocrine system utilizes chemical signals and the signal molecules are called hormones. The endocrine system refers to the set of glands, tissues, and cells that secrete hormones directly into bodily fluids.

Hormones regulate a wide variety of metabolic functions and transport functions, as well as development, growth, and reproduction. Hormones are structurally diverse and can exert physiological effects on their target tissues at very low concentrations. Hormones can be peptides, proteins, glycoproteins, biological amines, or steroids.

PROBLEM

Hormones do *not* regulate

a) growth.　　　　b) reproduction.

c) digestion.　　　　d) temperature.

Solution

b) Temperature regulation is controlled by the autonomic nervous system. Growth, reproduction, and digestion are controlled, in part, by hormones.

Cellular Mechanisms of Hormone Action

Many hormones exert their action by first binding to specific receptors on the cell surface. Some of the overall molecular characteristics of this type of signalling process has been discussed above.

Many hormones activate a cyclic AMP system. The first event is the binding of a hormone to a surface receptor on the plasma membrane of the target cell. The receptor-hormone complex then activates adenyl cyclase, which produces cAMP from ATP in the cytoplasm. The cAMP is a "second messenger" that relays the initial extracellular signal from the hormone (the first messenger) to an intracellular signal (increased levels of cAMP). The increased cAMP then, in turn, can activate a wide variety of physiological responses. Often a protein kinase is activated that can phosphorylate specific enzymes and thereby regulate their enzymatic activity.

Steroid hormones do not utilize a cAMP system. The steroid hormones are freely permeable to the plasma membrane and do not have surface receptors. Instead, they bind to cytoplasmic receptors, and the receptor-hormone complex then initiates a series of events leading to the activation of specific genes in the cells' nucleus.

PROBLEM

Steroid hormones are unlike other hormones because they

a) utilize the cyclic AMP system.

b) have cytoplasmic receptors.

c) have surface receptors on the plasma membrane.

d) are secreted directly into bodily fluids.

Solution

b) Steroids have cytoplasmic receptors. Most other hormones have surface receptors on the plasma membrane and utilize the cyclic AMP system. All hormones are secreted directly into bodily fluids.

Control of Hormone Secretion

The secretion of hormones into circulating blood is a very regulated process. Both negative and positive feedback loops help regulate this process (see Figure 4.39 a, b). In a negative feedback loop, gland A secretes hormone A, which stimulates gland B to produce hormone B, which then can inhibit the secretion of hormone A by gland A. In a positive feedback loop, gland A produces hormone A, which stimulates gland B to produce hormone B, which further stimulates gland A to secrete hormone A.

Hormonal secretions can also be controlled by the nervous system. For example, the adrenal medulla secretes catecholamines in response to nerve impulses and not by the influence of other hormones or any other stimulus.

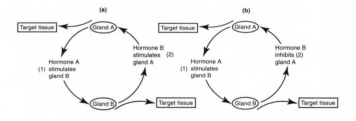

Figure 4.39.
Positive (a) and negative (b) feedback regulation.

PROBLEM

Hormone secretion is controlled by

a)   negative and positive feedback.

b)   the nervous system.

c)   negative feedback only.

d)   both a and b.

Solution

d) Hormone secretion can be controlled by positive feedback, negative feedback, or the nervous system.

Major Endocrine Glands

The major endocrine glands, their hormone products, their target tissues, and their functions are detailed below.

Pituitary Gland

The pituitary gland lies at the base of the brain, is connected to the hypothalamus and is controlled by the hypothalamus. The pituitary gland is divided into the anterior and posterior pituitary glands.

Anterior Pituitary Gland

All the major hormones produced by the anterior pituitary gland influence other glands, such as the thyroid, adrenals, gonads, and mammary glands.

Adrenocorticotropic hormone (ACTH)—a protein hormone whose target tissue is the adrenal cortex. ACTH controls the secretion of some adrenocortical hormones and thereby influences the metabolism of glucose, fats, and proteins.

Follicle stimulating hormone (FSH)—a protein hormone whose target tissue is the ovary. FSH stimulates the growth and reproductive activities of the gonads.

Growth hormone (GH)—a protein hormone that promotes body growth and has a major impact upon formation of body protein. It increases the transport of amino acids through cell membranes and the synthesis of proteins by ribosomes. It also decreases the rate of protein catabolism.

Luteinizing hormone (LH)—a protein hormone whose target tissue is the ovary. LH stimulates the growth and reproductive activities of the gonads.

Prolactin (PRL)—a protein hormone that stimulates growth of the mammary gland and production of milk.

Thyroid stimulating hormone (TSH)—a protein hormone whose target tissue is the thyroid gland. TSH controls the synthesis of thyroxine in the thyroid gland. Thyroxine, in turn, controls many metabolic reactions.

Posterior Pituitary Gland

Antidiuretic hormone (ADH)—this peptide hormone, also called vasopressin, causes a decreased secretion of water by the kidneys (i.e., antidiuresis).

Oxytocin—a peptide hormone whose target tissues include the uterus and the mammary gland. Oxytocin is thought to play a key role in the birthing process, causing contraction of the uterus. Oxytocin also stimulates the secretion of milk from the mammary gland in response to suckling.

Thyroid Gland

The thyroid gland is located below the larynx and on both sides of the trachea. It secretes hormones that control metabolism.

Thyroxine (T4)—an iodinated amino acid derivative that increases the overall metabolic rate and, in children, promotes growth. In particular, T4 increases protein synthesis, increases the number and size of mitochondria, and stimulates both carbohydrate and fat metabolism. Secretion of T4 is controlled by TSH from the anterior pituitary gland.

Triiodothyronine (T3)—an iodinated amino acid derivative whose functions are similar to those detailed for thyroxine.

Parathyroid Glands

The parathyroid glands are located on the posterior surface of the thyroid gland. The parathyroid hormone is the only hormone secreted by the parathyroid gland.

Parathyroid hormone (PTH)—this protein hormone causes an absorption of calcium and phosphate from bone. Moreover, the PTH causes a dramatic increase in the secretion of phosphate by the kidney. The overall result of increased levels of PTH in plasma is an increase in calcium, but a decrease in phosphate levels. PTH promotes the conversion of vitamin D into 1,25-Dihydroxycholecalciferol which, in turn, helps promote calcium transport through cell membranes. 1,25-Dihydroxycholecalciferol is the active form of vitamin D. High levels of plasma calcium decrease the secretion of PTH.

Adrenal Glands

The adrenal glands lie at the top of the kidney. The adrenal consists of two distinct glands that secrete different hormones. The exterior part of the adrenal is called the cortex and the central region the medulla. The cells of the medulla are modified postganglionic cells. The cells of the adrenal medulla are in contact with the sympathetic division of the autonomic nervous system.

Adrenal medulla

The adrenal medulla sends nerve impulses from the sympathetic nerve fibers to create the stimulus for the secretion of epinephrine and norepinephrine.

Epinephrine—this hormone, also called adrenalin, is a biological amine. Both epinephrine and norepinephrine are catecholamines. Epinephrine prepares the body for a "fight or flight" response (i.e., heart rate, metabolic rate, and systemic blood pressure increase). The liver converts glycogen into glucose, the airways dilate, and the force of cardiac muscle contraction increases.

Norepinephrine—this biological amine has a structure similar to that of epinephrine and its biological effects are very similar.

Adrenal Cortex

The adrenal cortex is a gland that secretes a group of hormones called corticosteroids that are all synthesized from cholesterol. Corticosteroids are further divided into glucocorticoids, mineralocorticoids, and androgenic hormones. The glucocorticoids increase blood glucose, the mineralocorticoids affect electrolytes. The androgenic hormones are similar to testosterone.

Aldosterone is the primary mineralocorticoid and causes sodium ions to be retained and potassium ions to be excreted. This hormone also reduces urinary output, promotes water retention, and increases extracellular fluid volume. Aldosterone exerts its effects on the tubules of the kidney. The secretion of aldosterone is controlled by many factors such as the potassium concentration in extracellular fluid, the renin-angiotensin system, body sodium, and adrenocorticotropic hormone.

Cortisol is the primary glucocorticoid and this hormone has the liver as its primary target. Cortisol influences carbohydrate, protein, and fat metabolism. One effect of cortisol is to stimulate gluconeogenesis (i.e., the synthesis of glucose from noncarbohydrates, particularly from amino acids). Increased gluconeogenesis, in turn, causes an increased formation of glycogen in the liver. In addition, cortisol causes an increased release of fatty acids from fat cells (adipocytes). The secretion of cortisol is first stimulated by the hypothalamus (of the brain), which secretes *corticotropin-releasing hormone* (CRH). CRH causes the anterior pituitary to secrete ACTH and ACTH causes the adrenal cortex to release cortisol. Stress of almost any kind will cause the release of ACTH, which is rapidly followed by secretion of cortisol. Cortisol also exerts an anti-inflammatory effect on tissues damaged by injury.

The Pancreas

The pancreas lies behind the stomach and is connected to the duodenum. The secretory cells of the pancreas play a role in both the endocrine and exocrine systems. The exocrine part of the pancreas secretes digestive enzymes into the small intestine (duodenum). The role of the pancreas in digestion is discussed below. The endocrine part of the pancreas is due to the islets of Langerhans, which contain alpha-, beta-, and delta-cells. These cells secrete their products directly into the blood stream. The alpha-cells secrete glucagon, the beta-cells secrete insulin, and the delta-cells secrete somatostatin. Humans with diabetes have beta-cells that are incapable of secreting insulin. Insulin and glucagon work in concert to control many metabolic activities.

Insulin—a protein hormone that influences carbohydrate, fat, and amino acid metabolism. Insulin decreases the release of fatty acids and fat cells and promotes the utilization of glucose. The high levels of blood glucose (e.g., after a meal) stimulates the secretion of insulin. Insulin promotes the uptake and storage of glucose by almost all tissues in the body, particularly those of the liver and muscles. In liver and muscle tissue, glucose is stored as glycogen. The glycogen in the liver is used to supply the blood with glucose when the dietary supply of glucose decreases. Insulin also causes the liver to convert glucose into fatty acids, which are subsequently stored in fat cells as triglycerides. Insulin also promotes the transport of amino acids into many tissues. Low levels of insulin cause fatty acids and glycerol to be released from adipocytes into plasma. The increased plasma levels of nonesterified fatty acids stimulate the liver to synthesize triglycerides, cholesterol esters, phospholipids, and cholesterol. These lipids are secreted by the liver in the form of very low-density lipoprotein. In addition, high levels of plasma fatty acids also stimulates liver mitochondrial fatty acid oxidation producing ketone bodies (i.e., betahydroxybutyrate and acetoacetate). Humans with the inability to secrete insulin often have very high levels of very low-density lipoprotein (LDL) and also develop premature atherosclerosis.

Glucagon—this protein hormone counteracts many of the metabolic effects of insulin. In particular, glucagon promotes an increase in blood glucose levels by causing a breakdown in glycogen (i.e., glycogenolysis). The secretion of glucagon is regulated by blood glucose levels (i.e., low levels of blood glucose stimulate glucagon secretion).

PROBLEM

The adrenal medulla is most closely associated with

a) insulin.   b) epinephrine.

c) chorionic gonadotropin.   d) vasopressin.

Solution

b) Epinephrine is a secretory product of the adrenal medulla. It causes a breakdown of glycogen to glucose in the liver and skeletal muscle with a consequent rise in blood glucose levels. Epinephrine elevates the blood pressure and heart rate. It also constricts cutaneous blood vessels and dilates skeletal muscle vessels. In addition, it causes the organs of the digestive tract to experience vasoconstriction.

5. Metabolic Rates

Metabolic rate is the amount of energy an organism uses over some time interval. The basal metabolic rate (BMR) is the amount of energy required for basic, resting functions (i.e., no growth, no activity except normal breathing, heartbeat, etc). The BMR for humans and other mammals varies with age, gender, size, thyroid health, diet, body/habitat temperature, activity levels, time of day, season and overall health. Adult males have a BMR of ~1,600–1,800 kcal (1 Calorie = kcal = 1,000 calories; note two "calorie" spellings) and for females it varies between 1,300–1,500 kcal per day. Any excess energy above BMR can be used for repair, defence against infection, growth, storage of reserves and reproduction. In general, the greater the activity, the greater the amount of energy required. The thyroid gland regulates the BMR by secreting a hormone called thyroxine. Hence, measuring thyroxine levels gives an estimate of how BMR is being regulated.

Differences in Adaptive Bioenergetics

Important terms associated with animal adaptation to their environments include these:

1. Endothermic: animals in which their body is warmed by the heat generated from metabolism (e.g., mammals and birds); these animals can maintain extended periods of activity but also require high energy input.

2. Ectothermic: these animals require much less energy than endotherms but their metabolism does not produce heat enough to maintain constant temperature, they must obtain most of their heat from their surroundings (e.g., fish, amphibians, invertebrates). The standard metabolic rate is used to define the base for these animals because it must be measured at some prespecified temperature and conditions since this type animal's rate varies with temperature.

3. Poikilothermic: body temperature varies with environment.

4. Homeothermic: body temperature remains relatively constant across a range of environments.

The four terms above can be used in combination:

1. Endothermic Homeothermic: old terminology called these "warm blooded"; they produce their own heat and maintain a fairly constant temperature regardless their environment (i.e., humans [96.8–100.4°F; 36–38°C], birds [100.4–107.6°F; 40–42°C]).

2. Endothermic Poikilothermic: generate their own heat, but allow some fluctuation among environments.

3. Ectothermic Poikilothermic: old terminology called these "cold blooded"; body temperature varies depending on environment temperature.

4. Ectothermic Homeothermic: body temperature remains fairly constant and the heat comes from the surroundings.

Measuring BMR

BMR can be measured by:

1. Energy loss from respiration can be measured as heat in a chamber called a calorimeter. Works best with smaller animals.

2. Cellular respiration rates can be measured by O_2 intake and/or CO_2 output per unit of time.

3. Food intake and energy content of food; must account for heat loss from urine and feces.

4. Assay the thyroxine in blood, i.e. high thyroxine would indicate a high BMR.

Body Size: Weight, Surface Area and Volume

It is well established, although not well understood, that there is a relationship between metabolic rate and body size. The relationship is inverse: as body size increase, metabolic rate decreases. Put another way, the amount of energy required to maintain a gram of body weight increases as the size of the animal decreases. Some estimates indicate that to maintain a gram of mouse requires about twenty times the amount of energy required to maintain a gram of elephant. The hypothesis is that smaller endothermic animals have a higher energy cost in order to maintain a constant temperature. This idea flows from surface to volume relationships. As the size of animals decrease the ratio of surface area to body volume increases. This would result in a greater loss of heat (or gain to the surroundings) and thus require more energy to maintain a constant temperature.

Metabolism and Activity

More Calories (1 Calorie = 1,000 calories, as in food energy) are required when activity levels go up, and vice versa. Increases in activity increase ATP demand, O_2 intake and CO_2 output. Likewise, animals that are living/working in cold environments expend more Calories.

Humans and many other animals are more active (generally) in the day time hours, while others like mice and bats are more active during the night; torpor describes when endothermic animals may allow their temperature to lower during a 24-hour period in order to conserve energy. Whether endotherms or ectotherms, the highest circadian activity level is 2–4 times the basic level (i.e., BMR or SMR). Hibernation is a seasonally low BMR strategy having low activity for some animals; bears, compared to some species, are not true hibernators because their body temperature remains fairly normal while other animals lower theirs, thus conserving energy.

Animal Reproduction and Development

1. Reproductive Structures

The successful production of offspring in higher organisms is complex and reserved for mature, fully developed organisms. The organs related to the reproductive process are called genitalia. The specific organs responsible for producing the sex cells (i.e., sperm cells in males and ova in females) are called gonads (the testes in males and the ovaries in females).

The Male Reproductive System

In human males (see Figure 5.1) the reproductive organs include two testes, the epididymides, the vas deferentia accessory organs (i.e., the seminal vesicles, ejaculatory ducts, the prostate gland, the urethra, and the bulbourethral glands) and accessory external organs (i.e., the scrotum, penis). These accessory organs primarily serve to store and deliver the sperm (or spermatozoa) to the female genitalia. Various glands secrete fluids to aid in this process.

The testes respond to luteinizing hormone (LH) secreted by cells in the anterior pituitary gland (called gonadotrophs; LH acts in both males and females) in the process of synthesizing and secreting testosterone following LH binding with receptors called Leydig cells. The anterior pituitary, in turn, is stimulated by the hypothalamus (see the section "The Endocrine System", Chapter 4). During puberty, testosterone stimulates testicular growth as well as the growth of the accessory male reproductive organs. Testosterone also helps to maintain secondary masculine sex characteristics such as facial hair and a deepened voice. Follicle stimulating hormone (FSH, another gonadotropin hormone) is required for successful reproduction in both human males and females. In males, FSH helps insure the maturation of sperm via its action on Sertoli cells.

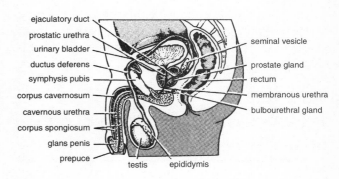

Figure 5.1.
The Male Reproductive System

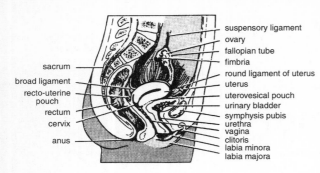

Figure 5.2.
The Female Reproductive System

Both LH and FSH are under the control of gonadotropin-releasing hormone (GnRH), which is secreted by the hypothalamus.

The Female Reproductive System

The human female reproductive system (Figure 5.2) produces, transports, and temporarily stores ova. Ova are produced in the ovaries, which are the primary reproductive organ. FSH in females, also secreted by the anterior pituitary, controls the maturation of ovarian follicles containing the ovum. LH stimulates the ovaries to produce testosterone, which is then converted to estrogen by adjacent cells called granulosa. Furthermore, the female reproductive system is specialized to accept sperm, aid in the process of fertilization, provide a highly controlled environment for long-term fetal development, and deliver the newborn from an intrauterine to an extrauterine environment. Hormones are also secreted by the ovaries. The accessory internal organs of the female reproductive tract include the fallopian tubes and the vagina. The external female accessory organs include the labia majora, the labia minora, the clitoris, and the vestibule.

Sequence of Sexual Arousal and Ejaculation in Males

The reproductive sequence in human males is (a) psychic stimulation, (b) erection of the penis, (c) lu-brication, (d) emission, and (e) ejaculation of semen. Erection is primarily a vascular event caused by dilation of arteries and constriction of veins in response to parasympathetic impulses from the sacral portion of the spinal column. The resulting high arterial blood pressure fills the erectile tissues of the penis. There are three cylindrical masses of erectile tissues (venous sinusoids) in the penis (i.e., two corpora cavernosa and one corpus spongiosum) (see Figure 5.1).

The parasympathetic impulses that promote erection also stimulate the secretion of a lubricating fluid from the bulbourethral gland. This fluid lubricates the end of the penis facilitating coitus (intercourse) and neutralizes any remaining acidity in the urethra.

Emission is the movement of sperm cells (from the testes) and secretions (from the prostate and seminal vesicles) to the urethra to form seminal fluid. Emission is a reflex that occurs in response to sympathetic impulses from the spinal cord that result in peristaltic contractions of the smooth muscles of the epididymis and the vas deferens. Sympathetic impulses also trigger the contraction of smooth muscles in the prostate glands and the seminal vesicles, which force the sperm down the urethra. Ejaculation is the expulsion of seminal fluid from the urethra by a reflexive contraction of the bulbocavernosus muscle (a skeletal muscle).

Sequence of Sexual Arousal in Females

The reproductive sequence in human females involves, in part, a monthly (28 days is normal) sexual cycle that is under hormonal regulation. Follicle-stimulating hormone (FSH) and luteinizing hormone (LH) are secreted by the anterior pituitary at the beginning of the sexual cycle and these hormones stimulate the process of ovulation. Both FSH and LH bind to cellular receptors that, in turn, activate adenylcyclase.

About two days before ovulation there is a marked increase in the secretion of LH, which causes the mature follicle to rupture and release an oocyte that enters the uterine tube (at about day 14). Following ovulation, the follicular cells turn into the corpus luteum, which releases increased amounts of estrogen and progesterone. The increased estrogen causes a thickening of the uterine endometrium in preparation for the potential implanting of a fertilized ovum. Similarly, progesterone causes an increased vascularization, swelling, and secretory activity of the endometrium, which is the innermost layer of tissue forming the uterine wall. Both estrogen and progesterone also inhibit the production of LH and FSH by the anterior pituitary gland. If fertilization does not occur, then the corpus luteum stops secreting estrogen and progesterone causing the disintegration of the uterine lining (i.e., menstrual flow). If fertilization does occur, then the placenta will secrete chorionic gonadotropin, which extends the life of the corpus luteum to the first 3–4 months of pregnancy.

Coitus in the female also involves psychic stimulation, erection, lubrication, and orgasm. The clitoris (Figure 5.2) contains two columns of erectile tissue called the corpus cavernosa that respond to parasympathetic impulses just as the penis does, i.e., the clitoris becomes erect. Concurrently, the Bartholin's glands located beneath the labia minor secrete a lubricating mucus.

2. Meiosis, Gametogenesis, and Fertilization

Meiosis

Refer back briefly to Chapter 3 and those sections discussing the formation of gametes and why genetic recombination is so important to sexually reproducing organisms. For sake of discussion, this section will describe meiosis and the steps leading to gametes (gametogenesis) in humans, and, potentially, to subsequent fertilization.

Recall that both mitosis (resulting in two identical diploid daughter cells, barring mutation) and meiosis must first have an Interphase when the DNA within each chromosome replicates (doubles). Following replication, each chromosome contains two identical sister chromatids joined by the centromere. Therefore, in humans, prior to the first division in meiosis, or mitosis, the sex (germ) cell contains $46 \times 2 = 92$ chromatids or 4 times the haploid DNA content. This is necessary, of course, because in meiosis in order to obtain a haploid number of 23 chromosomes in the mature gamete (1N=23) there must be two divisions. That is why in Figure 5.3 we note that gametogenesis requires that the original mother cell go though two series, meiosis I and II (Interphase, and Metaphase I not shown in Figure 5.3). In this illustration having 2N=6, note that each gamete at the end of Telophase II has N=3 chromosomes.

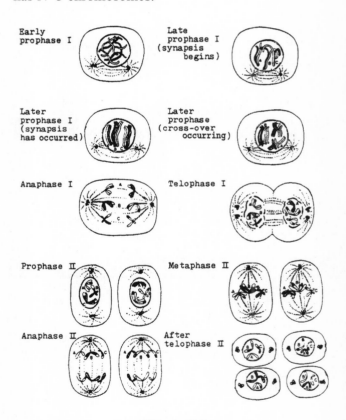

Figure 5.3.
The Stages of Meiosis in a Species with a Diploid Number of 6 Chromosomes (i.e., 2N=6).

Here we review the stages in meiosis in a diploid eukaryote, disregarding early and late phases:

— Interphase: DNA replication, chromosomes not visible.

— Prophase I: nonsister chromatids exchange genetic material (i.e., synapsis, formation of bivalents or tetrads by homologous pairs of chromosomes).

— Metaphase I: tetrads align along the equatorial plate of the dividing mother cell, centrioles at poles attached to centromeres by spindle fibers.

— Anaphase I: paired homologues separate, but centromeres do not, while moving to opposite poles.

— Telophase I: cytoplasm divides, no interphase, no more replication, meiosis II begins.

— Prophase II: centrioles divide, new spindles form, recombined chromatids still attached, align at second planar plate, perpendicular to the first.

— Anaphase II: the centromeres divide, the chromatids (now chromosomes) move to opposite poles.

— Telophase II: cytoplasm divides, chromosomes return to their dispersed form (i.e., naked DNA, no condensed coils, no heterochromatin).

The end result is four haploid cells. If successful fertilization occurs, the next diploid generation begins.

Spermatogenesis (Sperm Production)

Spermatogenesis (see Figure 5.4) in humans occurs in the coiled seminiferous tubules of the testes (Figure 5.1). The seminiferous tubules have supporting Sertoli cells, stimulated by FSH to promote sperm maturation, as well as spermatogenic cells. Sperm differentiation begins with spermatogonia, primitive unspecialized germ cells lining the walls of the tubules in the testes. During growth pre-puberty, spermatogonia increase mitotically concurrent with the growth of the testes. At puberty, some of the spermatogonia enlarge and become primary spermatocytes (see Figure 5.4). These undergo the first meiotic division, thus becoming secondary oocytes. After the second meiotic division, the resulting spermatids

are haploid (N=23 chromosomes in humans) and undergo further maturation (e.g., secretion of FSH by the Sertoli cells). Mature sperm are notably much smaller than the mature egg in the female and are composed of a head (with anterior acrosome, posterior nucleus with DNA and very small cytoplasm), carbohydrates (supply of energy for motility) and lysosomal enzymes (for enzymatic penetration of the egg), and a neck (mid-piece containing many mitochondria) that transitions to the flagellum or tail, which provides motility via the oviduct (upper Fallopian tube) following ejaculation of semen.

Oogenesis (Production of Egg Cells)

Oogenesis (ovum or gamete formation) in humans occurs in the in the ovary beginning with immature germ cells called oogonia that go through several mitiotic divisons during early development. The ovaries lie on each side of the pelvic cavity (Figure 5.2) and mainly contain two types of cells: germ cells and germinal epithelial cells. The germ cells give rise to the oogonia ovaries and the germinal epithelial line the ovary wall and eventually become the follicle. At birth in a human female the oogonia, or immature eggs, number approximately 400,000, but only about 400 will reach full maturity during an average human female reproductive life.

During oogenesis (see Figure 5.4), a fraction of the oogonia undergo further differentiation and become primary oocytes (these are arrested at Prophase I of meiosis). At maturation the primary oocyte undergoes the first meiotic division. The nuclear events (division) are the same as in spermatogenesis, but a notable difference is that in females the cytoplasm divides unequally resulting in two cells of very different sizes. The smaller cell is the first polar body and the large cell is the secondary oocyte (now arrested at Metaphase II). This cell undergoes the second meiotic division (again, the cytoplasm does not divide) that is also a result of an unequal division of cytoplasm and forms a second polar body and a large ootid. The first polar body divides to form two additional second polar cells. The polar cells disintegrate and do not become functional gametes. Note

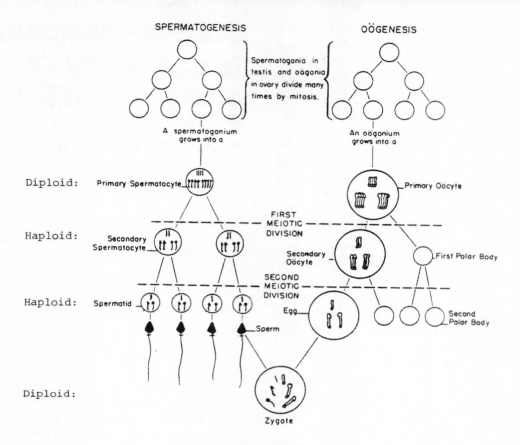

Figure 5.4.
Comparison of the Formation of Sperm and Eggs

that at this point the egg cell is much larger than the sperm and while four sperm resulted from meiosis, only one egg cell was formed in the female. It is important to understand that in the female, meiosis II is *not* completed unless the sperm contacts the plasma membrane of the egg.

Events leading to fertilization or sloughing of cells containing the mature egg: beware that that as the oocytes form and become surrounded by germinal epithelial cells they become part of what are the follicles on the surface of the ovaries. After about two weeks into oogenesis, ovulation occurs which means that the secondary oocyte is released from the follicle (i.e., rupture of part of the ovary wall and expulsion of the oocyte from the ovary). During ovulation, the oocyte is discharged into the body cavity from the ma-

ture follicle and then travels to the opening of the Fallopian tube (entry to the oviduct). The corpus leuteum (follicular material remaining after ovulation) secretes progesterone and estrogen that stimulate the uterine lining to prepare for a possible embryo implantation. If the ovum is fertilized, the corpus leuteum enlarges and forms the corpus leuteum of pregnancy, if there is no fertilization, it becomes the corpus leuteum of menstruation.

Fertilization (in Humans)

Following ejaculation of semen into the vagina, sperm enter the cervical canal and within about 30 minutes reach the intersection of the uterus and oviducts (uterine tubes or Fallopian). In the upper third of the uterine tubes, sperm (only about 1 in 3 make it this far) may contact an oocyte that has been released from an ovarian follicle during ovulation. Sperm motility is dependent upon not only the motility of the sperm tail but contractions within the uterine tubes that are stimulated by prostaglandins in the semen (semen is composed of sperm, and secretions from sex accessory glands, which are the prostate, two seminal vesicles and two Cowper's glands; only about 1% of semen is sperm). After about 6–7 hours in the oviduct fluids, the sperm has sufficiently undergone enzymatic changes to the protective layer surrounding the head such that the acrosome's enzymes can leak out through tiny openings allowing the sperm to dissolve its way (other sperm also contribute acrosome enzymes to sperm nearest the plasma membrane of the oocyte) through the follicle cells that surround the oocyte. Once inside the outer wall, sperm must digest through another layer called the zona pellucida that surrounds the oocyte. Generally only one sperm is allowed once it has made contact with the plasma membrane of the oocyte. All others are excluded, mainly by two mechanisms: fast block polyspermy (dipolarization via action potential) and slow block polyspermy (enzymatic hardening of the zona pallucida). Contact with the plasma membrane triggers the second meiotic division thus the now secondary oocyte becomes the ovum. Once inside the ovum, the cell becomes the zygote (two pronuclei) and after a series of events, including DNA replication, condensation of chromosomes, then spindle fibers form, the nuclear envelopes of the pronuclei disintegrate and the first of billions of cell divisions begins.

There are four results of fertilization:

1. fusion of the haploid sex cells' nuclei creating a single celled zygote which restores the diploid number of chromosomes;

2. fertilization results in the initial cleavage of the zygote, followed by rapid cell division leading to an embryo;

3. fertilization results in sex determination, a 46XX zygote is female, a 46XY zygote is male;

4. finally, fertilization in sexually reproducing organisms is an important factor in maintaining species genetic variation.

3. Early Development (e.g., Polarity, Cleavage, and Gastrulation)

Embryonic Development

Embryonic development, in humans, and other sexually reproducing, placental animals, begins when an ovum (1N=23 chromosomes) is fertilized by a sperm (1N=23 chromosomes) and ends at parturition (birth) of the fetus. It is a process of change and growth that transforms a single cell, the diploid zygote (2N=46 chromosomes), into a multicellular organism following a species-dependent gestation period.

While the timescale may differ among species, the processes are very similar. Because good ethical practices prevent us doing certain types of experimental research in human embryology, models from animals as distant as *Drosophila*, mice and birds (e.g., domestic fowl) have provided much information. It is instructive to recall that the so-called *Hox* gene complex that has been identified to regulate the very early organization of cells that become the adult organism have been found to be conserved in *Drosophila,* and also in humans.

For sake of illustration, but realizing that organismal development in *humans* is not a discrete process, the time from fertilization until birth can be divided into these phases:

— Pre-embryonic: fertilization to approximately 2 weeks.

— Embryonic: 3 weeks to approximately 7 weeks.

— Fetal: 7 weeks to approximately 38 weeks.

Pre-Embryonic Development

The earliest stage of embryonic development (Figure 5.5) is the diploid zygote. Next is a period called cleavage (dividing), in which mitotic division of the zygote results in the formation of daughter cells called blastomeres. At each succeeding division, the blasto-meres become smaller and smaller. There are three primary types of cleavage among animals:

1. radial, found in echinoderms and chordata (humans, of course, fall here).

2. spiral, found in annelids, molluscs and some flat-worms.

3. superficial, associated with arthropods.

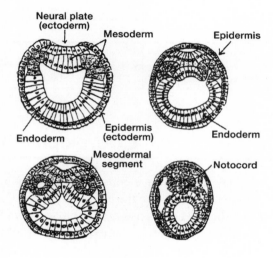

**Figure 5.5.
Early Embryo Development**

Radial cleavage is characterized by a mitotic spindle forming following fertilization that is at right angle to the egg's axis; the second spindle forms transverse to the first cleavage; the third and final division is parallel to the egg's axis. therefore, $2^3=8$ cells are the result. Additional mitotic divisions result in 16 or so cells (blastomeres), forming a solid ball of cells that is called a morula. As the morula divides further, a fluid-filled cavity is formed in the center of the sphere, converting the morula into a hollow ball of cells called a blastula or blastocyst. The blastocyst contains two groups of cells, one called the inner cell mass (ICM), which later becomes the embryo, and the other a ring of cells called the trophoblast, which is the precursor to the placenta.

The blastocyst may remain unattached in the uterine lumen for 2–3 days. If it remains intact and unharmed, it may then attach itself to the endometrium of the uterine wall, most often the back, upper wall. Lytic enzymes aid the blastocyst to digest into the endometrium and this becomes what is called implantation. Failure to implant can be due to any number of causes, including genetic mutations, maternal health, intrauterine device (IUD), "birth control pills," teratogens and so forth. Further development *in utero* is regulated by several hormones, including human chorionic gonadotropin (HCG), estrogen and progesterone.

Although research continues regarding the specific timing, it is generally accepted that during implantation polarity is established for the developing embryo. Polarity sets the pattern for further organogenesis and morphogenesis by establishing the orientation of axes, including right–left structures. These are the axes:

1. Anterior (head)–Posterior (feet).

2. Dorsal (back)–Ventral (face, or front, including position of limbs in the correct direction).

3. Left–Right: proper orientation of the left and right limbs, hands, feet and so forth.

It is believed that the A-P axis is also correlated with orientation of the primitive streak, midline of the blastocyst, which at it's anterior lies Hensen's Node and the anterior visceral endoderm (AVE), thought to be the two most important regions in the embryo during organogenesis; Hensen's appears to direct the overall body form and the AVE directs head development.

Once implanted, portions of the endometrial and embryonic tissues combine to form a placenta. It serves as a source for nourishment, oxygen and removal of wastes. As the placenta forms the ICM in the blastocyst undergoes changes and cells are laid down that form an amnion cavity, which fills with fluid and that acts as cushion for the developing fetus. It is about then that the blastocyst undergoes a sequence of changes called gastrulation. The cells of the ICM (below) begin to differentiate into three cell layers called the primary germ layers and this is the formal beginning of embryonic development (Figure 5.6).

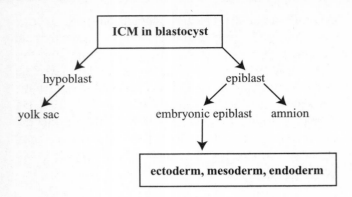

Figure 5.6.
Derivation of the Three Germ Cell Lines

These layers include the ectoderm (epiderm), mesoderm, and endoderm (see Figure 5.5).

The gastrular period generally extends until the early forms of all major structures (for example, the heart) are created. After this period, the developing organism is called a fetus. During the gestation of the fetus (the duration of which varies with different placental species (see Table 5.1), the various systems develop further. Though developmental changes in the late fetal period are not as dramatic as those occurring during the earlier embryonic periods (12 weeks and less), they are extremely important. Congenital defects may result from abnormal development during this period.

Table 5.1.
Length of Gestation in
Eight Different Placental Animals

Animal Species	Average length of gestation, in days
Human	280
American possum	12
Mouse	21
Rat	21
Dog	60
Cow	270
Giraffe	420
Elephant	>600

Formal Embryonic Development

Formal embryonic development begins with implantation of the blastocyst into the uterine wall. You learned, above, that the ICM gives rise to the primary germ layers of cells during the gastrular period. Forms of all major structures are constructed during the gastrular period. These primary germ layers begin to differentiate rapidly during the embryonic and through the fetal stages.

Table 5.2.
Fate of the Cells Arising from the Three Germ Layers[1]

Ectoderm	Mesoderm	Endoderm
epidermis	dermis	lining of digestive system
hair, nails, sweat glands	all muscles of body	lining of respiratory system
brain and spinal cord	cartilage	urethra and urinary bladder
cranial and spinal nerves	bone	gallbladder
retina, lens, and cornea of eye	blood	liver and pancreas
inner ear	all connective tissues	thyroid gland
epithelium of nose, mouth, and anus	blood vessels	parathyroid gland
enamel of teeth	reproductive organs	thymus
	kidneys	

[1] Source: page 467 in Chiras, D.D. (2001) *Human Biology*, 4th ed., Jones & Bartlett Pub., Boston, 625 pp.

The primary germ layers establish the lineages of what become tissues and organs (i.e., the fate of cells is determined and fate maps constructed from lineage analyses). Table 5.2 describe the fate of cells arising from the three primary, embryonic germ layers.

During embryonic development (organogenesis) the organism is very sensitive to chemical and physical agents that may cause birth defects. Such agents are called teratogens. They are said to be teratogenic. Some of the confirmed agents include these: progesterone, thalidomide, rubella (German measles), alcohol, and irradiation. Others that are suspected include, but are not limited to, these: tobacco, aspirin, insulin, antacids, dioxin, cortisone, excess vitamins A and D, and more. Table 5.3 lists nine organ systems or structures and their periods of greatest sensitivity to teratogens during development.

Table 5.3.
Development Time (38 Weeks Total) and Period of Sensitivity of Human Fetal Structures to Teratogens

Organ System or structure	Weeks needed to develop	Interval of greatest sensitivity to teratogens (beginning–ending)
Central nervous system	38.0	3rd to 6th week
Heart	9.5	3rd to 6th week
Arms	8.5	4th to 8th week
Eyes	4.5	4th to 9th week
Legs	8.5	4th to 8th week
Teeth	14.0	6th to 12th week
Palate	9.0	6th to 8th week
External genitalia	31.5	7th to 12th week
Ears	16.5	3rd to 12th week

[1] Source: adapted from Figure 20-8 in Chiras, D.D. (2001) *Human Biology*, 4th ed., Jones & Bartlett Pub., Boston, 625 pp.

The information in Table 5.3 underscores the vulnerability of the developing fetus to teratogens between the 3rd and 12th weeks *in utero*.

Reproduction and Development in Other Animals

A major difference between sexual and asexual reproduction is the number of parents involved. Sexual reproduction requires two sexes (male producing sperm, female producing eggs).

Insects—Insects are distinctly male and female and reproduce sexually. The male deposits sperm into the sperm receptacle of the female where the sperm are stored. When the female produces eggs, the eggs pass over the receptacle while they are being laid and are fertilized. In complete metamorphosis (e.g., beetles, the Coleoptera) an insect goes through the egg, larva, pupa, and adult stages. In incomplete metamorphosis (e.g., grasshoppers and crickets, the Orthoptera) the egg becomes a nymph (larval stage) and then an adult.

Sea Star (Starfish)—Usually sexual reproduction with a larval stage. Cleavage is radial in the sea star egg (Figure 5.7). The egg is split into two equal daughter cells. The second cleavage plane is vertical but at right angles to the first, separating the two cells into four. With further divisions, a blastula forms. It is a simple layered blastula, which is later converted into a double-layered sphere, the gastrula, by the invagination of a section of a wall of the blastula.

Sea stars can regenerate arms; one species can reproduce a complete animal from a body part.

Fish—Most fish use external fertilization, in which the female lays eggs in the water and the male deposits sperm-containing fluid (milt) over the eggs, resulting in fertilization without the two parents making physical contact. Pheromones released into the water by individuals may help to ensure that the release of eggs and sperm are properly timed. After fertilization, the zygotes develop into small fish (fry) that are nourished by an attached yolk sac. In the process of external fertilization, they develop outside the female's body. In some fish, fertilization occurs in an oviduct within the female's body and develops within the female's body.

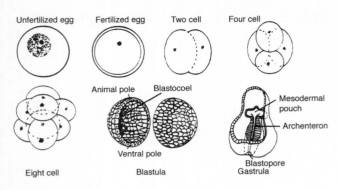

Figure 5.7.
Early Development Stages in the Sea Star

minal space separates the hypoblast from the underlying yolk. The gastrula is formed by cell migration.

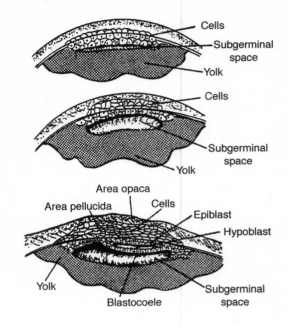

Figure 5.9.
Successive Stages of the Cleavage of a Hen's Egg

Amphibians—The cleavage pattern in amphibians is radial but because of the unequal distribution of yolk in the fertilized egg, the blastomeres are not of equal size (Figure 5.8). At the eight-celled stage, the egg consists of four smaller animal pole blastomeres (micromeres) and four larger, yolk-laden vegetal pole blastomeres (macromeres). The egg then becomes a gastrula. To form a gastrula, a groove appears on one side of the blastula and cells at the bottom of the groove stream into the interior of the embryo. This groove extends transversely until it is ring-shaped, at which point the yolk-filled cells remain as a yolk plug.

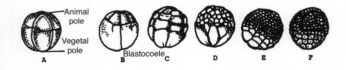

Figure 5.8.
Early Embryology in a Frog

Birds—Avian eggs display radial cleavage (Figure 5.9). Cleavage occurs in a small disc of cytoplasm in the animal pole. Horizontal cleavage separates upper blastomeres and lower blastomeres. The blastomeres at the margin of the disc and the lower cells in contact with the yolk lose the furrows that partially separated them and fuse into the periblast. The free blastomeres become incorporated into two layers: an upper epiblast and a lower hypoblast. Between them is a blastocoele. The subger-

Asexual reproduction usually requires only one individual that reproduces a genetic clone via budding, splitting or fragmenting. Examples of *asexually* reproducing animals include: Euglena, Paramecium, Amoeba, Hydra and flatworms.

Hermaphroditism is the ability to sexually reproduce by only one individual having both male and female sex organs. Examples are the Hydra and Earthworm:

Hydra—Hydra may reproduce asexually by budding, but may also reproduce sexually. An ovary containing one egg and a testis containing many sperm are formed along the sides of the hydra. Because hydras have organs of both sexes, they are known as hermaphrodites. Sperm are released into the water. One unites with the egg, and a zygote is created.

Earthworm—An earthworm is a hermaphrodite, but fertilization takes place when two worms transfer sperm cells to the other's sperm receptacle. A mucous ring around each worm's body becomes a cocoon, which is deposited into the soil and within which fertilization and the development of the zygote occurs.

Hymenoptera—Reproduction by parthenogenesis is when an egg develops into an adult individual without having been fertilized. This is similar to parthenocarpy used for commercial purposes in many parthenocarpic food crops (e.g., seedless grapes, fruit is produced without pollination) but it similarly occurs in animals. Examples are in the Hymenoptera (ants, bees, wasps). The queen bee has the ability to produce both fertilized and unfertilized eggs: the fertilized emerge to become queens and workers (females), the unfertilized become drones (males).

4. Developmental Processes

Organisms are composed of cells all having the same genetic information. Morphogenesis is a process under genetic control that causes genetically identical cells to give rise to new types of cells, tissues, organs and the anatomy or "phenotype." This is cellular differentiation. The resulting phenotype (refer back to Chapter 3) is the whole of the genotype, environmental factors and the effects of their interactions. This section describes some of the developmental processes that are believed important during embryo-fetal growth and differentiation.

Induction and Determination

Before cells develop into tissues and organs they must first be induced by receiving a chemical signal and this sets a process into motion whereby a cell is determined to be one type or another (its fate).

An organizer is a chemical secreted by certain cells in the embryo, which diffuses into and influences surrounding cells. The presence of organizers is the whole basis for the concept of embryonic induction. Induction is dependent upon competent cells (i.e., ones able to respond to signals).

There are two primary types of induction:

1. Permissive induction: a responding cell or tissue differentiates in the presence of an inducing signal, but does *not* differentiate if the signal is absent.

2. Instructive induction: a cell or tissue responds to a signal by differentiating into one specific cell type, however, in the absence of the signal, it *does* differentiate, but into a *different kind* of cell type than would have resulted with the signal.

Types of signalling molecules fall into three classes: ones at the cell surface and only affect neighbouring cells, and those localized but in the extracellular matrix and others that that can act from a distance because they are diffusible.

Transforming growth factor β (TGFβ) superfamily of signalling molecules is a major regulator of development and differentiation in both vertebrates and invertebrates. Transduction of the signal requires receptors (e.g., Types I, II and III are at the cell surface; I and II are serine/threonine kinases; III is β-glycan).

For example, in the frog and salamander, it was found that the dorsal tip of the blastopore was important in organizing the development of the body parts. In vertebrates (*Xenopus* [frog]), Spemann found that organizers in the optic vesicle in the embryo stimulate the adjacent ectoderm to form the lens of the eye. Spemann's transplantation experiments demonstrated that if the optic vesicle were removed and inserted anywhere under the skin of the embryo that the overlying ectoderm formed into a lens.

One of most studied models in developmental biology is that of morphogenes that code for proteins that activate/deactivate genes during the early development of *Drosophila*. These proteins act based upon concentration gradients, when high the protein acts a transcription factor (regulating the amount of a protein that is being expressed form a particular gene), when low, the transcription is "off." Examples are the *bicoid* proteins (transcription factors) that are involved in organizing the cells that become the head/anterior of the insect; the *hunchback* gene acts in concert with bicoid and is integral to anterior/posterior development in *Drosophila*.

Morphogenesis and Differentiation

The first sequences of differentiation in vertebrates have been described above and include the laying down of the three types of germ cells. From there, the cells change in structure and function based upon the determinants of genetic regulation and cell fate.

Early differentiation in the human embryo can be illustrated by the following: The first two organs to appear in the developing embryo are the brain and the spinal cord. They form during a process called neurulation. Soon after pre-embryo differentiation is done (Figure 5.7), the ectoderm in the primitive streak (not shown) develops a thickened plate of cells called the neural plate and a neural groove in the middle. Two longitudinal neural folds converge at the anterior pole (refer above to the anterior–posterior axes), later forming a tube whose anterior cavity becomes the brain. The posterior of the neural tube at the same time develops into the spinal cord. By the fifth week in human embryos the forebrain, midbrain and hindbrain are formed. The primitive streak gives rise to connective tissue precursor cells that later develop into the skeleton.

This section has discussed the development of the whole animal: this process is under genetic regulation via special regulatory genes called *Hox genes*. These are groups of genes (homeoboxes) that are common across species and apparently regulate the expression of other genes during development; these DNA sequences appear in species as diverse as sponges and humans, but to date, found only among animals.

Metamorphosis

Previous review has focused upon placental animals, mainly humans and other mammals. Not all animals follow continuum from zygote to adult. Many animals go through larval stages. The larva is a sexually immature form of the animal. The larva may live in different habitats than does the adult form, may eat different food and is morphologically distinct form the adult. Metamorphosis is represented by a new surge of development leading to the adult, sexually mature form. This type is called *complete* metamorphosis. Insects that have this type enter a pupal stage prior to becoming reproductive adults. It is during the adult stage that both dispersal and reproduction occur.

Some animals, such as grasshoppers, go through what is called *incomplete* metamorphosis. This means that the young resemble the adult (but smaller in size) and go through a series of molts and each time the animal more closely approaches the adult size and morphology.

Amphibians ("two lives"), such as frogs, have a swimming larval stage that appears more fish-like: has gills, is an herbivore and long finned tail and has a lateral line as in fish. During metamorphosis legs develop, the gills are lost, and the lateral line disappears.

Aquatic crustaceans (shrimp, crabs, lobsters) go through molts and also usually have a swimming larval stage.

5. External Control Mechanisms

Homeostasis is a balance among internal controls issued by the nervous and endocrine systems and the external environment. These are closely aligned with the concept of phenotype reviewed in Chapter 3 (i.e., expression of the genotype is modified by the environment and interactions between it and the genotype).

External factors, such as sound, humidity, light, nutrition and temperature, work in concert with internal controls mainly through negative feedback loops. It is becoming more clearly understood that light, including intensity, quality, and duration, is one of the most important factors affecting how organisms adapt both in the short and long term. These refer mainly to our 24-hour clock, the so-called light-dark photoperiods or *circadian rhythms*, and over longer periods are seasonal changes. Thus, while homeostasis is a state of constancy that the organism targets, it does not mean a static being. Although these short and long-term external cycles may vary over time, they are also predictable. Examples in humans include: 1) variation in our alertness, usually lowest around one o'clock in the afternoon; 2) male testosterone increases at night and maximizes during REM sleep (*rapid eye movement*, also called "dream sleep"); and 3) in women, estrogen is highest at about day 14 in the menstrual cycle when ovulation occurs and lowest at the beginning of their cycle.

Photoperiodism is defined as the biological response to a change in the proportions of light and darkness in a 24-hour daily cycle; short-term circadian responses. It has been shown to control a wide range of biological activities, including the induction of flowering in plants, stimulation of germination in certain seeds and the initiation of mating in certain insects, birds, fish and mammals.

Many of the biological cycles in humans are controlled by the interaction between external stimuli and a region in the brain called the suprachiasmatic nucleus (SCN); activation occurs via the eyes. The SCN is a group of nerve cells in the brain within the region of the hypothalamus. It is thought to operate analogously to a clock, constantly tracking time. The pineal gland, also in the brain, helps regulate the SCN, secretes a hormone called melatonin. Recent evidence indicates that in mammals upward of a dozen "clock" genes may be expressed in the SCN and in tissues that are control by melatonin. These genes, and melatonin, are also indicated to be involved in longer-term seasonal rhythms in mammals, not just circadian, thereby influencing reproduction, growth, appetite and so forth.

Actually, in humans, it is now accepted that we operate, on average, a 25-hour cycle, but that varies, as do most biological phenomena: some people are on a 24-hour wake-sleep cycle, others on a 28-hour cycle. "Jet lag" is one the common examples of the fatigue that can be associated with interrupting circadian rhythms in humans.

Normal growth of the fetus, and that occurring until adulthood, come under control of the anterior of the pituitary gland, also controlled by the region of the brain called the hypothalamus, and its secretion of growth hormone (GH). The primary targets of GH are the muscle and bone tissues. Concentrations of GH vary over the 24-hour cycle and with physical activity: highest blood levels are during sleep and following exercise.

Plant Structure, Function, and Organization

With Emphasis on Flowering Plants

Overview

Some Distinguishing Characteristics Among Higher Plants

Land plants are classified into two broad taxonomic divisions (refer to Chapter 8), nonvascular and vascular:

1) **Nonvascular Plants:** Bryophytes (mosses, liverworts)—Mosses have little or no organized (vascular) tissue for conducting water, minerals, and metabolites.

2) **Vascular plants** have conducting tissues and are further divided into seedless and seeded.

— **seedless** (Pteridophytes), include ferns and club mosses.

— **seeded**, include Gymnosperms (conifer) and Angiosperms (flowering).

Plants have three main components: root, stem, and leaf; reproduction organs like flower structures arise from buds originating in the stems, and in some, asexually via adventitious sprouting from stems and/or roots.

Alternation of Generations in Plants

The alternation of generations in plants involves two phases in the life of a single individual. These include a mature, diploid spore-producing phase (sporophyte) and a haploid gamete-producing phase (gametophyte).

Vascular Plants with Seeds

Most of this chapter will focus upon angiosperms, the so-called flowering plant species, and the most prevalent of the seed-bearing plants. Angiosperms and gymnosperms include both woody and nonwoody (herbaceous) species. Our discussion will focus mainly on the nonwoody.

Before leaving gymnosperms, the other vascular, seeded group, it is worth identifying a few of the distinguishing characteristics separating these two great subdivisions. These are shown in Table 6.2.

Table 6.1.
Summary of Alternation of Generations in Plants

Plant	Alternation of Generation
Mosses (nonvascular)	Separate male and female plants, gametophytes are the dominate visible structures, sperm from antheridia, eggs in archegonia, have swimming sperm, and zygote is protected within archegonia until new sporophyte develops.
Ferns (vascular)	Separate generations, dominant sporophyte, minute gametophyte, swimming sperm
Gymnosperms (vascular)	Gametophytes (male and female cones) on sporophyte, wind-pollinated, sperm moves in pollen tube.
Angiosperms (vascular)	Gametophyte in flower, on sporophyte, wind-pollinated and animal-pollinated (e.g., insects, bats, birds)

As noted in Table 6.2, Angiosperms are further divided into two major subclasses, monocots and dicots (*eudicot* is now often used, but this review will use dicot). Table 6.3 lists the primary traits that distinguish these groups.

Life Cycle

Plants that complete their life cycle from seed germination to reproduction to seed then death in one year are called *annuals* (e.g., most temperate flowering crop plants, many wildflowers). *Biennials* are those that live through a winter season, the vegetative growth in the spring/summer, and flower in the next spring/summer, thus requiring two years to complete their cycle (e.g., beets and carrots are biennials but we intervene by harvesting the root, rather than wanting their seed). Plants that live many years are called *perennials* (e.g., trees, shrubs, bulbs and many grasses). Once reproductively

Table 6.2.
Comparisons Between Angiosperms and Gymnosperms

Trait	Angiosperms	Gymnosperms
Seed	Seed produced in ovule encased in ovary, base of flower (i.e., "seed enclosed in vessel" per name, angiosperm).	Seed exposed upon sporophylls (i.e., "naked seed" per name, gymnosperm).
Flowers	Have flowers that produce seed that are formed from modified leaves and most, not all, are showy which attracts pollinators.	Do not have true flowers, seed produced in female (ovulate) strobili or cones; male (staminate) cones produce pollen.
Seed endosperm	Seed endosperm is triploid.	Seed endosperm is haploid.
Habitat	Adapted to more diverse habitats	Most species in cooler drier habitats
Reproduction (structures)	Male and female reproductive structures may/may not be in the same flower	Reproductive structures on the same plant but different structures
Reproduction (location of structures)	Some species have separate "male" and "female" plants (i.e., dioecious).	Most often both sex gametes produced on the same plant, different structures (i.e., monoecious).
Pollination	Animal pollination often the case	Wind pollination most common
Fertilization	Double fertilization	Single fertilization
Wood anatomy	Woody species have more complex wood structure	Wood anatomy less complex
Cotyledons (first leaves)	One (monocots) or two (dicots or eudicots)	Whorl of needle-like leaves, number varies with species

Table 6.3.
Contrasting Angiosperm Dicots and Monocots

Trait	Dicots (eudicots)	Monocots
Examples of each type	Trees (poplar, maples, lilac) and crops like peas, beans, and most annual herbs	Grasses, including crop species (e.g., wheat, corn), and ornamentals like lilies, irises
Flower structures	Usually in 4s and 5s, e.g. petals.	Usually in 3s.
Pollen grain	Usually 3 furrows in the grain	Usually having one furrow
Cotyledons	Two	One
Veins (vascular) in the leaves	Often "net-like"	Usually parallel
Vascular bundles in the stem	In a ring	No pattern, scattered
Secondary growth from vascular cambium	Often present	Absent

mature, they may produce seed crops every year, and it might take several years from germination to the time when the first seeds are produced (e.g., fruit and forest trees).

1. Tissues, Tissue Systems, and Organs

Higher Plant Cells Versus Animal Cells

Although eukaryotic cells share many similarities in their biochemistry, DNA-RNA processing and en-

ergy transfer processes, plant and animal cells do have important differences. A few of those follow:

1. Plant cell growth differs from animal cell growth: animal cells increase size via protoplasm, plants increase by water uptake via the vacuoles.

2. Mature plant cells are usually much larger than animal cells.

3. Plant cells need sturdy mechanical characteristics in order to withstand the osmotic pressure differences between a cell's interior and extracellular spaces.

4. Plant cells need vacuoles so that waste can be deposited (or, into the cell wall) whereas animals have the circulatory and excretory systems for managing wastes.

5. Plants are autotrophs, animals are heterotrophs: Plant cells (containing chloroplasts) capture energy from sunlight, uptake water, minerals and synthesize their own metabolites by fixing a carbon source (CO_2) used for energy and storage; these products (sugars and starches) are utilized by animals. Some animals, of course, ingest other animals that are linked with plant products.

6. Plants utilize CO_2 (fixed into sugar, stored as starch) and respire O_2, while animals utilize the O_2 catabolize the plant metabolites and respire as CO_2.

7. Plant cells contain three genomes (one nuclear, mitochondria and chloroplast), but animals have two (nuclear and mitochondria).

8. Generally, mature terrestrial plants are sessile (fixed in location) but are motile via spores/pollen and seeds, but animals rely upon their mobility for survival (i.e., gathering food and water and avoiding predators).

Seeds and Seedlings

Before reviewing the cells and tissues of mature plants, it is important to review some of the important structural characteristics of dicot and monocot seeds and seedlings. Keep in mind that there are many variations among the thousands of species in the two groups, so these are for the "typical" cases. Also, keep in mind that the embryo within the seed is made up of either one

Table 6.4.
Structures of Plant Seed and Embryo in Dicots and Monocots

Structure	Dicot	Monocot
Cotyledon	2-seed leaves, energy storage, provides nourishment to the emerging seedling	1-seed leaf (see scutellum)
Endosperm	Usually small to absent, but large in some species in which serves same function as in monocots	Often large proportion of the seed volume; provides nourishment to the emerging seedling
Plumule	Apical meristem or epicotyl above the cotyledons (+ any young leaves); absorbs nutrients from cotyledons or endosperm	See Coleorhiza.
Hypocotyl	Embryo axis [to become the shoot/stem] between the cotyledons and the root tip; absorbs nutrients from cotyledons or endosperm	Quite diminished in many species, particularly grasses; more identifiable in species, for example, such as the Onion family
Radical	Name of root-region if obvious in the embryo; absorbs nutrients from cotyledons or endosperm, water from surrounding media.	See Coleorhiza.
Scutellum	None	Same as cotyledon, closely aligned beside the endosperm; absorbs nutrition from the endosperm.
Coleorhiza	None	Sheath that encloses the radicl
Coleoptile	None	Sheath that encloses the plumule
Seed coat/pericarp	Hardness varies	Hardness varies

(monocot) or two (dicot) cotyledons joined to a small stem-like structure, and the basal portion that emerges during germination gives rise to the seedling root. Table 6.4 displays the main points.

Seeds imbibe (absorb) water, which together with optimum temperature and light (or absence of), that sets into action enzymatic pathways that result in the radicle then the emerging cotyledons to become established and develop into what is called a seedling. Depending upon the species, these events continue for days to weeks and are driven by very active mitotic cell divisions. Stems, branches, leaves and roots (primary, secondary and whether taproot or fibrous) develop into a mature plant (apical meristem [primary growth] yielding height and length and the vascular [lateral] or cambium meristem adding stem/branch diameter [secondary growth]) and depending upon the species and the growth environment (habitat for wild populations), the plant may be-

come reproductively mature within a matter of days or years (i.e., many woody species).

Tissues, Organs, and Systems

There are three main types of plant cells (parenchyma, collenchyma, and sclerenchyma) that are organized into various organs and three tissue systems: dermal, vascular, and ground or fundamental tissue). These are described in Table 6.4.

Simple tissues are those composed of only one cell type. Tissues of more than one type are called complex; these are noted in Table 6.5.

Meristematic cells, within the actively dividing apical region of the shoot and root, are the physiologi-

Table 6.5.
Cellular Organization in Angiosperms

Tissue System	Cell and Types	Structure and Function
Fundamental or ground tissue system	Parenchyma (simple)	Important in the primary plant body (i.e., tissues prior to secondary growth arising from the vascular cambium); maintain ability to divide and important in healing and regeneration; active in photosynthesis, storage, secretion, and movement of water and food; found in reproductive structures and glands (e.g., salt glands).
	Collenchyma (simple)	Like parenchyma, these are living tissues at maturity and allow for cell elongation in stems and leaves; aligned as strands or cylinders and with veins in dicots, and add support to young growing shoot.
	Sclerenchyma (simple)	These lack protoplasts at maturity, do not elongate and are often hard, thickened and highly lignified which gives the plant stem (fibers like hemp) or fruit (walnut shell) structural strength.
Vascular tissue system	Xylem (complex)	Principal water conducing tissue in plants, continuous along with phloem through the plant; conducts minerals, food and adds structural support; lack protoplast; conducting cells of the xylem are tracheids (with pits) and vessel elements (with pits and perforations for transport).
	Phloem (complex)	Phloem is the primary food conductor in the plant along side xylem; sieve elements are the main conducting cells in angiosperms and have protoplast at maturity, but most lack nuclei and do not have clear boundary between the cytoplasm and vacuoles; usually associated with parenchyma cell called companion cells.
Dermal tissue system	Epidermis (complex)	This is the outer layer of the plant body, it is dermal system for flower, fruits, seeds, leave roots, etc.; the epidermis (upper and lower leaf) may contain stomata cells and their guard cells which function in regulating CO_2 and O_2 exchange and water vapor; epidermis may have hairs or trichomes, which may become spiny; some species may have multiple epidermal layers, thought to be adapted to water storage and prevent loss; epidermis is covered by cutin a protective cuticle film and in some species may be waxy; epidermis covers and protects the young parts of the plant, specialized epidermal tissue includes roots hairs that increase the surface area for absorption.
	Periderm (complex)	Formed from secondary growth via the lateral vascular cambium that is augmented by activity of the cork cambium from which the periderm arises composed mainly of cork tissue.

cally young cells that give rise to all other cell-tissues upon differentiation; these are somewhat analogous to stem cells in animals. Monocots, like grasses, contain an apical meristem near the soil surface at the base of leaf whorls (recall adventitious shoots from base of the plant of many grass species) but most dicots have the apical meristem near the tips of the main stem and branches. The degree of dominance (due to distribution of plant hormones, see Chapter 7) of the apical meristem in mature dicots, particularly woody species, is demonstrated by the shape of the crown (i.e., Lombardy poplar columnar habit versus that of a widely branched, spreading oak).

2. Water Transport, Including Absorption and Transpiration

Water and dissolved minerals are absorbed from the soil solution into the plant via root hairs. There are two primary methods by which water enters the plant and reaches the xylem for transport up the stem to the leaves. One is by the *symplast* route that is across the root hair membrane then cell to cell until entering the xylem cells in the root core (stele). The other path is called *apoplast* whereby after entering the root hair it moves along the cell walls via the intercellular spaces. Plant cells enable these communications via plasmodesmata, open channels from the cell to the extracellular matrix.

Water (dissolved minerals, such as potassium [K^+], calcium [Ca^{2+}], nitrate [NO_3^-], and phosphate [PO_4^{3-}]) moves by simple diffusion and osmosis (see below) from the soil into the plant along a water potential gradient because of the higher solute concentration in the cell cytoplasm. Once it reaches the nonliving xylem cells (by apoplast and symplast pathways), it moves by TATC (transpiration-adhesion-tension-cohesion) up the stem. Water vapour loss by transpiration (evaporation at the leaf surface via stomata openings) sets up a pressure differential pulling the column of water up through tubular-like vessels, the column kept in-tact by adhesion, cohesion and tension forces. Some upward, positive pressure comes from the root, up the stem and into the leaves, mainly at night when there is no transpiration. Evidence of this can be seen in the morning when free water exudes from the tips/edges of leaves; this is called guttation.

The vascular system is continuous throughout the plant and in the stem xylem and phloem remain largely in separate streams having different functions; water is moved via xylem into leaf tissues actively photosynthesizing sugars by fixing CO_2 and it is here that the two vascular tissues tend to interact so that the metabolites (nutrients) can be moved throughout and down the plant via phloem.

Most of the water absorbed by the roots is lost to the atmosphere via evapo-transpiration; only approximately 1% is used during photosynthesis. The path is referred to as the soil, plant, atmosphere continuum (SPAC).

When the dynamic moisture equilibrium between the plant's interior and the external medium is disrupted the survival of a plant is threatened. The uptake and loss of water is dependent upon osmosis, the movement of water across a membrane. Plant cells have a cell wall that represents some different challenges compared with animal cells. Water potential (represented by the Greek letter psi, ψ) status in a plant is the key to whether a plant is flaccid (i.e., wilted), or in contrast, the leaves and stem are healthy and fully extended indicating proper turgor (i.e., osmotic pressure against the cell walls). Water moves across a membrane from a solution with higher water potential to one having lower water potential, which incorporates solute concentration. Potential in this sense refers to potential energy, or the ability to perform "work." The general formula for water potential is as follows:

$$\psi = \psi_p + \psi_s,$$

where ψ is the water potential and ψ_p is physical pressure (pressure potential) on a solution and ψ_s is the solute potential (osmotic potential); pressures are measured in MPa (megapascals; 1 MPa~10 atmospheres of pressure, or 1 kg/cm). Distilled water has a $\psi=0$, because $\psi_p + \psi_s = + 0 = 0$.

A flaccid plant has a $\psi_p = 0$ and $\psi_s < 0$, thus the plant is losing water to the surrounding medium. When to cells reach the same as the outside medium, plasmolysis occurs (i.e., the cytoplasm shrinks), the plasma membrane separates from the cell wall, and the plant proceeds to die. Conversely, if there is an initial relationship like this, $\psi = \psi_p + \psi_s = 0 + -0.5 = -0.5$, the plant cells have an initial lower water potential than the surrounding medium, but if water is added to the root environment, the relationship approaches $\psi = \psi_p + \psi_s = 0.5 + -0.5 = 0$. This results in turgor pressure and the plant cells become turgid, leaves and stems extend as expected of a normal healthy plant. Therefore, the influx of water causes the cell to swell and this exerts hydrostatic pressure against the cell wall (ψ_p) producing turgidity. This same pressure is required during gaseous exchange in plants. Turgor pressure helps guard cells to properly regulate the activity of stomata in the upper and lower leaf surfaces.

3. Phloem Transport and Storage

Sugars (mainly sucrose), amino acids, hormones and other organic molecules are transported via the phloem. These materials enter the transport system into the sieve tubes via plasmodesmata. They are distributed throughout the plant and the sugars, not used immediately for energy to aid growth and repair, are stored as starch in fruits and the roots

Current belief is that dissolved organics and "food" produced by photosynthesis move from the leaves (by active transport) through the phloem (via sieve elements) by a method called the "pressure flow hypothesis" (i.e., water containing the foods moves via pressure gradients set up by the differences in concentrations between water solution of the phloem and the lower solute concentration of the water in the xylem). As the solute concentration increases in the phloem, water enters the solution by osmotic pressure. Turgor pressure aids the transport similarly to the positive root pressure (see above) and pushes solute down the stem and sugars are removed by cortex cells in the root and stem/shoot (referred to as *sinks* by some) where they are either used or stored as starch (recall that starch is an insoluble carbohydrate [mixture of amylose and pectin compounds] and exerts no pressure). As the solute concentration changes down the stem, the water in the phloem begins to move back into the adjacent xylem channels and back up to the leaves again. Therefore, the pressure gradi-

ent between the leaves (high) and sinks (low) act as the driver of the transport of phloem contents.

4. Plant Mineral Nutrition

Plants require water and minerals (primarily carbon, oxygen, hydrogen, nitrogen, sulfur, and phosphorus) to form organic compounds. See Table 6.6. Micronutrients are also required in relatively small quantities. There are currently nine recognized macronutrients: carbon, oxygen, and hydrogen are available from carbon dioxide and water; the rest must be obtained from the soil or growing medium; recall that ~90–95% of plant biomass [dry weight basis] is C, H, O.

Plants obtain the nitrogen that they require from the soil after nitrogen-fixing bacteria convert it from atmospheric nitrogen gas (N_2), to a usable form such as ammonia NH_3 then to either NO_3 or NH_4. Some plants (e.g., legumes, alders) have nodules on their roots that contain nitrogen-fixing bacteria.

Nutritional adaptations by plants are exhibited by some plants that have developed ways to obtain supplemental nutrition. Plants may be carnivorous (kill and digest insects and small animals), parasitic (obtaining nourishment by attaching to other organisms), or participate in mutualistic relationships with fungi or bacteria.

5. Plant Energetics: Respiration and Photosynthesis

Respiration in plants occurs in all cells, whether they contain chlorophyll or not, thus in roots, stems and leaves. Photosynthesis occurs only in chlorophyll containing tissues (mainly in the mesophyll layer of leaf cells). Photosynthesis uses water, CO_2, and energy

Table 6.6.
Mineral Composition and Utilization in Angiosperms (and Most Higher Plants)

Element	Content (% of dry matter)	Absorbed in this form	Primary Function
Macronutrients (CHO, plus six)	C, H, O are all part of the plant organic molecules.		
Nitrogen	1–3%	NO_3 or NH_4	All amino acids, nucleotides and their macromolecules (i.e., proteins, DNA, RNA, some coenzymes) crucial for growth
Potassium	0.3–6%	K^+	Important in protein synthesis, regulating osmotic pressure/ Turgor and regulating leaf stomata
Calcium	0.1–3.5%	Ca^{2+}	Acts with Mg to maintain pectin, cell wall integrity, involved in cell osmotic pressure/turgor and maintenance of biomembranes (i.e., cell permeability)
Magnesium	0.05–0.7%	Mg^{2+}	Cofactor in chlorophyll, associated with pectin in cell wall and activator of enzymes
Phosphorus	0.05–1.0%	H_2PO_4 or HPO_4^{2-}	Important in flower/reproduction, involved in protein, nucleotide synthesis and energy transport systems (i.e., ATP); coenzymes, phosphorylation of sugars, phospholipids
Sulfur	0.05–1.5%	SO_4^{2-}	Important in amino acids and protein synthesis; coenzyme A
Micronutrients			
Chlorine	100–10,000 ppm	Cl^-	Osmotic and ionic balance, photosynthesis reactions
Boron	2–75 ppm	BO_3- or $B_4O_7^{2-}$	Inhibits high rates of pentose phosphate cycle, has influence upon calcium metabolism
Iron	10–1,500 ppm	Fe^{2+}, Fe^{3+}	Chlorophyll synthesis, ferrodoxin, cytochromes
Manganese	5–1,500 ppm	Mn^{2+}	Catalyst in photosystem II in photosynthesis
Zinc	3–150 ppm	Zn^{2+}	Enzyme activation
Copper	2–75 ppm	Cu^{2+}	Enzyme activation
Molybdenum	0.1–5.0 ppm	MoO_4^{2-}	Enzyme activation, nitrogen metabolism
Sodium	trace	Na^+	For some plants, but not essential for all: osmotic pressure and ionic balance
Cobalt	trace	Co^{2+}	Required by nitrogen fixing organisms, plants utilize the nitrogen

(sunlight) to synthesize 6-carbon sugars (i.e., glucose). Cellular respiration consumes oxygen and organic molecules and releases carbon dioxide. Photosynthesis is an energy absorbing process, the sun being the ultimate source. Respiration, on the other hand, is an energy releasing process, some is lost as heat and some is stored as ATP. While active photosynthesis occurs only when there is light, respiration occurs in both light and dark.

Photosynthesis occurs in the chlorophyll-containing organelle called a chloroplast (~30–40 per cell) that contains the enzymes and electron carriers to synthesize sugar, respiration occurs in another organelle, the mitochondrion, which contains the necessary oxidative enzymes of the tricarboxylic acid cycle (also known as the Krebs cycle; refer to Chapter 2 for details) and components of an electron transport system. Photosynthesis results in growth (increases in weight or biomass) while respiration reduces the weight of a plant because it is breaking down sugars, starch and fast for biosynthesis and release of energy. The glucose formed in photosynthesis may be used to derive energy, stored as starch, or converted to fats and oils, proteins, or vitamins

Gas Exchange

The exchange of gases in the leaf for photosynthesis, respiration and transpiration occurs through pores in the surface of the leaf known as the stomata. The stomata open in the presence of light (CO_2 uptake) and close in its absence, they open to release respiration of oxygen and open-close depending upon turgor pressure that reflects the plant's internal water balance with the external environment. The most direct cause of stomata activity this is the change of turgor in the guard cells that is a function of water potential and solute concentration mediated largely by K^+ ions (Figure 6.1).

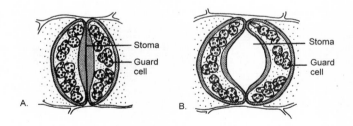

Figure 6.1.
(a) Stomata Closed (b) Stomata Opened
(When Turgor Builds in Guard Cells)

Chemical Reactions of Photosynthesis (C3 Plants)

The overall summary of reactions in photosynthesis is the following:

$$6CO_2 + 12H_2O + \text{Light energy} \rightarrow C_6H_{12}O_6 + 6O_2 + H_2O$$

The carbon (via CO_2) enters the cell through the stomata. Hydrogen is obtained from water, it is incorporated along with carbon into sugar and the waste, O_2, is returned to the atmosphere to be used in cellular respiration. The electron chain increases in potential as it moves from splitting water to sugar synthesis, the boost in energy required to reduce the sugar comes from the sunlight.

Photosynthesis is a complicated set of biochemistry. For discussion purposes we have simplified them to the two major steps: the light reaction and the Calvin cycle.

Light Reaction (Photolysis)

The light reactions convert solar energy into chemical energy; these occur in the thylakoid in the chloroplast; thylakoid is a membrane bound sac inside the chloroplast. A first step in photosynthesis is the decomposition of water molecules to separate hydrogen and oxygen components by photolysis. This decomposition is associated with processes involving chlorophyll and light and is thus known as the light reaction. Oxygen gas is formed, the released hydrogen is picked up by $NADP^+$ to form NADPH and phosphate is added to ADP to produce ATP via photophosphorylation. No sugar is synthesized in the light reaction that occurs in the next step, the "dark reaction." The ATP and NADPH feed into the dark reaction; two pathways called *cyclic and noncyclic photophosphorylation* generate these and are summarized next.

Cyclic and Noncyclic Photophosphorylation

Both these pathways require light and each is a photosystem (i.e., method of trapping energy from excited chlorophyll molecules and converted to electron streams down the transport chain). Photosystem I is called P700 (chlorophyll a absorbs light at this wavelength). Photosystem II is P680, the wavelength that excites chlorophyll b.

Cyclic photophosphorylation—light strikes chlorophyll a (Photosystem I) that causes an electron to be ejected from chlorophyll a, that is accepted by ferrodoxin and passed along a transport chain. As the electron is passed along, two ATP molecules are produced. Thus light energy is converted into chemical energy and not lost as heat.

Noncyclic photophosphorylation—oxygen is produced by the photolysis of water, and subsequently, ATP and NADPH are formed. When light hits chlorophyll b molecules (Photosystem II), high-energy electrons are released, as in the cyclic path above, these are accepted by "Q," and then through a series of steps they restore chlorophyll a to its original state. During these steps ADP is phosphorylated to ATP. Chlorophyll b is restored to its original state by the electrons released by the splitting of water → protons, electrons and oxygen. The oxygen combines with another oxygen and is released as O_2 to the atmosphere. The two electrons are gathered by chlorophyll b and the hydrogen ions are used to convert $NADP^+$ to NADPH.

Dark Reaction (CO_2 Fixation → Synthesis of Sugar)

In this second phase (dark reaction, also known as: light independent reactions, or the Calvin cycle, named for Melvin Calvin), the hydrogen is released from NADPH. Carbon fixation is the main end-point: NADPH (above) reacts with CO_2 obtained from the atmosphere and via the stomata and reacts with organic molecules (it is fixed) in the stroma (solute surrounding the thylakoid) in the chloroplast where carbohydrate forms as glucose, a 6-carbon sugar. The energy is to do this is obtained from ATP that is produced in the light reaction (above). CO_2 fixation does not require light. Thus, while the sugar is produced in the dark reaction, the ATP (energy source) and NADPH (reducing power) come from the light reaction.

C_4 Plants

There are species of plants (*Zea mays*, corn, is a good example) that incorporate CO_2 and first fix it in a 4-carbon molecule before entering into the "dark" or Calvin cycle. Given equal CO_2 concentrations, C4 plants assimilate at a faster rate than C3 plants. This means they are generally more efficient at converting CO_2 into a carbohydrate than are C_3 plants.

"Give and Take": Balancing Photosynthesis and Transpiration

Guard cells adjacent to the stomata help control the balance between the need for a plant to conserve water and its need to survive by photosynthesis. It has been estimated that a leaf can transpire its weight in water per day. This is due to both metabolism and the physical structures of tissues in which chloroplasts are kept. The parenchyma of the mesophyll has a surface area estimated to be 10 to 30 times the leaf's surface. During the course of opening stomata to release O_2 and uptake CO_2, water is loss as vapour; it is estimated that ~90% of water loss is through stomata. Transpiration acts also to cool the leaf surface and that can reduce the risk to enzymes needed in photosynthesis: temperatures may be lowered by 10 –15°C.

A method used to estimate the efficiency of water use is the *transpiration-to-photosynthesis ratio*. This represents the water loss per gram of CO_2 assimilated by photosynthesis. This may be as high as 600:1 in some C_3 plants, but less than 300:1 in C_4 plants.

Plant Reproduction, Growth, and Development

With Emphasis on Flowering Plants

1. Reproductive Structures

A "typical flowering plant" (i.e., the angiosperm flower, formed from the whorl of four modified leaves) is specially designed for plant reproduction (see Table 7.1 and Figure 7.1). A flower has brightly-colored petals (corolla), which attract pollinators (i.e., birds, insects) and are arranged in a circle. Note also that not all angiosperms have colourful floral structures, for example trees like the oak, crop plants like corn and wheat. However, regardless how "showy" they may appear, all have comparable reproductive structures that function in the same way, even if they may look quite different.

A *perfect* flower contains both sexes: male stamens and female structures enclosed in the carpel. If any of those structures are missing, it is *imperfect*. If both sexes are on the same plant, it is called *monoecious*. If they are on separate plants, the species is said to be *dioecious*. Therefore, monoecious species can have flowers that may be either perfect (roses, wheat) or imperfect (corn, oak). Dioecious species by definition are im-

perfect because they have "male" plants and "female" plants: examples include aspen and willow trees. Refer to Table 7.1 and Figure 7.1, for more details.

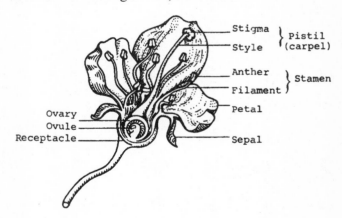

Figure 7.1.
Parts of a Flower

Male structures, the stamens, consist of a stalk (the filament) with an anther on the top containing the pollen sacs; the microgametophyte develops within the anther. The pistil contains an ovary with one or more ovules; the megagametophyte develops within the ovary. The pistil narrows at the top to form the style, the topmost region of which is known as the stigma and functions in pollen collection.

Table 7.1.
Components of an Angiosperm Flower (*Perfect*, Having Both Sexes)

Structure	Substructure/components	Function
Inflorescence	One or more flowers	Contains the seed producing structures and may attract pollinators and dispersing agents
Peduncle	Stalk that hold the inflorescence	Keeps the flower above soil surface and presents for pollinators
Pedicel	Stalk of a single flower in an inflorescence	Same as peduncle; offers support to individual flowers in a cluster
Receptacle	Held by the pedicel and to which floral parts are attached	Base of the reproductive structure
Petals, also called the corolla (sepals outside petals, at base)	Nonreproductive parts, attached to the receptacle below the reproductive structures, the carpels and stamens	"Showy" part of the flower and/or modified to protect the reproductive structures
Stamens	Male reproductive structure, the microsporophylls; stamen is composed of a stalk called the filament and two-lobed anther at the tip containing the microsporangia (pollen sacs)	Male gametes (following meiosis) are produced and dispersed from this organ
Carpels (Pistle)	Called the megasporophylls; composed of the stigma connected below to the ovary by a stalk-like structure called the style; may be one or more carpels, if more than one, may or may not be fused, contain one or more ovules	Female gametes (following meiosis) are produced in the ovary and seed(s) formed in ovules postfertilization (in some species the ovary later forms the fruit)
Placenta	Ovary is attached to the ovule by the placenta	Role is to nourish the flower

2. Meiosis and Gametogenesis

Meiosis

First, review the appropriate sections on meiosis in Chapter 3. Angiosperms and other higher plants proceed through the same two series of cell divisions described in that chapter: Interphase (DNA-chromosome replication), Prophase I (homologues synapse, genetic exchange between nonsister chromatids in the bivalent), Metaphase I (bivalents align at the cell equator) Anaphase I (division of cytoplasm, migration of recombined chromatids), Telophase I (Separation of homologues, two nuclei), Prophase II, Metaphase II, Anaphase II (reductive division, separation of centromeres), and Telophase II (cytokinesis, four haploid nuclei). Microspores (male gametes) and megaspores (female gametes) give rise to the gametophytes (next section).

The sequences and reproductive structures may vary among species, but the outcomes are the same following pollination and fertilization: haploid sex gametes unite to form a diploid zygote followed by repeated mitotic divisions giving rise to a new, diploid sporophyll generation.

3. Gametogenesis and Fertilization

Reproduction in the flowering plants begins with the development of the gametes. These processes are described here beginning with the female gamete. Refer to Figures 7.2–7.3 for more details.

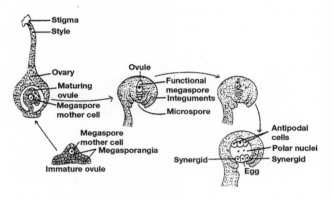

**Figure 7.2.
Megagametophyte Development in Angiosperms**

Megaspores. The female gametes, the megaspores, develop within the ovules of the ovary. Each ovule is attached to the ovary by a stalk. Embedded deep within the ovule is one cell called a megsporangium that enlarges to become the megaspore mother cell (see Figure 7.2) from which the gametes will be formed. The megaspore mother cell undergoes meiosis and four haploid megaspores result. In most angiosperm species, three of the four disintegrate leaving one functioning haploid megaspore. This surviving cell then undergoes three mitotic divisions producing eight haploid nuclei which migrate so that three position themselves at the far end of the now enlarged megaspore becoming the antipodal cells (these are short-lived). Three more of the eight move towards the micropyle (opening through which the pollen enters the ovule) and form the egg and two synergid cells. Like the antipodal cells, the synergid cells are short-lived. The remaining two of eight cells are the polar nuclei. The entire structure now is called the embryo sac and becomes enclosed by layers of cells called the integument (later giving rise to the seedcoat).

Microspores. The development of sperm or microgametophytes begins within the tissues of the anther. Each anther (numbers vary with species and whether monocot or dicot) typically contains four pollen sacs or microsporangia (see Figure 7.3). During early development, accelerated cell division in the microsporangia produces numerous microspore mother cells. Each of these microspore mother cells undergoes meiosis that results in a tetrad of four haploid microspores. Each microspore undergoes mitosis that results in a tube nucleus and a generative nucleus. It is now that it is called a pollen grain. When the pollen is mature, the anthers split open and shed the pollen. Pollination may then occur and the mode of dispersal varies among species (i.e., wind, insects, birds, bats, etc.).

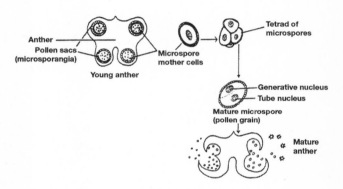

**Figure 7.3.
Microgametophyte Development in Angiosperms**

Pollination and Fertilization

Pollination occurs when pollen is transferred from the stamen to the stigma. Pollination may occur within the same flower (self-pollination) or among different flowers (cross-pollination). After pollination occurs, the pollen grains germinate and begin to grow down through the stigma, in the process forming a pollen tube (Figure 7.4). The pollen tube, carrying two sperm nuclei, enters the ovule through a pore called the micropyle. One sperm combines with the egg cell within the enlarged embryo sac to form the zygote (2N). The second sperm combines with the two polar nuclei to form a triploid nucleus (3N), which develops into the endosperm. The fertilization by two sperm, one to form the zygote (2N) and the other to form the endosperm (3N), is called *double fertilization*. Recall that gymnosperms do not have double fertilization and the endosperm is haploid and derived from the maternal plant; thus, in a conifer, if a gene is heterozygous the alleles will segregate 1:1 among the seed endosperm from an individual plant.

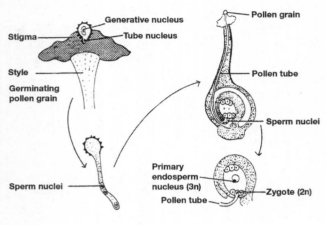

Figure 7.4.
Fertilization in Angiosperms

Embryogeny, Seed and Fruit Development

Mitosis and differentiation in the primary endosperm (formed by triple fusion, when the second sperm unites with the two polar nuclei; see Figure 7.4) along with the zygote (one of the two sperm fuses with the egg cell) will produce the embryo, all these enclosed in the seed coat, make up the seed. The embryo develops from the single cell, diploid zygote and is followed by several mitotic cell divisions creating a multicellular structure. Meanwhile, the primary endosperm (3N) is also dividing mitotically creating a mass of nutritive tissue (high starch content) surrounding the embryo.

Embryo development is quite similar in monocots and dicots (reminder: we are using dicot in lieu of *eudicot*). In most flowering plants the first division of the zygote is transverse (perpendicular to the long axis of the zygote cell) and this establishes polarity of the embryo. In other words, the "upper" cell from the first division will form the upper portion of the embryo; the "lower" cell will form the bottom. Therefore, shoot and root poles, respectively. Recall that animal embryos must also establish polarity. In fact, this is a phenomenon essential for all higher organisms so that a central axis, or backbone, is laid down from which appendages emerge.

Following a number of orderly mitotic divisions, the proembryo develops into the embryo proper upon a stalk of cells called the suspensor. It is thought that the suspensor acts both to push the developing embryo into the endosperm and also aids in absorbing nutrition from the endosperm to be used by the embryo. Periclinal cell divisions (parallel) to the surface of the of the embryo form the protoderm that will give rise to the epidermis. The protoderm, together with the emerging ground meristem and procambium cells, are called the primary meristems. These are continuous between the cotyledon and axis of the embryo. Cotyledon(s) develop concurrently with, or shortly subsequent to, the primary meristems. The cotyledons and axis elongate (some refer to this as the *torpedo* stage) and as it continues to develop, most of the new cell divisions result in apical meristem cells in both the shoot and root. The apical meritems eventually give rise, following germination and maturity, to all the cells contained in the new adult plant (these are somewhat analogous to stem cells in animals).

The seed is an interesting structure in that it is composed of tissues from three generations: (1) the previous sporophytes, (2) the current sporophytes, and (3) the next sporophytic generation. The embryo, consisting of diploid cells from the fusion of egg and sperm, is the new sporophytic generation and contributes to the continuation of the species by developing into the new reproducing plant. The seed coat, having developed from the integument surrounding the ovule, and of maternal origin, is from the previous sporophytic generation; this hard covering protects the seed against heat, cold, desiccation, pathogens, insects and other animals. See Figure 7.5.

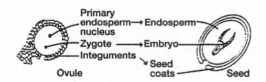

Figure 7.5.
Development of an Ovule into a Seed

In many plants, after fertilization, the ovary enlarges to become fruit, which provides protection and a mechanism for seed dispersal, as the fruit and seeds are eaten by animals and disposed of elsewhere. Thus, a fruit is a mature ovary. A pea pod is a fruit containing mature ovules, called seeds. Other mechanisms of seed dispersal include scattering by wind (e.g., dandelion), expulsion (e.g., legumes such as soy bean), attachment to fur (e.g., cocklebur), and floatation (e.g., coconut).

It is possible to grow mature fruit without pollination and fertilization. By applying a plant hormone like auxin to the carpel of a floral structure, seedless fruit may develop. This has been capitalized upon in commercial horticulture and agriculture (e.g., seedless grapes and oranges). Treated fruits grow larger, ripen faster and remain better on the plant until harvested.

Growth and Differentiation

Once the seed is formed, the embryo usually goes into a resting state. Some species enter what is referred to as dormancy. Growth does not normally begin until germination of the seed. Germination of mature seeds is dependent upon a combination of external and internal events. There are three factors that are important to successful germination: (1) water, (2) oxygen and (3) temperature. The limiting factor is usually water. Imbibation of moisture, together with optimum temperature and proper aeration, promote the ensuing activation of enzymes that hydrolyse the starch stored in the endosperm that affords energy and nutrition to the emerging cotyledon(s) and root.

Most species require that mature seed be maintained in a very dry state until germination. And, many require some cooler temperatures to help with after-ripening: note that the seeds of most temperate species mature in later summer or fall and do not germinate until the following spring-summer after having survived a winter. Most everyone is familiar with the importance of maintaining dry conditions for seeds and bulbs sold at the local garden store. Moisture, by weight, is usually kept at 5–20%, thus ensuring moisture potential within in the seed coat for imbibition. Germination is first under the control of anaerobic respiration and the aerobic, following the seed coat splitting. Optimum temperature for germination varies between 25–30°C, although some species germinate at lower or higher temperatures.

Some species have evolved dormancy. They will not germinate until certain requisite pregermination events have occurred. These include preconditioning by stratification (pretreatment with above-freezing cold, plus moisture; often with conifers) or scarification (physically or chemically scarring the seed coat to permit imbibition; e.g., mesquite which must pass through the gut of animals). Species associated with closed-cone, fire-dependent conifers (e.g., lodgepole pine) may require high temperatures caused by forest fires in order to germinate.

Upon germination, the radicle, or embryonic root, is the first structure to emerge. This event obviously fills the seedling's first requirements for moisture and mineral uptake. The first root, the primary root (tap root), produces lateral roots, which may produce even further branching. In monocots, the primary root withers and soon gives way to adventitious roots, which emerge from the leaf node above the apical meristem: these produce lateral roots. These systems are more fibrous.

Growth and elongation of the young seedling is first secured by moisture uptake form the primary root system, plus nutrition from the endosperm tissues, mainly in the monocots, or the cotyledons of the dicots. The apical meristem soon becomes more organized and functions to produce leaves, nodes and internodes. The plant life cycle to this stage, and up to and including mature stature (before flowering), is the vegetative phase. Internal (hormones, water, nutrition) and external (light, temperature) factors cause one or more of the vegetative meristems to switch over to becoming reproductive apical meristem cells. These give rise to flowers (inflorescence) and the reproductive cycle begins again.

Control Mechanisms: Plant Hormones and Tropisms

Plant hormones (Table 7.2), like animal hormones, are organic molecules that can produce very dramatic effects on cell metabolism and growth. They are produced mainly in actively growing tissue, especially the apical meristem. Like animal hormones, plant hormones usually produce their effects at sites removed from where they are produced. Movement throughout the plant is via phloem tissues.

Kinds of Hormones

Table 7.2.
Plant Hormones[1]

Hormone	Production site or where usually found	Functions
Auxin (e.g., indole acetic acid); synthetic is 2,4-D.	Embryo of seed, apical meristems, young leaves.	Low concentration will stimulate stem elongation, root growth, cell differentiation, branching, enhances apical dominance (pyramidal or branched crown), involved with fruit development and exerts influence during phototropism and gravitropism. 2,4-D is a herbicide on dicots, can stimulate monocots.
Cytokinin (e.g., zeatin)	Produced in roots then transported to other structures.	Cytokinins affect root growth and differentiation, stimulate cell division and growth, branching and stimulate germination and delay senescence. Used to promote cell division in culture.
Gibberellins (e.g., gibberellic acid)	Apical and root meristems, young leaves and the embryo.	Promotes seed germination, bud break, stem elongation, leaf growth, promotes flowering and development of fruit, affects root growth and development.
Abscisic acid	Roots, stems green fruit and leaf.	Inhibits growth, closes stomata during water stress, counters break in dormancy.
Ethylene (only gaseous molecule)	Tissues of ripening fruit, stem nodes, aging leaves and flowers.	Promotes ripening of fruit, counters some of auxin effects, in some species ethylene can promote or inhibit growth of roots, leaves, flowers.
Brassinosteriods (e.g., brassinolide)	Seeds, fruit, shoots, leaves, and floral buds.	Inhibits root growth, retards abscission of leaves, promotes xylem differentiation.
Florigen	Produced in leaves.	Stimulates flowering.

[1] This table is based upon Table 39.1 (p. 808) in Campbell, N. and Reece, J. (2002) *Biology*, 6th ed., Benjamin Cummings Pub., San Francisco, 1247 pp.

Tropisms

Auxin is involved in tropisms, or growth responses to specific stimuli. They include positive phototropism (bending or growing toward light), negative phototropism (bending or growing away from light), geotropism (influenced by gravity), and thigmotropism (mechanically or by touch, as exhibited by tendrils and vines).

Photoperiodism is any response to changing lengths of night or day (Table 7.3).

The stages of plant development are regulated, at least in part, by the plant's responsiveness to the relative lengths of light and darkness. Plant photoperiodism is defined in terms of day length, though it usually depends on critical night length. Flowering occurs

Table 7.3.
Examples of Different Types of Plants with Regard to Photoperiod

Type of photoperiod response	Normal time of flowering	Examples
Long-night (short-day)	Early spring, late summer or fall	Asters, poinsettias, dahlia, cosmos, potato
Short-night (long-day)	Late spring, early summer	Beets, coreopsis, gladiolus, delphinium, clover
Indeterminate (night-neutral)	Spring to fall	Carnations, cotton, tomato, sunflower

following a maximum (short-day plants; > 9 hr dark) and minimum (long-day plants, < 9 hr dark) number of hours of uninterrupted darkness. Day-neutral plants flower without regard to photoperiod. *Phytochrome* is a plant pigment that detects specific wavelengths of light and plays a role in the measurement of the length of darkness in a photoperiod, thus regulating time of flowering.

Chapter 8

Diversity of Life

1. Classification of Living Organisms

As humans we like to have things organized into manageable information. Early on humans recognized that we live with diversity, including biological, ecological, and climatic. Other living organisms were among the first things that we attempted to name and classify into similar groupings. The science became known as taxonomy. Until most recently, emphasis (referring to C. Linnaeus [1707–1778]) was on classification based on morphology, reproductive structures, common function, and sexual compatibility. Prior to Linnaeus, classification was based mainly on the habitat in which organisms lived (i.e., air, land, or water). Obviously, lumping elephants and lizards together based on their both being terrestrial animals is considerably lacking. Recent years have seen more efforts put toward identifying evolutionary and genetic relationships based on Charles Darwin's notion of "common descent": these disciplines are referred to as biosystematics and/or cladistics, the latter being what some scientists believe may replace the Linnaean system. Cladistics incorpo-rates molecular and genomic analyses coupled with comparative anatomy, fossils, and modern computer algorithms to develop phylogenetic trees, referred to in some texts as phylogenetic systematics based on cladistics analysis (see *Biology*, 6th edition, by Campbell and Reece, 2002).

While certainly not the first person to develop a system for classification, Carolus Linnaeus (1707–1778) published the *Systema Naturae* that ran through 12 editions during his lifetime (first edition, 1735). Linnaeus' system has survived until the present. Because common names can mean different things to different people, Linnaeus's method gave each organism a scientific Latin name having two parts (binomial nomenclature): the genus (always capitalized) and the *specific epithet* indicates a particular species (not capitalized) within the genus (plural, genera). You may also see in the literature a third name when a species is further divided into subspecies, varieties, ecotypes, forms, and so forth. The author (or abbreviation) who assigned the scientific name to an organism will also be shown at the very end of the name in scientific literature so that proper credit is given: not surprisingly, many times that is an "L.," which stands for Linnaeus.

Systematics Is Very Dynamic Today

Taxonomy used to be considered pretty staid (not to a taxonomist, of course). But today, phylogenetic relationships are changing at an even greater rate because of molecular technology, new computer algorithms, and the nearly immediate dissemination of information. We include here a few URLs (addresses) of Internet sites where you can view the current status of classification and find lots of other helpful links:

http://www.worldhistory.com/wiki/D/Three-domain-system.htm

http://www.geocities.com/RainForest/6243/diversity

http://tolweb.org/tree?group=Life_on_Earth&contgroup=

2. Current Classification Methods: Three Domains

Organisms are now first classified into three broad domains: Bacteria, Archaea, and Eukarya (or Eukaryota); viruses are not included since, among other traits, they require another organism in order to reproduce. These three groups were first recommended by Carl Woese at the University of Illinois ~30 years ago. The first two comprise all prokaryotes and were formerly called Eubacteria and Archaebacteria, respectively. The third comprises all eukaryotes. These are then divided into kingdoms that contain six major subgroupings or clades when cladistic analyses are done (see below). These taxa (groupings [taxon, singular]) are shown here:

Kingdom

Phylum (animals) or Division (plants)

Class

Order

Family

Genus

Species

Becoming acquainted with taxa suffixes can help identify where in the hierarchy an organism may be located. Table 8.1 displays these suffixes and then the rules are applied to an example.

Table 8.1.
Taxa Suffixes Organized for Type of Organism

Taxon	Plants	Algae	Fungi	Animals
Division/ Phylum	-phyta	-phyta	-mycota	Varies
Subdivision/ Subphylum	-phytina	-phytina	-mycotina	Varies
Class	-opsida	-phyceae	-mycetes	Varies
Subclass	-idea	-phycidae	-mycetidae	Varies
Order	-ales	-ales	-ales	Varies
Suborder	-ineae	-ineae	-ineae	Varies
Superfamily	-acea	-acea	-acea	-oidea
Family	-aceae	-aceae	-aceae	-idae
Subfamily	-oideae	-oideae	-oideae	-inae
Tribe	-eae	-eae	-eae	-ini
Subtribe	-inae	-inae	-inae	-ina

Table 8.2 provides an example of how the classification system is applied to humans.

3. Phylogenetic Systematics: Analyzing Clades

This section is a brief review of some of the underlying methodology that is used to classify organisms today. First, understand the molecular premise for

Table 8.2.
Heirarchy for Classifying Humans:
Domain—Eukarya

Kingdom	Animalia
Phylum	Chordata
Subphylum	Vertebrata
Class	Mammalia
Subclass	Eutheria
Order	Primates
Suborder	Catarrhini
Family	Hominidae
Genus	*Homo*
Species	*sapiens*

grouping (i.e., include organisms that are descended from a common ancestor and that are genetically more alike, though, maybe not phenotypically as similar as others). You are referred to Campbell and Reece (*Biology*, 6th ed., 2002), which states that: "*Classification based on evolutionary history is called phylogenetic systematics.*" By combining morphological/anatomy, fossil, and molecular analyses, systematists conduct what is known as *cladistic analysis*. This method (first developed ~50 years ago by Hennig) is used to build phylogenetic trees or *cladograms*. Each branch point, node, represents a divergence between two species that have shared a common ancestor; decisions at each node are dichotomous (i.e., two branches based on one or more criteria). Each branch is called a clade; clades can be nested in hierarchies. There are three types of cladistic groupings: monophyletic, paraphyletic, and polyphyletic. An acceptable clade taxon should be monophyletic (i.e., comprised of an ancestral species and all it descendants, including family, genera, species, etc.).

How is similarity decided? This requires choosing between *homology* and *analogy*. Homology is similarity due to shared ancestry. Analogy is similarity that has arisen because organisms may have evolved in similar habitats and natural selection has resulted in two different taxa having evolved structures that serve analogous functions (e.g., the wings of a moth and bird both serve similar functions but they share little genetic homology). Both of these are convergent processes, but homology is called *convergent evolution*. That is the basis for constructing a monophyletic clade.

4. Previous Classification Scheme: Five Kingdoms

We include this section to remind the reviewer that you will still see references to the former method of naming, using five kingdoms. These are Monera (bacteria and archaebacteria, all the prokaryotes) and the four remaining groups of eukaryotes: Protista, Fungi, Plantae, and Animalia. The descending hierarchy of naming remains the same, or similar, but beware, systematics is very dynamic (i.e., it is changing almost daily because of new molecular technology and computer algorithms).

Defining the Three Domains

Table 8.3 describes a number of the traits that discriminate among the three domains of living organisms.

Domain: Archaea

Caveat: Classification is "in progress."

This domain contains the prokaryotes that are ancient bacteria. The Archae (ancient bacteria) are distinct from the more familiar Bacteria (true bacteria) by differences in their cell wall, cell membrane, and ribosome structure (see Table 8.3).

Table 8.3.
Differences Among the Three Domains[1]

Characteristic	Domain		
	Bacteria	Archaea	Eukarya
Nuclear envelope	Absent	Absent	Present
Membrane-bound organelles	Absent	Absent	Present
Peptidoglycan I cell wall	Present	Absent	Absent
Membrane lipids	Unbranched hydrocarbons	Some branched hydrocarbons	Unbranched hydrocarbons
RNA polymerase	One kind	Several kinds	Several kinds
Initiator amino acid for start of protein synthesis	Formyl-methionine	Methionine	Methionine
Contain Introns	Rare	Present in some genes	Present
Response to antibiotics: streptomycin and chloramphenicol	Growth inhibited	Growth not inhibited	Growth not inhibited
Histones with the DNA	Absent	Present	Present
Circular chromosome	Present	Present	Absent
Ability to grow at >100 degrees C	No	Some species	No

[1] This information was taken from table 27.2, p. 537, in Campbell, N. and Reece, J., 2002, *Biology*, 6th ed., Benjamin Cummings Pub., San Francisco, 1247 pp.

Archaea often live in extreme environments (e.g., water whose temperature exceeds 100°C; high salt, acid or alkaline water). They can be single-celled or form filaments or aggregates. They may be spherical, rod-shaped, spiral, or lobed. Size varies in diameter from 0.1 to more than 15 μm (filaments up to 200 μm). Multiplication modes include binary fission, budding, or fragmentation.

The two main groups (plus two putative) of the ancient bacteria are the following:

1. Kingdom—Euryarchaeota, including methanogens (methane generators), halophiles (having salt tolerance), and a few thermopiles (having high temperature tolerance).

   Phyla–Halobacteria, Methanobacteria, Methanococci, Methanopyri, Archeoglobi, Thermoplasmata, Thermococci.

2. Kingdom—Crenarchaeota (mostly thermophiles).

3. (putative) Korarchaeota (based on RNA analyses but poorly defined).

4. (putative) Nanoarchaeota (one putative species).

Archaea are compared to the other domains in Table 8.3, above.

Domain: Bacteria

Caveat: Classification is "in-progress"

True bacteria are the oldest organisms on Earth. They diverged from the Archaea at some very early time in earth's history. They are the most abundant living organisms on earth. They exist in most all habitats (e.g., soil, water, and as parasites of other organisms). They are infamous for causing a great proportion of nonhereditary diseases in animals. Many antibiotics have been developed to combat bacterial diseases.

They have been grouped into four main groups based on energy and nutrition sources (i.e., how and what they do in order to survive):

1. Photoautotrophs—similarly to plants, these harvest light and fix carbon via photosynthesis.

2. Chemoautotrophs—use only CO_2 as a carbon source, but rather than light for energy, they get energy via oxidation of inorganic substances (i.e., hydrogen sulfide, ammonia, ferrous ions).

3. Photohetrotrophs—use light to produce ATP but obtain carbon from organic molecules.

4. Chemoheterotrophs—use organic molecules for obtaining both energy and carbon. Recall that many eukaryotes, including mammals, are of this type.

Bacteria are compared to the other domains in Table 8.3, above.

Some people now suggest there are currently five phylogenetic kingdoms in the Bacteria. We have not attempted to further divide these into the many subclades that are in progress. These are as follows:

1. Proteobacteria—which include five Phyla: Alpha, Beta, Gamma, Delta, Epsilon. These are all gram-negative bacteria and include both anaerobic and aerobic species. Include photoautotrophs, chemoautotrophs, and heterotrophs.

2. Chlamydias—survive only within animal cells; lack peptidoglycan; one species in this clade (*C. trachomatis*) causes blindness, and another is infamous for causing a sexually transmitted disease, nongonococal urethritis.

3. Spirochetes—this clade contains helical heterotrophs; contains flagellar type structures that move in a corkscrew-like motion; includes nasty ones like *Treponema palladium* (a.k.a. syphilis) and Lyme disease (*Borrelia burgdorferi*).

4. Gram-positive—most diverse, next to Proteobacteria kingdom; actinomycetes once confused with fungi due to chains of cells in colonies; two nasty ones cause leprosy and another tuberculosis; many soil dwellers that decompose litter; there are spore forming types in the *Clostridium* (e.g., botulism) and *Bacillus* (e.g., anthrax); *Staphylococcus* and *Streptococcus* are two other genera that are notorious for disease; *Streptomyces* is cultured to obtain antibiotics (i.e., streptomycin); this clade also contains the mycoplasmas, smallest of bacteria and without cell walls, many in the soil, but a nasty one is responsible for so-called walking pneumonia in humans.

5. Cyanobacteria—only bacterial photoautotrophs; it is believed that plant chloroplasts represent endosymbionts that were free-living cyanobacteria; clade contains both single and colonial types; provide much nutrition to freshwater and marine aquatic organisms; some can fix nitrogen (N_2).

Domain: Eukarya[2]

The former five-domain scheme lumped all eukaryotes into one kingdom called Protista: included plants, animals and fungi. The three-domain system breaks the Eukarya into several clades that are still "in-progress."

Two of the distinguishing traits (Table 8.3, above) separating eukaryotes from the prokaryotes are a nucleus and membrane bound organelles. We can add also that cell division by mitosis and formation of sexual gametes by meiosis are also important indicators.

Eukarya currently contains eleven clades designated as Kingdoms. Within these are cladistic groups called phyla, classes and so forth (see section above). The list of Eukarya follows:

1. Kingdom—Parabasala (trichomonads)

2. Kingdom—Diplomadida (diplomonads)

---

[2] This information was taken from Appendix 3 in Campbell, N., and Reece, J. (2002) *Biology*, 6th ed., Benjamin Cummings Pub., San Francisco, 1247 pp.

3. Kingdom—Euglenozoa
   phylum—Euglenophyta (euglenoids)
   phylum—Kinetoplastida (trypanosomes)

4. Kingdom—Alveolata
   phylum—Dinoflagellata (dinoflagellates)
   phylum—Apicomplexa (apicomplexans)
   phylum—Ciliophora (ciliates)

5. Kingdom—Stramenophila
   phylum—Phaeophyta (brown algae)
   phylum—Oomycota (water molds)
   phylum—Chrsophyta (golden algae)
   phylum—Bacillariophyta (diatoms)

6. Kingdom—Rhodophyta (red algae)

7. Kingdom—Chlorophyta (green algae)

8. Kingdom—Mycetozoa
   phylum—Myxogastrida (plasmodial slime molds)
   phylum—Dictyostelida (cellular slime molds)
   phylum—Rhizopoda (amoebas)
   phylum—Actinopoda (heliozoans and radiolarians)
   phylum—Formaminifera (forams)

9. Kingdom—Plantae

   Non-vascular—Bryophytes
   phylum—Hepatophyta (liverworts)
   phylum—Anthocerophyta (hornworts)
   phylum—Bryophyta (mosses)

   Vascular, seedless plants—Pteridophytes
   phylum—Lycophyta (lycophytes)
   phylum—Pterophyta (ferns and horsetails)

Vascular, seeded plants (conifer)—Gymnosperms
phylum—Ginkgophyta (ginkgo)
phylum—Cycadophyta (cycads)
phylum—Gnetophyta (gnetae)
phylum—Coniferophyta (conifers)

Vascular, seeded plants (flowering)—Angiosperms
phylum—Anthophyta (flowering plants)

10. Kingdom—Fungi
    phylum—Chytridiomycota (chytrids)
    phylum—Zygomycota (zygomycetes)
    phylum—Ascomycota (sac fungi)
    phylum—Basidiomycota (club fungi)
    Subkingdom—Deuteromycetes (imperfect fungi)
    Subkingdom—Lichens (symbiotic: fungi and algae)

11. Kingdom—Animalia[3]
    phylum—Porifera (sponges)
    phylum—Cnidaria (cnidarians)
    phylum—Ctenophora (comb jellies)
    phylum—Platyhelminthes (flatworms)
    phylum—Bryozoa (bryozoans)
    phylum—Phoronida (phoronids)
    phylum—Brachiopoda (brachipods)
    phylum—Rotifera (rotifers)
    phylum—Nemertea (ribbon worms)
    phylum—Mollusca (mollusks)
    phylum—Annelida (segmented worms)
    phylum—Nematoda (roundworms)
    phylum—Arthropoda (insects)
    phylum—Echinodermata (echinoderms)
    phylum—Chordata (vertebrates)

[3] Some of these phyla contain sub-clades called "classes" that are omitted; consult the Internet and the text by Campbell and Reece (2002) for additional details.

Ecology

Ecology is the study of organisms and their relationships to each other and to their environment. The numbers and types of organisms present are important, but it is the interaction among and between the organisms and the environment that is the basis of ecology. The topics addressed in this summary are the distribution of organisms, behavioral ecology, population ecology, community ecology, and the structure and function of ecosystems.

1. The Distribution of Organisms

Biogeography

Biogeography is the study of the past and present distribution of organisms and the biotic (living) and abiotic (nonliving) factors that determine the patterns. The biotic factors include dispersal, habitat selection, behavior and intraspecific competition. The abiotic factors are the chemical and physical characteristics of the environment that determine the habitats of the organisms. The two main habitat types are terrestrial and aquatic.

Terrestrial Biomes

The interaction of climate, topography, and soil parent-material determines the abiotic conditions in terrestrial ecosystems. The specific abiotic factors are temperature, wind, moisture, sunlight (solar radiation), and soil-type. These factors affect the physiology of the organisms and are important selective agents in evolution. They determine the types of vegetation in different regions, ranging from the rainforests of the tropics to the tundra in the Arctic. These regions with similar abiotic conditions and plant and animal communities are the terrestrial biomes. There are nine major terrestrial biomes.

Tropical Forests

All biomes are defined primarily by temperature and photoperiod. In tropical forests, the mean annual temperature is 25°C and day length is 11 to 12 hours. There are three main types of tropical forests: tropical dry forests (also called tropical thorn forests), tropical deciduous forest (sometimes called tropical monsoon forests) and tropical rain forests. The type of tropical forest depends on the total and seasonal distribution of rain.

Savannas

Savannas are grasslands with some trees and shrubs. Frequent fires and grazing by herbivores maintain the short vegetation and limit the growth of trees. The rainfall is seasonal and animals must migrate to watering holes during periods of drought. Savannas are found scattered in North and South America, Africa, and Australia.

Deserts

Deserts have low and unpredictable rainfall, less than 30 cm per year. They occur on the leeward side of mountains where the dry air descends and warms. Most deserts are at about 30°, north and south of the Equator. Deserts may be warm or cold and can have extreme variations of temperature between day (over 50°C) and night (less than 0°C). Desert plants include many types of cacti, which are well adapted to obtain and store water. Desert animals include snakes, lizards, mammals such as fox, dingoes and kangaroo rats, and even some amphibians that stay in cocoons until the rains come.

Chaparral

Chaparral biomes have mild wet winters and long, hot, dry summers. The predominant vegetation consists of spiny evergreen shrubs, many of which are adapted to frequent fires, and some require hot fires for the seeds to germinate. The roots store food, which is available after fires and thus they can quickly resprout. Chaparrals are found in the Mediterranean, coastal Chile, southwestern Africa and Australia and California.

Temperate Grassland

Temperate grasslands are similar to savannas but have no trees. They have seasonal drought and frequent fires and large herbivore gazers maintain the vegetation. These grasslands are known by specific names: the pampas of Argentina, the veldts of South Africa, the steppes of Russia and the prairies of North America. The rainfall is low enough that there is little penetration into the soil and thus nutrients are maintained in the soil. The result is fertile areas used for agriculture.

Temperate Deciduous Forests

Temperate deciduous forests have warm summers and cold winters, and sufficient moisture to support the large trees. Deciduous plants lose their leaves in the colder winter months to prevent loss of water. The vegetation is diverse and includes large trees, an understory of shrubs, small flowering plants, and grasses.

Coniferous Forests

Conifers, cone-bearing trees such as pine, fir, spruce, and hemlock dominate this biome. Coniferous forests are found where the winters are cold and the summers short. The northern boreal forest or taiga, which extends across North America and Eurasia is the largest terrestrial biome. The conifer leaves (called needles) are shed only ever two or three years. They accumulate on the ground resulting in a thin and acidic soil.

Tundra

The tundra biome occurs where the winters are long and harsh. The surface soil supports short vegetation such as shrubs, grasses, moss and lichens. There are no deep roots because the subsoil is permanently frozen (called permafrost).

Polar

The polar biome is high mountain ice or which occurs on the tops of high mountain ranges such as the Rocky Mountains in North America, the Andes in South America, and the Urals and Himalayas in Eurasia. There is no vegetation.

Three concluding comments about terrestrial biomes warrant mention. First, there are no distinct boundaries between biomes. The biomes grade into each other in an area called the ecotone. Second, the plant and animal communities in the biomes have similar abiotic requirements and thus the plant communities appear similar. However, the same biome in different places has different species. Third, biomes are dynamic areas and there are significant temporal changes within parts of any biome. For instance, fires are an integral part of many biomes and help to maintain the plant communities of the biome. Overall, the biome maintains the defining plant communities but there are temporal changes in the communities present in areas after a fire.

Aquatic Biomes

One common feature of aquatic biomes is vertical stratification of sunlight, temperature, and organisms. Because the penetration of sunlight decreases with depth so do the temperature and extent of photosynthesis. The top area that receives sunlight is where photosynthesis occurs and is called the photic zone. The area without light and insignificant photosynthesis is called the aphotic zone. There is a narrow layer of water frequently within the photic layer where the temperature decreases rapidly. This band is called the thermocline. The bottom of aquatic biomes is called the benthic zone. Organisms in the benthic zone are called benthos. The dead organic material within the benthic zone is detritus.

Aquatic biomes are divided into freshwater and marine based primarily on salt concentration. Freshwater biomes have salt concentrations less than 1% and marine have salt concentrations about 3%. Estuaries occur where rivers empty into the marine waters.

Freshwater Biomes

Freshwater biomes include rivers, lakes, and wetlands.

Rivers vary from the headwaters to the mouth. In the headwaters, the velocity is rapid and the water is cool. The energy source is primarily detritus from the adjacent land, although there is usually some attached algae (periphyton) on the rocks. Immature stages of invertebrates, such as mayflies and caddisflies, and fish are prevalent. Near the mouth, rivers are larger and the velocity is generally less than in the headwaters. (Note: Although the velocity is less, the flow is greater than at the headwaters because flow equals velocity times cross-sectional area of the river.) The waters frequently carry sediment eroded from upstream areas and thus appear quite muddy. There are plants that grow in the shallow shoreline areas or literal areas, and algae that live in the open water, the limnetic zone. Invertebrates and fish are also present in these areas. Much of the dead material is carried downstream to the mouth or estuary where decomposers use the energy and recycle the nutrients.

Lakes are similar to large, slow-moving rivers, and in many cases they are the result of the widening of rivers. Rooted plants may occur in the littoral zone and phytoplankton are present in the photic zone of the deeper waters (limnetic zone). Nutrients enter lakes from the surrounding land and rivers (allochthonous input) and can be regenerated within the lake by decomposition of dead organisms (autochthonous input). As organisms die, they settle to the benthic zone at the bottom of the lake where decomposition occurs, releasing the nutrients. To return these nutrients to the water column, the lake water must be mixed. The fall season is important to lake turnover in two ways. First, dropping daily temperatures cause the surface water to cool; this dense water falls to the bottom, initiating mixing of the water by convection. Second, the strong winds associated with the seasonal transition from summer to winter help to create vertical circulation in the lake, which may be mediated by the movement of organisms. In very deep lakes and many tropical lakes this turnover of lake-water and the resulting recycling of the nutrients does not occur. The nutrients are "trapped" in the sediments and the lakes are nutrient poor. These nutrient poor lakes have low productivity and are called oligotrophic. Lakes with high nutrient levels and high productivity are usually more shallow. They are called eutrophic. Between oligotrophic and eutrophic lakes are those with moderate nutrient levels and productivity, called mesotrophic.

Wetlands include marshes, bogs, and swamps. They contain plants called hydrophytes that can grow in water or in saturated soil that may be anaerobic. Examples include cattails, sedges, pond lilies, and black spruce. There are three main types of wetlands, each associated with a topographic feature. Basin wetlands occur in shallow basins or potholes, riverine wetlands result from periodic flooding of rivers and streams, and fringe wetlands occur adjacent to large lakes and seas. Fringe wetlands include mangrove wetlands such as the Florida Everglades. The amount of water in the wetlands varies with season. Extensive drying may occur in some years in some wetlands. Wetlands are important recharge areas (sources) for groundwater, they provide surface water storage areas, they are the habitat for an incredible diversity of organisms, and they play an important role in filtering pollutants from surface water.

Estuaries

Estuaries are where rivers or streams meet the marine biome. The sediments carried by rivers are deposited in estuaries making the land very fertile. The estuaries are feeding/breeding areas for many organisms, and they are on the migration routes of both birds and fish. In general they are productive areas of high species diversity.

Marine Biome

The marine biome is divided into zones based on distance from shore (intertidal, neritic, and oceanic zones) and light penetration (photic and aphotic zones).

Intertidal Zone: The intertidal zone is the area between high and low tides. Intertidal areas include sandy areas, mudflats, and rocky extrusions with tide pools. The organisms are well adapted to the wave action and to the tidal cycle. For instance, large brown and red algae are attached to the substrate and have air-filled structures that allow them to stay anchored and protected from wave action and simultaneously stay afloat, organisms in tide pools are usually attached to the rocks and organisms in the sand can bury themselves.

Neritic Zone: The neritic zone is the part of the marine environment that extends outward from the intertidal area over the continental shelf. The continental shelf is so called because it is a submerged part of the edge of continents that slopes very gently (about 2 m per km). It varies in width but averages 80 km wide and 432 km deep at the outward edge. It may be as wide as 1,500 km. The continental shelf area of the marine biome is the most productive part of the ocean. It is the main source of fish and invertebrates harvested as food

for people. One special area of the continental shelf is coral reefs, which occur primarily in warm clear waters of the Pacific and Indian oceans.

Oceanic Zone: The oceanic zone is the open oceans which cover about 70% of the earth's surface. Most of the organisms live in the photic zone but when they die they sink through the pelagic (deep ocean) zone and become part of the benthos. Upwelling along the coasts of continents can bring the nutrient from the benthos to the surface. Thermal vents (also called smokers, chimneys, or hot vents) occur on the ocean floor where tectonic plates are moving apart (diverging). These thermal vents occur at depths greater than 2 to 3 km, which is in the aphotic zone well below the photic zone. They are a source of many metals (e.g., zinc) and have a unique assemblage of organisms present. The initial energy source for these organisms is inorganic compounds. Ideas about the origin of life are emerging from studying these organisms.

2. Behavioral Ecology

Behavior is primarily about animals. Animal behavior is what an animal does, how it does it, and the consequences of the action. Most specifically, animal behavior is a coordinated response to stimuli. The cause of the behavior may be proximal or ultimate. Proximate causes are the mechanisms that elicit the response whereas ultimate causes are the evolutionary basis for the response. Behavioral ecology deals with the ultimate causes, that is, why natural selection favors a particular behavior.

The response of the animal may be instinctive (innate) or learned. Instinctive responses are prewired within a newborn such that when the naïve newborn is exposed to a specific signal, it elicits a stereotypical response called a fixed action pattern. These depend on genes that code for the development and operation of nervous, endocrine and skeleto-muscular systems. That is, the stimulus is received by a sensory system and relayed within the body, ultimately producing an observable and predictable motor response. Learned behaviors are those which are modified after an experience in the environment. There are several types of learning but they all have an environmental as well as a genetic component.

In some cases, the animal receives and interprets the stimulus and then performs a particular behavior. That is, some animals can process information about their environment. This is a type of thinking. The relationship between the sensory input (nervous or endocrine systems) and behavior is cognition. There are several types of cognitive mechanisms that animals use to move through their environment. Some animals respond to a stimulus by changing their activity (kinesis) or moving to or from the stimulus (taxis). Some animals use landmarks or mental maps to relocate nests or food sources. Some animals move or migrate using the sun, stars, and the earth's magnetic field.

Habitat Selection

Habitat selection is a function of behavioral mechanisms (proximal causes) and the evolution of habitat preferences (ultimate causes).

Proximal Behavioral Mechanisms

Proximal behavioral mechanisms operate at the level of the individual animal and vary with respect to spatial scale. For instance, observations may show that a bird selects a forest vs. a grassland habitat and within the forest habitat it selects a specific species or size of tree. The selection is a result of stimuli about the environment: the general characteristics of the terrain or landscape, the feeding and mating areas present, food and water availability, cover objects and protection, and the presence of conspecifics and other organisms.

Ultimate Behavioral Mechanisms

Natural selection will favor animals that have a high evolutionary fitness (high contribution of genes to subsequent generations). For instance, animals that select poor quality or marginal habitats will not have as many

offspring as those in better quality habitats. Natural selection may act on the behavior that results in the habitat choice or it may select for individuals that learn which habitats are appropriate.

Social Behavior and Sociobiology

Social behavior is the interaction among organisms, usually of the same species. Sociobiology is the study of the behavior between individuals of the same species and how these behaviors are adaptive. There are two main types of intraspecific behaviors: those associated with competition for food or resources, and those associated with mate selection. In both cases the "winner" enhances his/her survival and fitness (contribution of genes to the next generation). However, individuals can proliferate their own genes by producing offspring and by helping close relatives. This is the concept of inclusive fitness. All of these behaviors require signals between organisms. The exchange of these signals is communication.

Intraspecific Competition

Not all competitive behavior is the same. Three general categories are recognized. *Agonistic* behaviour includes threatening and submissive gestures that are used to determine which individual gets the prize (e.g., food or a mate). Both individuals begin with threatening behaviors but usually one shows submissive behavior. In animals that live in groups, the competition frequently ends with reconciliation behavior. *Dominance hierarchies* are sometimes referred to as pecking orders. The top individual, called the alpha individual, acquires this status by fighting or intimidating, using agonistic gestures. The alpha individual has the first pick of resources and mates. A third type of intraspecifc competition is the *defense of territories*. The territory may be for an individual or a social group of individuals.

Mating Behaviors

Mating behaviors are those that enhance reproductive success and thus increase fitness. *Courtship* is a series of specific actions designed to attract a mate. It allows individuals to identify mates of the same species and confirm that the individuals are physiologically ready to mate. Some behaviors that were considered courtship are now being interpreted as forced mating (e.g., garter snakes). *Mating systems* vary among species. Mating in which there is no consistent pairing of individuals and both males and females have many mates is called promiscuous. Mating in which one of the sexes has several mates is called polygamous. If one male mates with several females, the mating is called polygamy. If one female mates with many males, it is called polyandry. The relationship between animals that form long-term bonds between one female and one male is called monogamous. The factors that influence the mating system include the extent of parental care and certainty of paternity. Investment in offspring increases the chance of their survival and certainty of paternity guarantees contribution of the males input to the next generation.

Animal Communication

The signals sent and received by animals may be visual (e.g., the dance of the honeybee or postural displays), auditory (male songbirds, particularly in spring), tactile (male snakes), electrical (some fish), or chemical. Pheremones are an example of chemical signals. They may be emitted into the air to attract mates (e.g., moths), included in urine to mark territories (e.g., wolves), and released from injuries (e.g., some fish).

Inclusive Fitness

Competition for food resources and mates, and use of different mating systems increases the survival and fitness of individuals. However, some animals are *altruistic*. They show behaviors that can decrease their fitness and increase the fitness of another individual. For instance, the slap of the tail of a beaver is an alarm call for the other beavers but can expose the one beaver to the predator. Altruistic behaviors usually occur between and among closely related individuals. Closely related individuals have more common genes than distantly related individuals and thus altruistic behaviors towards closely related individuals enhances the fitness of the altruist. Natural selection that favors altruistic behaviors among closely related individuals is called *kin selection*.

3. Population Ecology

A population is a group of individuals of the same species occupying the same area. Populations have basic characteristics or demographics, they have different life history strategies or life events between birth and death, and they are dynamic and grow in different ways. These three main topics, demographics, life history strategies, and population growth and limitation are outlined below.

Demographics

Demographics are the basic characteristics of populations. They include density, distribution (or dispersion) patterns, age structure, and survivorship.

Density

Density is the number of individuals per unit area or volume. It can be determined directly by counting the individuals, indirectly using mark-recapture studies or sample plots, and relatively by relative measures. Direct approaches give absolute numbers, indirect procedures provide estimates, and relative measures provide only relative changes (or trends). There are two important indirect methods, mark-recapture methods and sample plots. *Mark-recapture* methods are frequently used for animals. An example is the number of sunfish in a lake. In its simplest form, the procedure is to collect, mark, and release a sample of sunfish, and after a short period of time, obtain a second sample. The ratio of the number of marked sunfish in the second sample (m) to the total number of marked sunfish (M) is an estimate of the ratio of the number of sunfish in the sample (n) to the number of sunfish in the population (N). If 60 sunfish were marked (M) and 4 marked sunfish (m) were collected in a second sample of 80 sunfish (n), the population estimate is $4 / 60 \approx 80 / N$, or $N \approx 1200$. The second indirect approach, *sample plots*, is most appropriate for nonmotile individuals. The sample area is divided into quadrats, or some form of subsamples, and the information from the subsample is extrapolated to the whole population. For instance, if the average number of oak trees in five different plots 10 m by 10 m is 76, the esti-

Figure 9.1.
Age Distribution of the People in British Columbia, Canada, in 1989.

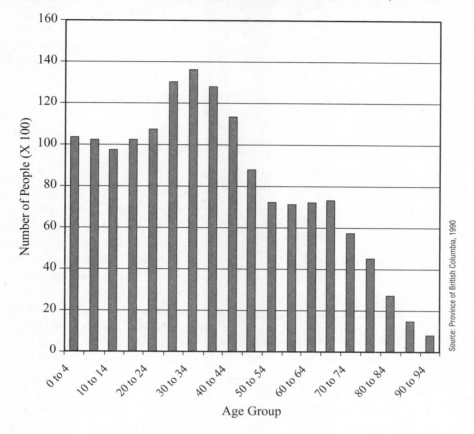

Source: Province of British Columbia, 1990

mated number of oak trees in a three hectare (1 hectare = 10,000 square meters) forest is 7,600. Relative measures include counting the number of nests or scats of a particular species. The results over several years show only relative increases or decreases.

Distribution Patterns

Dispersion patterns describe the spacing among individuals. They may be clumped, random, or uniform. Different patterns may exist simultaneously. For instance, a population may be uniformly distributed over a large area but as clumps within the area. The type of spacing pattern is determined by factors such as dispersal distance, competition for resources and ability to defend an area.

Age Structure

Plots of age distribution can show the potential for growth of future populations and the effect of past events on certain age groups. The age distribution of the population of British Columbia, Canada, in 1989 is given in Figure 9.1. Three points warrant mention. First, the large number of people that were 25 to 45 years old in 1989 are the baby boomers, the babies born in the prosperous years after the Second World War. A bulge representing the baby boomers is typical of the North American population. Second, the relatively high number of people over 65 years of age is due, in large part, to movement of people to this province to retire. Third, this is a stable or aging population. In a rapidly growing population the age distribution would decrease with age; there would be a very high number of young people.

Survivorship

Life tables and survivorship curves are used to track mortality and survival of a cohort. A cohort is a group of individuals born at the same time or in the same year. The number of individuals in the cohort alive at the beginning and end of defined time intervals is recorded in life tables. The information is graphed (number of survivors vs. percent of maximum life span) in survivorship curves. There are three basic types of survivorship curves. Type I shows low mortality in early and mid life but high mortality later in life. This is typical of humans. Type II shows a constant rate of mortality throughout life. This is typical of some annual plants and rodents. Type III shows high mortality in the early years and low mortality in later years. It is typical of organisms that produce large numbers of eggs directly into the environment. Large numbers of the eggs and larvae die.

Life History Strategies

The life history strategies are the life events of the individuals from birth to death that affect survival and reproduction. Courtship and mating behaviors affect the survival and fitness of organisms (see the section "Behavioral Ecology"). Fitness also depends on the age at sexual maturity (maturity), the number of offspring produced at each mating (parity), and the number of reproductive events (fecundity). Mature individuals may produce large numbers of offspring only once or infrequently (big-bang reproduction or semelparity) or they may produce offspring annually (repeated reproduction or iteroparity).

In the life of an individual there are also trade-offs observed between reproduction and survival. For instance, in years when resources are limited, some organisms either do not breed or produce small numbers of offspring. This increases the chance of survival and future breeding events. This type of trade-off is found in many species; it is particularly well studied in birds.

Population Growth and Regulation

Population dynamics is the study of how and why the numbers of individuals in a population change. The size of the change depends on the initial size (N_0) and the number of births (B), deaths (D), immigrants (I), and emigrants (E). Births and immigrations increase the population, and deaths and emigrations decrease the population. Mathematically, using the given terminology, the number of individuals at time t (N_t) is expressed as $N_t = N_0 + B - D + I - E$. If $(B + I)$ is greater than $(D + E)$, the population increases or grows; if $(B + I)$ is less than $(D + E)$, the population decreases. To simplify the equation, we can assume that $I = 0$ and $E = 0$, although they are likely to be greater than zero in real situations.

We can also assume that birth and death are continuous processes. This allows us to use simple calculus to describe population dynamics. The change in the size of a population is usually expressed in terms of rates (per individual per unit time); thus the birth rate (b) is B/N; similarly the death rate (d) is D/N. The change in size of the population (dN/dt) is $(b - d)N$ (note: dN/dt is the differential, it is not $d \times N$). The rate of change in population size in a given time period is $r = b - d$. If $r = 0$, the population stays the same size. If $r < 0$, the population declines in size. If $r > 0$, it increases. If r varies in value from one time period to the next then the size of the population fluctuates. However, if r is constant the population changes in size at a constant rate. When r is constant and greater than zero, the population is described as growing exponentially. Exponential growth is unlimited growth (Figure 9.2). Real populations rarely show exponential growth for long because resources or other factors become limiting; the value of r decreases, the growth decreases, and the population size hovers around a constant level, the carrying capacity. This type of population change is called logistic growth (Figure 9.3).

There are two types of factors that affect the change in the size of populations. These are called ***density-dependent and density-independent factors***. Density-dependent factors are those whose effect increases as the density increases, such as food or suitable nesting sites. Some density-independent factors are fires, hurricanes, droughts, floods or severe cold. However, density-dependent and -independent factors are not always mutually exclusive. For instance, severe weather (a density-independent factor) can affect the available food or space (density-dependent factors).

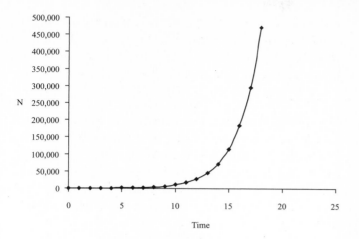

Figure 9.2.
The Change in Numbers of Individuals (N)
Over Time During Exponential Growth

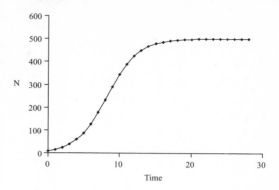

Figure 9.3.
Change in Numbers of Individuals (N)
Over Time During Logistic Growth

A continual decrease in a population can result in the extinction of a population. Mathematically this occurs because r is continually negative. This may result from a decrease in the birth rate or an increase in the death rate. Both the birth and the death rates are affected by natural and imposed (e.g., decreased space, harvesting) conditions.

4. Community Ecology

Communities are all of the populations in a defined area. The numbers and kinds of organisms present (biological diversity), the interactions among organisms, and the consequences of disturbances in communities (succession) are three basic ideas in community ecology. They are outlined below.

Biological Diversity (Biodiversity)

Biodiversity is a measure of the numbers and kinds of individuals present in a community. Although the information on biodiversity may be recorded and compared among communities in different locations, it is frequently summarized as a mathematical index. One common index calculated is the species diversity index, sometimes referred to only as species diversity. There are several ways of calculating this index, but the final number is a measure of both the species richness (number of species) and evenness (relative abundance of each species).

Biodiversity depends on factors such as geographic location, size of the area, and natural barriers to movement (e.g., mountain ranges and oceans). In general, there is a general decrease in species diversity with latitude from the equator to the poles. On islands or in isolated areas (called habitat islands), the diversity is related to the rate at which species immigrate to the island and the rate at which the species become extinct. These factors are in turn dependant on the distance of the island from the mainland (or other communities), the size of the island, and the number of species on the island. Models that predict the effects of the interaction of these factors on biodiversity are the basis of island biogeography theory. They are used in conservation biology to assess and study potential effects of habitat fragmentation.

Interactions Within Communities

Interactions occur between two species and among several species with immediate and long-term consequences. The types of interactions between two species and among several species, and some consequences of these interactions are outlined below.

Interactions Between Two Species

When two species interact, there are three possible outcomes for each species: it may benefit (positive effect), be harmed (negative effect), or not be affected (zero or neutral effect). The main types of interactions between two species and the types of outcomes for each are summarized in Table 9.1.

Table 9.1.
Types of Interactions That Occur
Between Two Species in a Community

Type of Interaction	Species 1	Species 2	Effect Using Symbols
Competition	Harmed	Harmed	$(-, -)$
Predation	Benefits	Harmed	$(+, -)$
Commensalism	Benefits	No effect	$(+, 0)$
Amenalism	No effect	Harmed	$(0, -)$
Mutualism	Benefits	Benefits	$(+, +)$

Competition: Competition has a negative effect on both species. It occurs between individuals of different species and is called interspecific competition. In an environment, the total space, physical conditions and resources required by a species is called the ecological niche. If two species have the same niche, interspecific competition can have two different outcomes: the eventual elimination of one species, called competitive exclusion, or the sharing of the resources by the two species, called resource partitioning.

Predators

Predators feed on prey and the predator/prey interaction benefits the predator but harms the prey. *Parasites* are predators that feed on their hosts; the parasite benefits and the host is harmed. Parasites do not usually kill their hosts. However, *parasitoids* are a specific kind of parasite that do kill their hosts. For instance, larvae of some wasps and flies consume the eggs, larvae and pupae of other insects. Predators include herbivores, carnivores, omnivores and filter feeders. Carnivores kill whole animals, whereas herbivores may only nibble and consume part of individual plants. Filter feeders consume plankton, or small (microscopic) plants and animals in water (ocean, lakes, etc.). A predator can also be a prey as is discussed in the section on ecosystems.

Commensalism is an interaction in which one species benefits but with no harm to the second species. The example frequently used to describe commensalism is the sitting of egrets on the backs of grazing cattle. The cattle's movement flushes insects from the vegetation and the egret, with its vantage point on the cattle's back, readily obtains the food. There is no effect on the cattle but the cattle helps the egret obtain the food.

Amenalism is an interaction that is not consistently included in the summary of the interactions between species in a community. Although one species is harmed there is no direct benefit for any other species from the harm to the first species. In a forest, dead branches on a tree fall to the forest floor and damage small plants or animals on the forest floor. The tree that loses the branch may benefit from the loss because it no longer has to support the branch and because other living branches are exposed to more sunlight. However, the tree does not benefit from the damage to the small plant or animal. Similarly, animals that trample plants may benefit from having a path but not from the damage to the plant.

Mutualism is the interaction in which both species benefit. It occurs between a diverse array of organism. Many flowering plants produce excess nectar, a sugar solution within the flower. Bees, moths, butterflies, birds, bats, and even some small rodents eat the nectar. The nectar is within flower and thus they must touch the stamens containing the pollen to obtain the nectar. These organism, referred to as *pollinators*, then move to other flowers carrying the pollen. The flowers benefit because the pollen is passed among the flowers, resulting in cross-pollination or mixing of the genetic material; the pollinators benefit because they obtain the high-energy sugar solution.

Lichens consist of two organisms, a modified fungus and an alga. The algae live on the fungi hyphae and receives water and nutrients from the fungi. The algae provide sugars, produced by photosynthesis, to the fungi. This close and usually obligatory mutualistic relationship is called *symbiosis*. Another example of symbiosis is the nitrogen-fixing bacteria that live in the soil or in nodules on the roots of plants such as legumes (peas) and alders. The bacteria obtain food and protection from the tree and provide nitrogen to the plant.

Interactions Among Species

Most of the interspecific interactions between two organisms involve feeding. The feeding relationships among organisms in a community is called the trophic structure and is described as a food chain or food web.

Parasitism can also be an interaction among species. Although most parasites have one definitive host, they usually require one or more intermediate hosts at different stages of their development. The various hosts are often quite specific and must be in close proximity for the parasite to complete its life cycle.

Consequences of the Interactions

Organisms interact and a new genetically determined trait that improves an organism's situation in the interaction is an adaptation. Important consequences of the interactions of species within a community include adaptations of predators, defenses of plants and animals against predators, and coevolution. The importance of keystone species is a consequence of the interactions among species.

Adaptations of Predators. In general, predators have well-developed sensory systems and special structures that assist in catching and subduing the prey.

Adaptations of Plants Against Predators. Plants can not escape from predators and thus have adaptations that make them unpalatable to animals. For instance, some plants have spines and thorns which deter predators and some contain toxic chemicals or distasteful chemicals.

Adaptations of Animals Against Predators. Animals have a variety of adaptations to avoid predators. Some move quickly and can hide or escape from predators, some have mechanical (porcupines), chemical (rattlesnakes) or postural (cobra) defenses that deter predators, and some produce toxins (newts and monarch butterflies). Many animals rely on adaptive coloration in which the animals are camouflaged because they resemble the habitat (cryptic coloration). Organisms that produce toxins are usually brightly colored (aposematic coloration) and nontoxic animals of similar species mimic the colors (Batesian mimicry). In some cases two unpalatable animals resemble each other (Mullerian mimicry), enhancing the protective advantage.

Coevolution. Because two species are involved in an interaction, an adaptation in one species results in selective pressure on the second. The result is like a see-saw with each adaptation in one effecting a change in the second. This reciprocal selection and the resulting adaptations is called *coevolution*. Coevolution can occur in predator-prey, parasite-host, and mutualistic interactions. Strict coevolution between two species is not widely known. General or diffuse coevolution is more common. General coevolution results when several species interact with several other species and the adaptations within each of the groups are similar. The following example is coevolution occurring in a mutualistic interaction. Flowers that attract bees are brightly colored, usually blue and yellow but not red because bees cannot see red. In addition, the flowers are flagrant to the attract bees, have a landing platform positioned so that stamen and pistil (reproductive structures) contact the bee's body, and have a narrow floral tube. The mouth parts of the bee are fused into a long tube that contains a tongue, an adapter for sucking the nectar. This tube fits into the flower's floral tube. Insects cannot fit into the floral tube. Several different flowers attract bees. In general, these flowers mature at different times of the year. Therefore the bees have a continuous food supply and there is reduced opportunity for bees to deliver pollen from species to another.

Keystone Species. The community structure is the sum result of all of the interactions. However, some species have a greater effect on the community structure than is expected on the basis of their numbers and the general types of interactions in which they are involved. One of these is a ***keystone predator***. A keystone predator consumes several different kinds of prey and keeps the population numbers of each of the prey populations at levels that reduce competition. This reduces the chance of competitive exclusion. The net result is that a keystone predator helps to maintain the diversity of a community. Keystone species do not all act as keystone predators. Woodpeckers, such as the red-naped sapsucker, drill holes in trees to obtain sap and insects, and to nest. Once the holes are drilled, other species can eat the sap and insects and use the holes for nesting. The red-naped sapsucker thus provides food and habitat for other species that would otherwise not be available; it enhances the diversity of the community more than is immediately apparent from its numbers or lifestyle.

Succession

The change in biodiversity of a community after a disturbance is called ***ecological succession***. Traditionally, succession is referred to as primary or secondary.

Primary Succession

Primary succession is the change in a community following a disturbance that virtually eliminates the existing life (e.g., on the volcanic ash of Mount St. Helens or on the asphalt of an abandoned road).

Secondary Succession

Secondary succession refers to changes following less dramatic disturbances (e.g., fires, floods, and hurricanes) that leave the soil intact.

In both primary and secondary succession there is a gradual recolonization of the area. The first species present, the *pioneer species*, are different than the dominant ones present before the disturbance. Furthermore, neither the sequence of species nor the species composition is consistent among sites during ecological suc-

cession. The species that colonize the area depends on the composition of the surrounding communities, the ability of a species to invade a disturbed area and on the response of the species to changes during succession. The ultimate association of species is called the *climax community*.

5. Ecosystems

Ecosystems are defined as the biotic (living) and abiotic (nonliving) components of an area or habitat that interact in the exchange of energy and nutrients. The biotic components are the different populations that comprise the community. The abiotic components are the physical habitats of the organisms and the conditions of the habitats. The two main topics in ecosystems are energy flow and biogeochemical cycles. It is important to remember that energy flows whereas nutrients cycle. This does not mean that all nutrients can be recaptured in a nutrient cycle but it does mean that energy cannot be recycled.

Energy Flow and Trophic Levels

The trophic levels show the feeding relationships among the species in an ecosystem. There are three main trophic levels: primary producer, consumers, and decomposers. The flow of energy through the trophic levels is called the food chain or food web.

Primary Producers

Primary producers begin the energy flow through an ecosystem. There are two types of primary producers: autotrophs or chemoautotrophs. Autotrophs use the sun's energy to fix atmospheric carbon into sugars during photosynthesis. Chemically, the sun's energy is contained in the organic sugar molecules and can be released and used by the plant for growth and maintenance or passed to a higher trophic level. Plants, algae and photosynthetic prokaryotes are the main autotrophs. Chemoautotrophs use inorganic compounds to produce energy. Most chemoautotrophs are prokaryotes.

Consumers

Consumers obtain their energy from food. There are several levels of consumers but because they all obtain energy from other organisms they are called heterotrophs. Consumers that feed on plants are the primary consumers and are referred to as herbivores. Consumers that feed on animals are called carnivores. Carnivores that feed on herbivores are secondary consumers and those that feed on other carnivores are tertiary or quaternary consumers.

Detrivores

Detritivores or decomposers obtain energy from dead organic material called detritus.

Food Chains and Food Webs

A linear series of organisms that transfer the energy is called a *food chain.* One of the food chains in Figure 4 (shown by a dashed line) is the consumption of plants by mice that are in turn consumed by foxes. The transfer of energy among the organisms is usually more complex than a food chain because one organism may consume several organisms or several organisms may consume the same one organism. The result is a *food web*. The relationship between a food chain and a food web is shown in Figure 9.4.

Trophic Efficiency and Ecological Pyramids

Trophic efficiency is the percentage of the production transferred from one trophic level to the next.

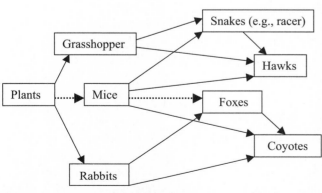

Figure 9.4.
An Example of a Food Chain (·······▶)
Within a Food Web (——▶)

That is a measure of the net production of one level divided by the net production at a lower level expressed as a percent. The trophic efficiency between trophic levels in an ecosystem is about 10% (range from 5% to 20%). This means that only 10% of the production is passed to each successive trophic level. In a food chain with a primary producer, a herbivore and a secondary consumer, 10% of the primary production (energy) would pass to the herbivore and 1% of the primary production energy (10% of 10%) would pass to the secondary consumer. This decrease in energy through the trophic levels is shown as a pyramid. There are several types of energy pyramids depending on the way the energy is measured at the different trophic levels. A production pyramid shows the energy content at each level. A biomass pyramid shows the biomass (e.g., dry weight); a numbers pyramid shows numbers of organisms. A generalized pyramid is shown in Figure 9.5.

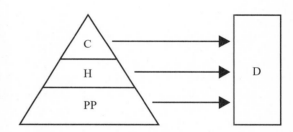

Figure 9.5.
The Energy Flow (kcal/m²/year)
Through the Primary Producers (PP),
Primary Consumers or Herbivores (H),
Secondary Consumers or Carnivores (C), and
the Decomposers in a Generalized Ecosystem

Specific terminology is used to describe the energy movement between trophic levels. The total energy produced by photosynthesis is called the *gross primary production (GPP)*. The total energy ingested by each consumer group is E_I. The energy used for maintenance in both plants and animals is the respiratory energy (E_R). The energy put to growth in animals is **net *production (NP) or growth*.** The net production or growth is the ingested energy (E_I) minus the respired energy (E_R). In plants it is called *net primary production (NPP)*. Calculation of production efficiencies for a primary producer and a primary consumer are summarized below.

Production efficiency of primary producers = NPP / GPP

But GPP = NPP + E_R,

Therefore, production efficiency for primary producers = NPP / (NPP + E_R)

AND

Production efficiency for primary consumer = NP / E_I

But E_I = NP + E_R

Therefore, production efficiency for primary consumers = NP / NP + E_R

Biogeochemical Cycles

Biogeochemical cycles define the movement of elements between the physical environment and the organisms. As the name implies, movement is in a cycle. The elements most commonly examined are the nutrients, carbon (C), nitrogen (N) and phosphorous (P) that are required in large quantities by the organisms. In addition, water, which is an important medium for the nutrients and a transporter of the nutrients, moves in a global cycle. The physical environment includes land, air (atmosphere) and water (fresh and marine). Within each of these environments the nutrients may be available for immediate uptake by the organisms, or unavailable. Unavailable nutrients include organic material buried as coal and other fossil fuels, and inorganic material such as phosphorus contained in minerals in rocks or carbon contained in limestone. The living organisms are the producers, consumers and decomposers. Recall that one important similarity between the flow of energy and the cycling of nutrients is that both occur through the same organisms.

A general biogeochemical cycle showing the three main components (available nutrients, unavailable nutrients, and nutrients contained in organic matter) and the general types of movement among the components are summarized in Figure 9.6. The following sections outline the global water cycle and generalized nitrogen, carbon and phosphorus cycles.

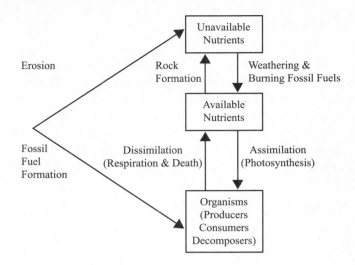

Figure 9.6.
An Outline of a General Biogeochemical Cycle

Hydrological Cycle

The global hydrological cycle is the movement of water between the land and the oceans. On the oceans and land there is both precipitation as rain and snow, and evaporation. Over the oceans, evaporation exceeds precipitation. The excess moisture condenses in clouds and moves toward the land. On the land, precipitation exceeds evaporation. The excess water moves into the ground (groundwater), through the soil (subsurface flow) and over the land (surface flow) to the oceans. Therefore, the overall cycle is evaporation from the oceans, movement via clouds to the land, precipitation on the land and runoff to the oceans. It must be remembered that this is a global cycle; there are many natural and unnatural local factors that affect the cycle and alter the movement of water. In addition, in the 1980s it was realized that fresh water, as huge chunks of ice from comets, was being added to the earth. Therefore, there is a small net gain of water on the earth.

Nitrogen Cycle

In living organisms, nitrogen is a main component of proteins. As an inorganic nutrient, nitrogen exists in several forms. Interconversions among these forms are a major part of the nitrogen cycle. The chemical changes are outlined in Figure 9.7, and each step is summarized in Table 9.2.

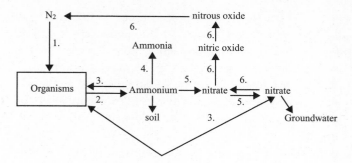

Figure 9.7.
Outline of the Chemical Reactions
of the Nitrogen Cycle

The atmosphere is an important part of the cycle because it is the initial source of nitrogen. Some nitrogen is also returned to the atmosphere during denitrification and volatilization. Mineralization and nitrification occur in water (lakes, rivers, and oceans) and in the soil. They are the processes that are responsible for the decomposition of organic nitrogen in organisms. Uptake of nitrate or ammonium by organisms is immobilization or assimilation, and like mineralization and nitrification, it occurs in water or soil. Some organisms are selective and use only nitrate or only ammonium, but many organisms use both. Nitrogen is unavailable when it is lost as nitrate to groundwater or adsorbed (tied up) as ammonium to clay particles in the soil. In summary the atmosphere, water and land are all important environments for the nitrogen cycle and much of the nitrogen cycle are the interconversions among the different forms of nitrogen. The rate of the many steps is controlled by the organisms present and physical features (pH, temperature, etc.) of the environment.

Carbon Cycle

The movement of carbon among the main components of the general biogeochemical cycle is shown in Figure 9.8. The numbers are explained in the following text.

The carbon cycle is closely allied with energy flow and it is important to understand both the similarities and the differences. In energy flow, the first step is the fixation of carbon, using energy from the sun in the

Table 9.2.
Summary of Steps in the Nitrogen Cycle

Step #	Step Name	Comments
1.	Nitrogen fixation	• Bacteria fix atmospheric into nodules on roots of some plants such as legumes (e.g., peas) and alders.
2.	Mineralization	• Release of ammonium during decomposition.
3.	Immobilization (assimilation)	• Assimilation of ammonium or nitrate by organisms (plants).
4.	Volatilization	• Release of ammonia, a gas.
5.	Nitrification	• Two step conversion of ammonium to nitrate. • Both steps require bacteria.
6.	Denitrification	• Conversion of nitrate via nitrite to gaseous nitrogen. • Occurs under anaerobic (no oxygen) conditions. • Involves soil organisms.
7.	Ammonium adsorption	• Adsorption of ammonium to soil clay particles.

process of photosynthesis. Similarly, a major part of the carbon cycle is the uptake of carbon by the organisms during photosynthesis. The carbon is from carbon dioxide in the air or bicarbonate ions (HCO_3^-) in water. Carbon is present as bicarbonate in the water because carbon dioxide and water combine to form carbonic acid and, in water, the acid dissociates giving bicarbonate and sometimes carbonate (CO_3^-). Whether it is carbonate or bicarbonate depends primarily on the pH of the water. In energy flow, the energy moves through the trophic levels. As a nutrient, the carbon moves through the organisms and is returned to the atmosphere or water during respiration and when the organisms die. The land (soil) is not given as a source of available carbon because available carbon is not present in the soil particles but among the particles in water and air. Overall, the exchange of carbon between

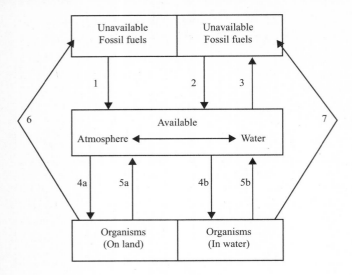

Figure 9.8.
Summary of the Carbon Cycle

the organisms and the available nutrient source is by photosynthesis, respiration, and death and decomposition of the organisms.

Some of the organisms may be buried, or at least were buried in geological time (6). These buried organisms are the present source of fossil fuels. In addition, the shells of marine organisms contain calcium carbonate. Death and accumulation of these shells is the source of limestone deposits (7). However, limestone takes millions of years to form from the shells. Before limestone is produced, the carbon in shells is made unavailable (3) simply as shells on the bottom of water bodies. Burning of fossil fuels (1) and removal of limestone from the ground for use as dimension stone in construction, or release of carbon from limestone in the water body (2) releases the unavailable carbon.

Phosphorus Cycle

The chemistry of phosphorus in extremely complex and thus a chemical explanation of the phosphorus cycle is not realistic. A summary of the movement of phosphorus among the three parts of the biogeochemical cycle is summarized in Figure 9.9.

In organisms, phosphorus is contained primarily in high-energy molecules of adenosine triphosphate (ATP) and in deoxyribonucleic acid (DNA). In the soil, phosphorus is present in numerous compounds that are not available to the organisms, and as the ion phosphate that is available and used by organisms. Unavailable phosphorus is also found as an element in minerals of numerous rocks. Unavailable phosphorus is released and made available by weathering and erosion of rocks, and by chemical reactions in the soil and in the sediments of water bodies. In addition to its presence in the soil, available phosphorus is present in marine and fresh water. In marine water phosphate is part of the salt that makes the ocean "salty." There is a rare phosphorus gas, but in reality there is not an atmospheric component of the phosphorus cycle. Therefore the movement of phosphorus from land to water is by runoff from the soil to the water but movement from the water (oceans) to land requires animals such as birds that obtain phosphorus from marine sources and then die or leave feces on the land. As with all of the cycles, death and decomposition of the organisms returns the phosphorus to the available pool of nutrients, or to the unavailable pool if the organisms are buried or the chemical form of the phosphorus cannot be used by the organisms.

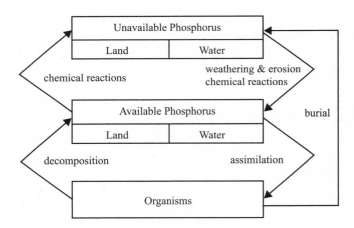

Figure 9.9.
A Summary of the Movement of
Phosphorus Among the Three Parts
of the Biogeochemical Cycle

In fresh water, but not in the soil, phosphorus is the *limiting nutrient* to primary production. This means that the addition of available phosphorus, usually as phosphate, will result in increased primary production. This is because all of the other nutrients are present in larger quantities than are being used. The visible result is algal and phytoplankton blooms and eutrophication of previously oligotrophic or mesotrophic waters. This change in the numbers (and kinds) of primary producers affects the numbers and kinds of consumers. The decreased oxygen decreases the water quality and limits the kinds of organisms that can survive. This is the reason for using phosphate-free detergents.

Evolution

Charles Darwin defined evolution as "descent with modification." Evolution is the change in a lineage of populations between generations. It is an ongoing process that accounts for the changes in living organisms from their onset over about 3.8 billion years ago to the present day. The topics addressed in this summary are genetic variability, evolutionary processes, evolutionary consequences, and the history of life.

1. Genetic Variability

Several processes give gene pools the ability to vary; the amount that the genes vary is genetic variation. Genetic variation is the diversity of the gene pool both within populations and among populations. The gene pool is all of the alleles at all gene loci in all individuals of a population. The locus is the position in the chromosome where a gene is located, and alleles are the alternative versions of the gene. A gene is a specific nucleotide sequence in a DNA molecule. Alternate or identical versions of genes (i.e., alleles) are paired on homologous chromosomes of diploid individuals. If the alleles are the same, the pair is homozygous; if the alleles are different, the pair is heterozygous. During meiosis and mitosis, homologous chromosomes duplicate themselves into two sister chromatids.

Genetic variation is assessed at both the gene (gene diversity) and molecular (nucleotide diversity) levels. Nucleotide diversity is a measure of the similarity of nucleotide sequences. Gene diversity is the percent of loci that are heterozygous and thus will contribute different alleles to different gametes.

Diversity of a gene pool results from genetic recombinations, specific relationships among genes (e.g., linkages), mutations, polymorphism, heterozygous advantage, geographic variation of the gene pool within populations, and neutral variation. How each of these contributes to genetic variation within populations is outlined below.

Genetic Recombination

Genetic recombination, usually through meiosis, produces offspring with new combinations of traits and thus enhances genetic variability. In general, it requires two sets of alleles and occurs between individuals with at least two (diploid) sets of chromosomes. That is, it results from sexual reproduction. Traits may be both discrete and quantitative. Discrete traits are determined by a single gene with different alleles. Quantitative traits vary along a continuum because their expression is due to several genes.

Recombination does not always require sexual reproduction. Bacteria have only one set of chromosomes but can undergo sexual reproduction due to transfer of F factors during conjugation (see Chapter 3, Genetics and Molecular Biology). Bacteria can also pick up naked DNA from the environment (transformation) or pick up DNA from viruses (transduction). The former is used in synthesizing recombinant DNA.

Linkage

Linked genes are on the same chromosome and move together during meiosis and fertilization. The result is offspring that should have both traits of one of the parents. That is, there should be no recombination of the individual linked traits. However, during meiosis crossing-over occurs. Breakage occurs in sister chromatids of both homologous chromosomes, and the fragments exchange places. In general, both genes are more likely to be retained in the gene pool because they are linked.

Mutations

A mutation is a change in the DNA of an organism or RNA of a virus due to rearrangement of sections of chromosomes (DNA) during meiosis (chromosome mutations) or to changes in specific nucleotides (point mutations).

Chromosome Mutations

Chromosome mutations can change the structure and number of chromosomes. Because they occur during meiosis, the gametes have the aberrations that can be passed to the offspring. Change in chromosome structure occurs due to breakage and incorrect repositioning of the broken fragment. The results are deletion, duplication, inversion, and translocation. In *deletion*,

a fragment of DNA is lost and the chromosome lacks certain genes. If the deleted fragment becomes attached to a sister chromatid (replicated chromosome), the recipient has a duplicated or repeated segment. This is *duplication*. If the fragment reattaches to the chromosome but in reverse order, the result is an *inversion*. If the fragment attaches to another chromosome (nonhomologous chromosome), it is called a *translocation*. A reciprocal translocation occurs when nonhomologous chromosomes exchange fragments. Inversions and reciprocal translocations produce a normal number of genes, but because the position of a gene is important in gene expression, the phenotype of the offspring may be affected. The altered genes can be inherited and can be important in evolutionary change.

Change in chromosome numbers occurs if the sister chromatids or homologous chromosomes do not separate correctly during meiosis. This is called *nondisjunction*. The result is gametes with either two or no copies of the chromosome. Combination of these gametes with normal gametes produces offspring with abnormal numbers of chromosomes. This is called *aneuploidy*. During mitosis, the aneuploidy condition is transferred to all cells. Nondisjunction and subsequent mating can result in offspring with more than the normal two sets of chromosomes. This is called *polyploidy*. It is common in plants and important in their evolution. It occurs in only a few animals.

Point Mutations

There are two main types of point mutations: base-pair substitution and base-pair deletions and insertions (Figure 10.1). Substitution occurs during DNA replication. Because there are several different nucleotide sequences for the same amino acid, many substitutions have no effect on the resulting protein. These are called *silent* mutations. Base-pair substitutions that result in a code for a different amino acid are called *missense* mutations. Those that change the code to a stop signal are called *nonsense* mutations. The addition or loss of nucleotide pairs results in insertions or deletions. Because the code for each amino acid is three nucleotides, additions or deletions that are not in multiples of three affect all of the codes downstream from the deletion with serious, usually lethal consequences. This is called a *frameshift*. If the insertion or deletion is in multiples of three, new amino acids will be contained in the protein, which may be retained and passed to future generations.

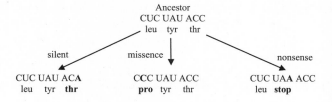

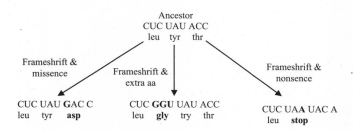

Figure 10.1.
Consequences of Point Mutations
Due to Substitutions and Insertions
(Deletions can have the same consequences as substitutions. See text for details.)

Polymorphism

In some populations, two or more specific phenotypes from the same set of alleles are maintained. That is, there is a stable frequency of two or more phenotypes instead of the predominance of one phenotype. These distinct phenotypes are called morphs, and the population that contains the morphs is called polymorphic. Polymorphism applies only to discrete traits. Because all of the alleles of the set are represented in high frequencies, they are retained in a population.

E. Brodie studied the evolutionary aspects of pronounced polymorphism in color pattern found in *Thamnophis ordinoides* (northwestern garter snake). There are two main morphs, striped and unstriped. Individuals that are striped are more likely to attempt to evade predators by flight, relying on the optical illusion created by the moving stripe to make it difficult for visually hunting predators to focus on them. Unstriped animals, by contrast, are more likely to exhibit reversals and then

use crypsis to evade a predator. Individual snakes are consistent over time in both color pattern and escape behavior, and both morphs are maintained in the population, thus maintaining genetic diversity

Heterozygous Advantage

In some cases, the heterozygote is better suited to the environment (has a higher fitness) than the homozygote. Therefore, the two alleles will be maintained, thus enhancing genetic variability. An example is sickle-cell anemia alleles, which provide an advantage against malaria.

Geographic Variation

The environment is not perfectly consistent over the distribution of a species, and thus most species exhibit some geographic variation in the gene pools. That is, there are differences in the gene pool between populations or subgroups of populations over the range of the species. The resulting differences in an observable trait may be discrete or vary along a continuum. The latter is called a *cline*. These environmental differences contribute to the genetic variation of the population.

Neutral Variation

Some traits do not affect the fitness of an individual, and the alleles are maintained in the gene pool. Their presence increases the variation.

2. Evolutionary Processes

Evolutionary processes are the mechanisms that cause a population to evolve. In order to understand what causes change, it is important to identify the characteristics of a non-evolving gene pool. This is done by the Hardy-Weinberg theorem. In a non-evolving gene pool, the gene frequencies in the gene pool must remain constant from one generation to the next. Conversely, in an evolving population, the gene frequencies in the gene pool change. Therefore this topic on evolutionary process begins with the Hardy-Weinberg theorem and then outlines how populations evolve—the causes of microevolution.

Hardy-Weinberg Theorem and Equation

For a non-evolving population to exist, five conditions must be met.

1. The population must be infinitely large.

2. There can be no natural selection within the population.

3. The can be no net mutations within the population.

4. There can be no migration or gene flow into or out of the population.

5. Mating within the population must be entirely random.

If these conditions are met, the population is said to be in Hardy-Weinberg equilibrium, and the gene pool is described using the Hardy-Weinberg theorem and quantified using the Hardy-Weinberg equation.

The simplest form of the equation is for a gene with two alleles A and a. The frequency of A = p, and the frequency of a = q. There are only two alleles, and thus the frequency for A and a equals 1. Therefore, $p + q = 1$. From this it is possible to obtain the frequencies of AA (homozygous dominant), Aa (heterozygous dominant), and aa (homozygous recessive). The frequency of AA is pp or p^2, and the frequency of aa is qq or q^2. The frequency of Aa is pq, but the heterozygote may be Aa or aA, and therefore the frequency of the heterozygote is 2 pq. The frequencies of the homozygous dominant plus the heterozygotes plus homozygous recessive must be 1 (or 100%). This gives the Hardy-Weinberg equation, $p^2 + 2pq + q^2 = 1$.

If the frequency of the genotypes in a population is known, it is possible to use the Hardy-Weinberg equation to calculate the frequency of alleles in a gene pool. Similarly, if the frequencies of the alleles are known, it is possible to calculate the frequencies of the genotypes.

How Populations Evolve— Causes of Microevolution

If the frequency of alleles or genotypes calculated using the Hardy-Weinberg equation is different than the predicted values, it suggests that the population is evolving. Therefore, the factors that contribute to microevolution are those that negate the conditions required for the Hardy-Weinberg theorem. These are genetic drift, natural selection, gene flow, mutations and nonrandom mating. In general, genetic drift and natural selection are the two dominant factors.

Genetic Drift

Genetic drift is a change in the frequency of the alleles in a population due to chance. In the same way that the chance of the observed result being close to the expected result in a coin toss decreases with decreasing sample size (number of coin tosses), the chance of an allele being represented in a population decreases with decreasing population size. Therefore, drift is most pronounced in small populations. Two main situations can shrink populations so that genetic drift may affect the gene frequencies. These are the founder effect and the bottleneck effect.

Founder Effect: If a small number of individuals of a population colonize a new area, the new population is small and there is a good chance that alleles from the source population are not fully represented. Genetic drift in the new colony is called the founder effect.

Bottleneck Effect: If the size of a population is reduced due to a natural disaster such as an earthquake, fire, or flood, there is a good chance that the frequency of the genes in the new gene pool is different than the original population. Genetic drift in the new population is called the bottleneck effect.

Natural Selection

Natural selection is a process by which forms of organisms in a population that are best adapted to the environment increase in frequency relative to the forms of less well-adapted organisms over a number of generations. In other words, the organisms that are best suited to the given environmental conditions are more likely to survive. The characteristic that allows an organism to survive and reproduce better than if it lacked the characteristic is called an adaptation. Selection is for these advantageous traits of organisms, in a given environment, so that the frequency of these alleles increases. Overall, there is a change in the frequencies of the alleles in the population. There are three ways that natural selection can act on the frequency of an allele: directional selection, stabilizing selection, and diversifying selection. These are shown in Figure 10.2 and are explained below. Another type of selection is sexual selection, which is discussed below as well.

Directional Selection: Directional selection shifts the overall makeup of the population by selecting against individuals at one end of the trait (phenotypes) and selecting for individuals at the other end. In the distribution of the frequencies, the mean value of the trait shifts, but the range remains constant (Figure 10.2A).

Stabilizing Selection: Stabilizing selection occurs when the intermediate variants are favored. This type of selection preserves the most common phenotypes. In the distribution the mean value of the traits remains constant and the range decreases (Figure 10.2B).

Diversifying Selection: Diversifying selection occurs when environemntal conditions favor the trait (phenotypes) at both ends of the range. The most common or intermediate forms are selected against. The result is a bimodal distribution of the range of the trait with the same range (Figure 10.2C).

Sexual Selection: The difference in appearance of males and females due to secondary sexual charac-

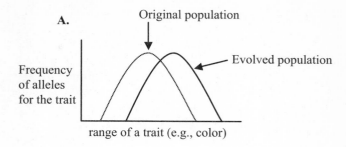

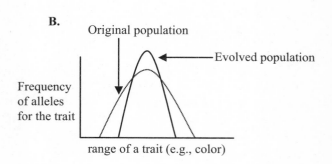

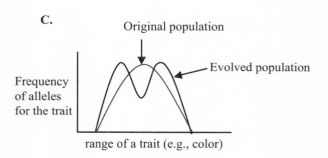

Figure 10.2.
Diagrammatic Representation
of the Three Ways in Which Selection
Can Act on the Frequency of Alleles

A = directional selection, B = stabilizing selection,
C = diversifying selection.

teristics is called sexual dimorphism. Sexual selection is selection on mating behavior, frequently associated with secondary sexual characteristics, for access to the other sex. It differs from natural selection in one important way. It acts on the fitness of a genotype relative to the other genotypes of the same sex. Natural selection works on the fitness of the genotype relative to the whole population.

Mutations

The types of mutations and their contribution to genetic diversity were outlined in the Genetic Variability. Mutations occur in about one in 10 million cell divisions. For most organisms this is a negligible contribution to microevolution. However, the cumulative effect can be important in organisms with high reproductive rates, such as bacteria. A single bacterium can produce a billion descendents in 12 hours. If the original bacterium had a mutation, the frequency of this mutation in the population would change considerably.

Gene Flow

Gene flow occurs due to migration of fertile individuals (or gametes) between populations. Migration decreases differences between populations because the alleles from the two populations become mixed. But the new mix of genes in the amalgamated population can result in microeveolution.

Nonrandom Mating

For most organisms, mating is not completely random. Individuals choose mates based on some trait, and they usually mate with individuals that are nearby. Nonrandom mating can result in inbreeding, which drastically affects the frequency of the alleles in the gene pool of the population.

3. Evolutionary Consequences

Microevolution occurs over generations within a population. Speciation is the production of new species largely due to isolation of populations. Millions of speciation events result in macroevolution, which is the dramatic changes in organisms found in the fossil record. This topic is divided into three parts. The first two are speciation and macroevolution. The third part is on phylogeny and systematics—or how the organisms are named and classified

Speciation

The definition of a species is somewhat controversial, largely due to new molecular ways of examining similarities and differences among individuals from different populations. In general, a species is a population or group of individuals that can interbreed and produce fertile offspring. An offspring produced from two different species is called a hybrid. In order to maintain a species, barriers to the production of viable and fertile hybrids must exist. These barriers act either before the formation of zygotes and fertilized eggs (prezygotic barrier) or during development of the zygote (postzygotic barriers). Speciation does occur and has occurred on millions of occasions through geological time. The three main modes of speciation are allopatric, sympatric, and parapatric. The final subject in speciation is an overview of the models that describe the rate of speciation.

Prezygotic Barriers

Five prezygotic barriers restrict mating or fertilization between different species by some type of isolation.

Habitat Isolation: If species live in different habitats, they are effectively isolated from each other and will not mate.

Behavioral Isolation: Mating for many species requires recognition of signals. If one species does not recognize the signals of another species, they will not mate.

Temporal Isolation: Species that are reproductively active at different times of year breed at different times of year and cannot interbreed.

Mechanical Isolation: Species can be anatomically incompatible. This is most prevalent in plants that have specific structures for specific pollinators.

Gametic Isolation: Some gametes can recognize appropriate gametes due to receptors or special molecules on the surface of the gamete. In plants, some flowers can distinguish between pollens from the same and different species.

Postzygotic Barriers

These are three barriers that prevent the development of a viable and fertile hybrid.

Reduced Hybrid Viability: The hybrid either is aborted or is very frail with a short life expectancy.

Reduced Hybrid Fertility: Two species mate and produce a viable offspring, but this hybrid is infertile. The most common example is the mule from a female horse and male donkey. When the female is a donkey and the male a horse, the offspring is called a hinny. Mules are more common than hinnies.

Hybrid Breakdown: The hybrid may survive and reproduce, but the second generation does not survive.

Allopatric Speciation

Allopatric speciation is the result of interrupted gene flow due to geographic separation of parts of a population. There is no gene flow between the two parts, but microevolution continues in each part. The separation may be in one area over geological time such as the formation of a mountain, or it may be due to dispersal to a new location. The result is not always a new species. In some cases, the two new populations can interbreed and are still a single species. However, the environmental conditions were different for the separated populations, and natural selection would have selected for different alleles in the two parts. Therefore, when the separated populations are reunited, the new population would have a greater genetic diversity than the original population. If the two new populations cannot interbreed, they are different species. A specific case of allopatric speciation is adaptive radiation.

Adaptive Radiation: Adaptive radiation is the evolution of several species from a common ancestor. It re-

quires dispersal from an original population to several new locations with different habitats (environments). The example most frequently used is the evolution of finches on the Galápagos Islands. The finches that initially dispersed to the islands were separated from the parent population. Because the environment of the island was different than the mainland, continued separation from the mainland population and microevolution resulted in a new species. However, the new species moved to other islands that had still different environments, and allopatric speciation occurred on these islands. The result was that several species evolved from the mainland ancestor.

Sympatric Speciation

In sympatric speciation, new species evolve within the habitat of the parent population. A reproductive barrier occurs within the population. This can happen due to polyploidy and ecological separation.

Polyploidy: Some chromosomal mutations change the number of chromosomes. One consequence is a condition called polyploidy in which the offspring can have more than two sets of chromosomes. If this organism can mate with itself (e.g., self-pollination in plants or another organism) and produce viable offspring, these offspring will have two times as many chromosomes as those in the original population. The polyploidy individuals cannot mate with the individuals in the original population because there would be an odd number of chromosomes and meiosis could not occur. Therefore, the polyploids are reproductively isolated from the original population.

There are two types of polyploidy. An autopolyploid is an individual that derives all of its chromosomes from the same species. An allopolyploid derives it chromosomes from different species. Allopolyploidy usually requires asexual reproduction because the hybrids are infertile, but additional chromosome alterations can produce fertile individuals.

Polyploidy is common in plants but rare in animals. In animals, an extra X or Y chromosome usually results in abnormalities and infertility. However, polyploidy does occur in some amphibians and reptiles and in at least one mammal.

Ecological Separation: Ecological separation occurs when a subset of the population is isolated within the original population. For instance, the subpopulation may exploit different resources and thus be physically separated from the original population but still live within the range of the original population. This is well studied in the fish of Lake Victoria in East Africa.

Parapatric Speciation

In parapatric speciation, the new species evolve from contiguous populations. This differs from allopatric speciation in which the new species are completely separate from the original population. It occurs in hybrid zones, which are contact areas between two species. However, because there is hybridization, either speciation is incomplete or allopatric species are coming together. It gives insight into how speciation may occur.

The three forms of speciation are not necessarily mutually exclusive and discrete during the evolution of species. For example, two or more populations could begin developing different characteristics while contiguous (parapatric) and then become isolated within the same area (sympatric) before separating into distinct habitats (allopatric). Conversely, they could begin speciation while separated and then merge into the same area while continuing their speciation.

Macroevolution

The fossil record shows large changes such as the origin of new organs and body plans, or of new taxa above the species level. These are macroevolutionary changes. The rate of these macroevolutionary changes and some proposed macroevolutionary events are discussed below.

Rates of Evolutionary Change

There are two models used to describe the rates of macroevolution observed in the fossil record: gradual-

ism and punctuated equilibrium. Although presented separately, it is more probable that these two models represent the extremes of a continuum.

Gradualism: This model suggests that new species arose slowly and evenly due to the transformations of an ancestral population into its descendants. The transformations involve the whole population, and the change occurs over most of the ancestral species' range. This model is not well supported by the fossil record where new and different species occur suddenly. However, this may be explained by allopatric speciation. That is, if the new species evolved in an area isolated from the original population and then invaded the area of the original population, a sudden new species would be recorded in the fossil record. But the greatest criticism of this model is the dearth of transitional forms in the fossil record.

Punctuated Equilibrium: This model proposes that new forms arose rapidly but the changes were separated by long periods with minimal change. In addition only a small subpopulation of the original population gave rise to the new species in an area separated, from the ancestral (original) species, due to a bottleneck effect, founder effect, directional selection or diversifying selection. (Remember that rapid change in geological time is usually thousands of years.)

Macroevolutionary Events

There are four main ways in which macroevolutionary events can occur: changes in the use of existing structures (exaptation), changes in rate of growth of different body parts (allometry), change in the rate or timing of development (heterochrony), and changes in genes that determine spatial organization (homeotic genes).

Exaptation: An exaptation has also been called a pre-adaptation. It is a characteristic that performs a function different from the reason it originally evolved. The structure or behavior possesses the necessary form before the new biological role arises. For example, feathers initially evolved in birds as insulation. Feathers are still important in conserving body heat, but flight is a conspicuous use of feathers.

Allometry: Allometry is the change in size or shape of one part of an organism relative to the change in size or shape of another. It is compared to *isometry* in which the two parts grow at the same rate. One way in which differences in the relative growth of different parts is used is to compare the appearance of new derived structures using transformation or Cartesian grids. These grids are horizontal and vertical lines. The coordinates of different parts of a structure (e.g., a chimpanzee skull) are put on a grid. The same coordinates are located on the second structure, and the grid is redrawn (stretched and skewed as necessary) to show how the shape has changed in the second structure.

Heterochrony: Heterochrony is a type of allometry. Heterochrony is a change in the rate or timing of a developmental event. There are two major evolutionary outcomes: paedomorphosis and peramorphosis. Paedomorphosis is the retention of embryonic or larval characteristics by adults. Both the adult and descendant are sexually mature. Peramorphosis is the development of entirely new adult shapes, beyond the typical adult form. The result is an ancestor with the adult body shape and a descendant that is significantly larger than the ancestor. Both paedomorphosis and peramorphosis occur due to changes in the rate of development, the time of onset of development, and the time of offset of development. Examples of how a change in the rate of development can change the shape of the descendant is given in Figure 10.3.

Homeotic Genes: Homeotic genes control the placement and spatial organization of body parts. For example, they dictate the position of wings or limbs in animals and the arrangement of the parts of flowers. One class of homeotic genes are the *Hox* genes that control the expression of genes that determine the position where structures develop. The *Hox* genes are considered to be important in the evolution of fins and limbs in vertebrates.

Phylogeny and Systematics

Systematics and phylogeny are interrelated. Systematics is the study and classification of biological diversity in an evolutionary context (taxonomy and no-

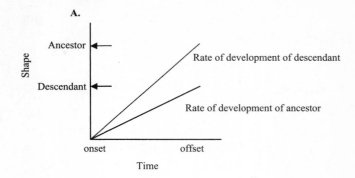

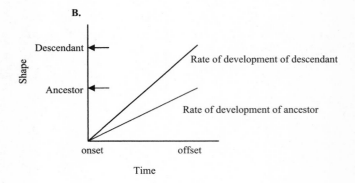

Figure 10.3.
Paedomorphosis and Peramorphosis
Due to a Change in the Rate of Development

A. Paedomorphosis due to a decrease in the rate of growth. The slope of the line (rate of development) decreases from ancestor to descendant. The development of the shape is truncated. B. Peramorphosis due to a increase in the rate of growth. The slope of the line (rate of development) increases from ancestor to descendant. The development of the shape is extended.

menclature), and phylogeny is a hypothesis about the evolutionary relations among members of a taxonomic group. Systematics is used to create a phylogeny.

Taxonomy and Nomenclature

The Linnaean system is used to name and classify organisms. It has two important characteristics: each organism has a two-part name, genus and species; and there is a hierarchy of taxonomic groups. In naming an organism, the genus name is followed by the species name, both are italicized, and the genus name is capitalized. The hierarchical groups are species, genus, family, order, class, phylum, kingdom and domain. Ascending the hierarchy results in broader groups.

Phylogeny

A phylogeny, the evolutionary history of a group of organisms, is represented in a phylogenetic tree. The tree may be constructed from bottom to top or left to right (Figure 10.4). The lines may be either horizontal or vertical or at an angle. Many phylogenetic trees are built using cladistics and are called cladograms. They have certain consistent characteristics, and their construction is based on specific principles.

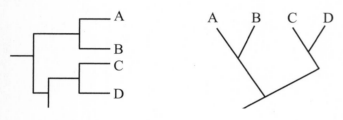

Figure 10.4.
Two Different Ways of Showing
the Same Phylogenetic Tree

Characteristics of Phylogenetic Trees. A phylogeny based on cladistics has a series of dichotomous branches that diverge from nodes. A node represents a hypothetical common ancestor. The oldest divergence is the root, and the terminal branches are the most recently evolved taxa. A taxon is a named taxonomic group such as a species, genus, or family. A taxon that includes the common ancestor and all of the descendants is called a monophyletic group or clade. If some of the descendants are missing, it is called a paraphyletic group. In a phylogenetic tree, it is the branching pattern, not the length of the lines that is important because the specific times of divergence are not usually known. However, the branching pattern shows the relative times of divergence.

Constructing a Phylogenetic Tree. A cladogram is constructed using homologous and shared derived characteristics based on information from paleontology, anatomy, molecular biology, behavior, and ecology.

Homologies are similarities due to a common ancestry. Homologous characteristics may be similar in appearance and obvious such as the bones of the forelimbs of a horse and seal, or they may be dissimilar and less obvious such as portions of the gill arches of fish and ear ossicles of mammals. However, structures that look similar and have the same function may not be homologous. These structures, such as the mouth parts of true bugs (order Hemiptera) and biting mosquitoes (order Diptera), which are both modified for biting and sucking, are not homologous. They have similar ecological roles, and natural selection has produced similar structures with similar functions. This is referred to as convergent evolution. The characteristics are analogous. The prevalence of analogous structures is why numerous types of information are used to identify homologous characteristics.

Shared derived characteristics are those that are unique to monophyletic groups. They must be distinguished from ancestral derived characteristics. A characteristic is both ancestral and derived depending on the taxonomic level. For example, the vertebral column is a derived characteristic of vertebrates, but the monophyletic group of vertebrates contains smaller monophyletic groups such as mammals. The derived characteristics of mammals include mammary glands and hair. For mammals, the vertebral column is an ancestral characteristic.

Out-groups are used to determine whether a characteristic is ancestral or derived for a particular monophyletic group. An out-group is closely related to the group of species being studied but less closely related than the study-group members (in-group).

4. History of Life

The history of life is contained in fossils and shown in the fossil record. Paleontology is the study of fossils and the fossil record and includes paleoecology, which is the study of how fossil organisms may have lived. Life is believed to have begun with prokaryotic cells about 3.8 billion years ago and has evolved to produce the diversity of organisms known today. The subjects of this topic are paleontology, the origin of prokaryotes and eukaryotes, and the evolution of eukaryotic cells.

Paleontology

Fossils

Fossils are parts, replicas, or imprints of organisms preserved in rocks and other materials. There are several kinds of fossils.

- Hard parts of organisms such as bones, teeth, and shells. These are the most common fossils.

- Petrified fossils in which the organic material is replaced by minerals. An example is large petrified trees.

- Remains of thin structures that are pressed between layers of rocks and may contain organic matter. Different parts of plants may be preserved in this way.

- Molds or casts of organisms that result when the organism dies and becomes impregnated with minerals. This differs from petrified organisms because there is only a cast of the original organism.

- Trace fossils such as footprints and burrows.

- Organisms preserved in resins.

- Organisms preserved in ice.

- Dung and stomach contents called coprolites.

The Fossil Record and Geological Time

The age of fossils is determined from the rocks in which they are found. Because most fossil are found in sedimentary rocks, the relative position of the fossils in the strata gives the relative age. Not all sedimentary rock strata are complete, and thus strata from different areas are used to get the complete record. Certain organisms were abundant and were present for a known period of geological time. These are called index fossils, a well-known one being trilobites. Using the relative positions of the strata, index fossils, and certain geological principles to attain the relative ages of fossils is called stratigraphic layering. It is also possible to add information on the absolute ages of rock using radiometric dating. This is based on known decay rates of radioactive elements.

The time series of fossils is called the fossil record and is given over geological time. Geological time is divided into eras, periods, and epochs. There are four eras: Precambrian, Paleozoic, Mesozoic and Cenozoic. Life began in the Precambrian about 3.8 billion years ago. Representatives of most of the living (extant) groups of organisms appeared early in the Paleozoic. This occurred in the Cambrian period and is called the Cambrian explosion. Three interacting conditions help explain this explosion: the evolution of *Hox* genes, which increased variation during embryonic development, an atmospheric oxygen level well suited for aerobic respiration, and evolution of prey/predator interactions resulting in new adaptations by prey and predators.

During the Paleozoic era insects, fish, amphibians, reptiles, and vascular plants appeared. During the Mesozoic era, gymnosperms (e.g., conifers), angiosperms (flowering plants), and dinosaurs dominated. During the Cenozoic era mammals, birds, and plants diversified. Humans appeared in the Pleistocene of the Cenozoic.

The fossil record shows the explosion and diversification of life, but there are also mass extinctions. Mass extinctions delineate the end of the Paleozoic (Permian extinction about 250 million years ago) and the end of the Mesozoic (Cretaceous extinction about 65 million years ago). The latter resulted in the extinction of the dinosaurs and many gymnosperms. In the Permian extinction, marine and terrestrial life crashed. This extinction spanned about 5 million years, and during this time, there was extensive tectonic plate movement with two important consequences: increased volcanic activity and destruction of marine habitats. The volcanic activity destroyed terrestrial and marine habitats and added carbon dioxide to the atmosphere. The increase in carbon dioxide contributed to global warming and further loss of habitats.

Origin of Prokaryotic and Eukaryotic Cells

The oldest known fossils are prokaryotes found in stromatolites and in sediments that were deposited near hydrothermal vents (at diverging tectonic plates) about 3.2 billion years ago. The explanation of the origin of life is therefore only a logical or testable sequence of events that could produce prokaryotes. Eukaryotes are believed to have evolved from prokaryotes.

The Origin of Life and Prokaryotes

Life is believed to have developed from nonliving material that became ordered into molecules that could replicate. One hypothesis suggests that there were four important steps.

1. The synthesis of small organic molecules such as amino acids and nucleotides.

2. The combination of simple single molecules into small chains or polymers (proteins and nucleic acids).

3. The origin of self-replicating molecules.

4. The production of protobionts containing the self-replicating molecules and surrounded by a membrane. These are the first prokaryotic cells.

There is some evidence for each of these steps. Of particular importance is the autocatalytic property of RNA. This means that the RNA could have acted as a catalyst and protein enzymes were not required. The RNA catalyst is called a ribozyme.

Evolution of Eukaryotic Cells

Eukaryotic cells have a membrane-bound nucleus and membrane-enclosed organelles (endoplasmic recticulum, Golgi apparatus, etc.). Three further events are suggested for the evolution of eukaryotic cells from the first prokaryotic cells.

1. The evolution of multicellular prokaryotes with different and specialized types of cells (e.g., cyanobacteria).

2. The evolution of complex communities so that there were with different and specialized kinds of prokaryotes. Each member benefited because of the metabolic specialties of the other members.

3. The evolution of membrane-bound compartments within single cells. This required two processes.

   (a) The first was to produce the endomembrane system. This may have occurred from infoldings of the plasma membrane of a prokaryote.

   (b) The second process is called endosymbiosis. Endosymbionts are cells that live within other cells. This is the process used to explain the presence of mitochondria and chloroplasts within eukaryotic cells. It assumes that mitochondria and chloroplasts were small prokaryotes living within larger cells. The ancestors of the mitochondria were aerobic heterotrophic prokaryotes that became endosymbionts. The proposed ancestors of the chloroplasts were photosynthetic prokaryotes that became endosymbionts. These prokaryotes were ingested, and the result was a mutually beneficial association that eventually became a single organism—a eukaryote.

Other Important Eukaryotic Organisms

Fungi

All fungi are eukaryotic organisms, having at least one nucleus with a nuclear membrane, an endoplasmic reticulum, and mitochondria. They lack chloroplasts and chlorophyll. Fungi are spore-bearing organisms with absorptive nutrition. They reproduce sexually and asexually. The primitive plant body formed by fungi is called a thallus, but it has no true roots, leaves, stems, or vascular tissue. Although there are over 100,000 species of fungi, only about 100 are important in human diseases. Ringworm (dermatophytoses) is, however, a very common infectious disease caused by a fungus. Mushrooms, yeasts, and molds are examples of fungi.

Fungi Life History. Yeasts are unicellular forms of fungi with a spherical shape. Yeast reproduce by budding or by binary fission. Molds grow in multicellular tubular colonies called hyphae. During growth, these hyphae bunch together to form a mycelium.

Fungi are also capable of sexual reproduction. In all sexual reproduction there is an alteration in chromosome number. At fertilization, two haploid nuclei join to form a diploid nucleus. The diploid cells eventually give rise to haploid cells by meiosis. In lower fungi, the visible organism often exists primarily in the haploid state (haplophase) and only transiently in the diploid state (diplophase). Sexual reproduction in fungi follows this sequence:

1. Compatible haploid nuclei are brought together in the same cell of the thallus.

2. Two genetically different nuclei fuse to form a diploid nucleus.

3. Meiosis occurs to form haploid nuclei, which develop into sexual spores.

Yeast Genetics

Nuclear Genes. Yeasts have become more than simple biological tools for the brewer and baker. Because yeasts are single-celled eukaryotes and contain mitochondria DNA (an organelle) like multicellular eukaryotes, they have been used extensively for recombinant DNA research, often acting as a conduit between prokaryotes and eukaryotes.

Saccharomyces cerevisiae (referred to by some as baker's yeast, but also long used for making beer and wine) is one of the often used organisms and is somewhat a standard for yeast genetic terminology. Alleles at a nuclear chromosome gene locus are defined by three italicized letters: dominant alleles are capitalized, recessives are lowercased (e.g., for a leucine gene, a mutant allele might be designated *LEU*, while the wildtype [normal expression] would carry a "+" as superscript, *LEU*$^+$). If two loci exist, the locus is followed by a number (i.e., *LEU2*).

Mitochondrial Genes. Mitochondria gene function is often tested using nonfermentable carbon substrate sources (Nfs). Inability to grow on Nfs may be designated Nfs-. If an Nfs- is found due to joint action, or lack of, by nuclear and mitochondrial genes, it may also have *pet* added to its designation, but if more specific substrate criteria are known, these would be used.

Yeast Chromosomal Genes

Transformation of Yeast Cells. At the time of preparing this document, these are the primary methods for moving genes from/to yeast cells:

1. spheroplasts (DNA contained within cells with some portion of the cell walls removed)

2. lithium salts (using single-strand carrier DNA)

3. electroporation (electrical pulse causes micropores to temporarily open in the plasma membrane)

4. for mitochondrial gene transformation, high-velocity microprojectiles containing DNA fragments can be used because methods 1–3 above are ineffective

Keep in mind that all transformation methods have as their primary goal the maximum number of cells receiving and incorporating the nucleic acid molecule that is being moved about between organisms.

One of the more interesting, novel methods for moving genes about (via a vector), especially between eukaryotes and prokaryotes, are yeast artificial chromosomes (YAC). These are fabricated and contain a centromere, two arms, and the necessary origin of replication. YAC have a distinct advantage also in that larger pieces of DNA can be cloned as compared to most prokaryotic plasmids.

For more in-depth information about yeast biology and genetics, go to this website: *http://dbb.urmc. rochester.edu/labs/sherman_f/yeast/6.html*.

Yeasts, along with *E. coli,* are powerful tools nowadays and they "collaborate" in many genetic ways via *shuttle vectors*. These are small plasmids that contain both origins of replication for yeast genes and origins of replication for bacterial genes. They are used for moving DNA fragments back and forth between yeast (transforming via spheroplasts) and *E. coli* in order to identify/isolate mutants with specific expression and increase the quantity of those genes of interest via cloning in the bacterium.

Modern genetics owes a lot to yeast organisms, the lowly unicellular eukaryotes. Recently, the entire DNA sequence of *Saccharomyces cerevisiae* has been published. There is evidence that the organism's nuclear genome contains 5,885 genes. Most interesting was the discovery that of those nearly 6,000 genes, approximately a third have almost identical homologous sequences in other eukaryotes, including humans (e.g., about 70 human diseases share gene sequences with yeast). These findings are only a part of what drives modern molecular genetics/biology, and of course, commercialization of recombinant technology for all sorts of gene-based products (i.e., pharmaceuticals). There are now artificial yeast "chromosomes" used as gene vectors (molecules for transferring genes).

GRE BIOLOGY

Practice Test 1

This test is also on CD-ROM in our special interactive GRE Biology TEST*ware*®. It is highly recommended that you first take this exam on computer. You will then have the additional study features and benefits of enforced timed conditions and instant, accurate scoring. See page xi for guidance on how to get the most out of our GRE Biology software.

Answer Sheet

1. Ⓐ Ⓑ Ⓒ Ⓓ Ⓔ
2. Ⓐ Ⓑ Ⓒ Ⓓ Ⓔ
3. Ⓐ Ⓑ Ⓒ Ⓓ Ⓔ
4. Ⓐ Ⓑ Ⓒ Ⓓ Ⓔ
5. Ⓐ Ⓑ Ⓒ Ⓓ Ⓔ
6. Ⓐ Ⓑ Ⓒ Ⓓ Ⓔ
7. Ⓐ Ⓑ Ⓒ Ⓓ Ⓔ
8. Ⓐ Ⓑ Ⓒ Ⓓ Ⓔ
9. Ⓐ Ⓑ Ⓒ Ⓓ Ⓔ
10. Ⓐ Ⓑ Ⓒ Ⓓ Ⓔ
11. Ⓐ Ⓑ Ⓒ Ⓓ Ⓔ
12. Ⓐ Ⓑ Ⓒ Ⓓ Ⓔ
13. Ⓐ Ⓑ Ⓒ Ⓓ Ⓔ
14. Ⓐ Ⓑ Ⓒ Ⓓ Ⓔ
15. Ⓐ Ⓑ Ⓒ Ⓓ Ⓔ
16. Ⓐ Ⓑ Ⓒ Ⓓ Ⓔ
17. Ⓐ Ⓑ Ⓒ Ⓓ Ⓔ
18. Ⓐ Ⓑ Ⓒ Ⓓ Ⓔ
19. Ⓐ Ⓑ Ⓒ Ⓓ Ⓔ
20. Ⓐ Ⓑ Ⓒ Ⓓ Ⓔ
21. Ⓐ Ⓑ Ⓒ Ⓓ Ⓔ
22. Ⓐ Ⓑ Ⓒ Ⓓ Ⓔ
23. Ⓐ Ⓑ Ⓒ Ⓓ Ⓔ
24. Ⓐ Ⓑ Ⓒ Ⓓ Ⓔ
25. Ⓐ Ⓑ Ⓒ Ⓓ Ⓔ
26. Ⓐ Ⓑ Ⓒ Ⓓ Ⓔ
27. Ⓐ Ⓑ Ⓒ Ⓓ Ⓔ
28. Ⓐ Ⓑ Ⓒ Ⓓ Ⓔ

29. Ⓐ Ⓑ Ⓒ Ⓓ Ⓔ
30. Ⓐ Ⓑ Ⓒ Ⓓ Ⓔ
31. Ⓐ Ⓑ Ⓒ Ⓓ Ⓔ
32. Ⓐ Ⓑ Ⓒ Ⓓ Ⓔ
33. Ⓐ Ⓑ Ⓒ Ⓓ Ⓔ
34. Ⓐ Ⓑ Ⓒ Ⓓ Ⓔ
35. Ⓐ Ⓑ Ⓒ Ⓓ Ⓔ
36. Ⓐ Ⓑ Ⓒ Ⓓ Ⓔ
37. Ⓐ Ⓑ Ⓒ Ⓓ Ⓔ
38. Ⓐ Ⓑ Ⓒ Ⓓ Ⓔ
39. Ⓐ Ⓑ Ⓒ Ⓓ Ⓔ
40. Ⓐ Ⓑ Ⓒ Ⓓ Ⓔ
41. Ⓐ Ⓑ Ⓒ Ⓓ Ⓔ
42. Ⓐ Ⓑ Ⓒ Ⓓ Ⓔ
43. Ⓐ Ⓑ Ⓒ Ⓓ Ⓔ
44. Ⓐ Ⓑ Ⓒ Ⓓ Ⓔ
45. Ⓐ Ⓑ Ⓒ Ⓓ Ⓔ
46. Ⓐ Ⓑ Ⓒ Ⓓ Ⓔ
47. Ⓐ Ⓑ Ⓒ Ⓓ Ⓔ
48. Ⓐ Ⓑ Ⓒ Ⓓ Ⓔ
49. Ⓐ Ⓑ Ⓒ Ⓓ Ⓔ
50. Ⓐ Ⓑ Ⓒ Ⓓ Ⓔ
51. Ⓐ Ⓑ Ⓒ Ⓓ Ⓔ
52. Ⓐ Ⓑ Ⓒ Ⓓ Ⓔ
53. Ⓐ Ⓑ Ⓒ Ⓓ Ⓔ
54. Ⓐ Ⓑ Ⓒ Ⓓ Ⓔ
55. Ⓐ Ⓑ Ⓒ Ⓓ Ⓔ
56. Ⓐ Ⓑ Ⓒ Ⓓ Ⓔ

57. Ⓐ Ⓑ Ⓒ Ⓓ Ⓔ
58. Ⓐ Ⓑ Ⓒ Ⓓ Ⓔ
59. Ⓐ Ⓑ Ⓒ Ⓓ Ⓔ
60. Ⓐ Ⓑ Ⓒ Ⓓ Ⓔ
61. Ⓐ Ⓑ Ⓒ Ⓓ Ⓔ
62. Ⓐ Ⓑ Ⓒ Ⓓ Ⓔ
63. Ⓐ Ⓑ Ⓒ Ⓓ Ⓔ
64. Ⓐ Ⓑ Ⓒ Ⓓ Ⓔ
65. Ⓐ Ⓑ Ⓒ Ⓓ Ⓔ
66. Ⓐ Ⓑ Ⓒ Ⓓ Ⓔ
67. Ⓐ Ⓑ Ⓒ Ⓓ Ⓔ
68. Ⓐ Ⓑ Ⓒ Ⓓ Ⓔ
69. Ⓐ Ⓑ Ⓒ Ⓓ Ⓔ
70. Ⓐ Ⓑ Ⓒ Ⓓ Ⓔ
71. Ⓐ Ⓑ Ⓒ Ⓓ Ⓔ
72. Ⓐ Ⓑ Ⓒ Ⓓ Ⓔ
73. Ⓐ Ⓑ Ⓒ Ⓓ Ⓔ
74. Ⓐ Ⓑ Ⓒ Ⓓ Ⓔ
75. Ⓐ Ⓑ Ⓒ Ⓓ Ⓔ
76. Ⓐ Ⓑ Ⓒ Ⓓ Ⓔ
77. Ⓐ Ⓑ Ⓒ Ⓓ Ⓔ
78. Ⓐ Ⓑ Ⓒ Ⓓ Ⓔ
79. Ⓐ Ⓑ Ⓒ Ⓓ Ⓔ
80. Ⓐ Ⓑ Ⓒ Ⓓ Ⓔ
81. Ⓐ Ⓑ Ⓒ Ⓓ Ⓔ
82. Ⓐ Ⓑ Ⓒ Ⓓ Ⓔ
83. Ⓐ Ⓑ Ⓒ Ⓓ Ⓔ
84. Ⓐ Ⓑ Ⓒ Ⓓ Ⓔ

85. Ⓐ Ⓑ Ⓒ Ⓓ Ⓔ
86. Ⓐ Ⓑ Ⓒ Ⓓ Ⓔ
87. Ⓐ Ⓑ Ⓒ Ⓓ Ⓔ
88. Ⓐ Ⓑ Ⓒ Ⓓ Ⓔ
89. Ⓐ Ⓑ Ⓒ Ⓓ Ⓔ
90. Ⓐ Ⓑ Ⓒ Ⓓ Ⓔ
91. Ⓐ Ⓑ Ⓒ Ⓓ Ⓔ
92. Ⓐ Ⓑ Ⓒ Ⓓ Ⓔ
93. Ⓐ Ⓑ Ⓒ Ⓓ Ⓔ
94. Ⓐ Ⓑ Ⓒ Ⓓ Ⓔ
95. Ⓐ Ⓑ Ⓒ Ⓓ Ⓔ
96. Ⓐ Ⓑ Ⓒ Ⓓ Ⓔ
97. Ⓐ Ⓑ Ⓒ Ⓓ Ⓔ
98. Ⓐ Ⓑ Ⓒ Ⓓ Ⓔ
99. Ⓐ Ⓑ Ⓒ Ⓓ Ⓔ
100. Ⓐ Ⓑ Ⓒ Ⓓ Ⓔ
101. Ⓐ Ⓑ Ⓒ Ⓓ Ⓔ
102. Ⓐ Ⓑ Ⓒ Ⓓ Ⓔ
103. Ⓐ Ⓑ Ⓒ Ⓓ Ⓔ
104. Ⓐ Ⓑ Ⓒ Ⓓ Ⓔ
105. Ⓐ Ⓑ Ⓒ Ⓓ Ⓔ
106. Ⓐ Ⓑ Ⓒ Ⓓ Ⓔ
107. Ⓐ Ⓑ Ⓒ Ⓓ Ⓔ
108. Ⓐ Ⓑ Ⓒ Ⓓ Ⓔ
109. Ⓐ Ⓑ Ⓒ Ⓓ Ⓔ
110. Ⓐ Ⓑ Ⓒ Ⓓ Ⓔ
111. Ⓐ Ⓑ Ⓒ Ⓓ Ⓔ
112. Ⓐ Ⓑ Ⓒ Ⓓ Ⓔ

Continued

Answer Sheet

Continued

113. Ⓐ Ⓑ Ⓒ Ⓓ Ⓔ   135. Ⓐ Ⓑ Ⓒ Ⓓ Ⓔ   157. Ⓐ Ⓑ Ⓒ Ⓓ Ⓔ   179. Ⓐ Ⓑ Ⓒ Ⓓ Ⓔ
114. Ⓐ Ⓑ Ⓒ Ⓓ Ⓔ   136. Ⓐ Ⓑ Ⓒ Ⓓ Ⓔ   158. Ⓐ Ⓑ Ⓒ Ⓓ Ⓔ   180. Ⓐ Ⓑ Ⓒ Ⓓ Ⓔ
115. Ⓐ Ⓑ Ⓒ Ⓓ Ⓔ   137. Ⓐ Ⓑ Ⓒ Ⓓ Ⓔ   159. Ⓐ Ⓑ Ⓒ Ⓓ Ⓔ   181. Ⓐ Ⓑ Ⓒ Ⓓ Ⓔ
116. Ⓐ Ⓑ Ⓒ Ⓓ Ⓔ   138. Ⓐ Ⓑ Ⓒ Ⓓ Ⓔ   160. Ⓐ Ⓑ Ⓒ Ⓓ Ⓔ   182. Ⓐ Ⓑ Ⓒ Ⓓ Ⓔ
117. Ⓐ Ⓑ Ⓒ Ⓓ Ⓔ   139. Ⓐ Ⓑ Ⓒ Ⓓ Ⓔ   161. Ⓐ Ⓑ Ⓒ Ⓓ Ⓔ   183. Ⓐ Ⓑ Ⓒ Ⓓ Ⓔ
118. Ⓐ Ⓑ Ⓒ Ⓓ Ⓔ   140. Ⓐ Ⓑ Ⓒ Ⓓ Ⓔ   162. Ⓐ Ⓑ Ⓒ Ⓓ Ⓔ   184. Ⓐ Ⓑ Ⓒ Ⓓ Ⓔ
119. Ⓐ Ⓑ Ⓒ Ⓓ Ⓔ   141. Ⓐ Ⓑ Ⓒ Ⓓ Ⓔ   163. Ⓐ Ⓑ Ⓒ Ⓓ Ⓔ   185. Ⓐ Ⓑ Ⓒ Ⓓ Ⓔ
120. Ⓐ Ⓑ Ⓒ Ⓓ Ⓔ   142. Ⓐ Ⓑ Ⓒ Ⓓ Ⓔ   164. Ⓐ Ⓑ Ⓒ Ⓓ Ⓔ   186. Ⓐ Ⓑ Ⓒ Ⓓ Ⓔ
121. Ⓐ Ⓑ Ⓒ Ⓓ Ⓔ   143. Ⓐ Ⓑ Ⓒ Ⓓ Ⓔ   165. Ⓐ Ⓑ Ⓒ Ⓓ Ⓔ   187. Ⓐ Ⓑ Ⓒ Ⓓ Ⓔ
122. Ⓐ Ⓑ Ⓒ Ⓓ Ⓔ   144. Ⓐ Ⓑ Ⓒ Ⓓ Ⓔ   166. Ⓐ Ⓑ Ⓒ Ⓓ Ⓔ   188. Ⓐ Ⓑ Ⓒ Ⓓ Ⓔ
123. Ⓐ Ⓑ Ⓒ Ⓓ Ⓔ   145. Ⓐ Ⓑ Ⓒ Ⓓ Ⓔ   167. Ⓐ Ⓑ Ⓒ Ⓓ Ⓔ   189. Ⓐ Ⓑ Ⓒ Ⓓ Ⓔ
124. Ⓐ Ⓑ Ⓒ Ⓓ Ⓔ   146. Ⓐ Ⓑ Ⓒ Ⓓ Ⓔ   168. Ⓐ Ⓑ Ⓒ Ⓓ Ⓔ   190. Ⓐ Ⓑ Ⓒ Ⓓ Ⓔ
125. Ⓐ Ⓑ Ⓒ Ⓓ Ⓔ   147. Ⓐ Ⓑ Ⓒ Ⓓ Ⓔ   169. Ⓐ Ⓑ Ⓒ Ⓓ Ⓔ   191. Ⓐ Ⓑ Ⓒ Ⓓ Ⓔ
126. Ⓐ Ⓑ Ⓒ Ⓓ Ⓔ   148. Ⓐ Ⓑ Ⓒ Ⓓ Ⓔ   170. Ⓐ Ⓑ Ⓒ Ⓓ Ⓔ   192. Ⓐ Ⓑ Ⓒ Ⓓ Ⓔ
127. Ⓐ Ⓑ Ⓒ Ⓓ Ⓔ   149. Ⓐ Ⓑ Ⓒ Ⓓ Ⓔ   171. Ⓐ Ⓑ Ⓒ Ⓓ Ⓔ   193. Ⓐ Ⓑ Ⓒ Ⓓ Ⓔ
128. Ⓐ Ⓑ Ⓒ Ⓓ Ⓔ   150. Ⓐ Ⓑ Ⓒ Ⓓ Ⓔ   172. Ⓐ Ⓑ Ⓒ Ⓓ Ⓔ   194. Ⓐ Ⓑ Ⓒ Ⓓ Ⓔ
129. Ⓐ Ⓑ Ⓒ Ⓓ Ⓔ   151. Ⓐ Ⓑ Ⓒ Ⓓ Ⓔ   173. Ⓐ Ⓑ Ⓒ Ⓓ Ⓔ   195. Ⓐ Ⓑ Ⓒ Ⓓ Ⓔ
130. Ⓐ Ⓑ Ⓒ Ⓓ Ⓔ   152. Ⓐ Ⓑ Ⓒ Ⓓ Ⓔ   174. Ⓐ Ⓑ Ⓒ Ⓓ Ⓔ   196. Ⓐ Ⓑ Ⓒ Ⓓ Ⓔ
131. Ⓐ Ⓑ Ⓒ Ⓓ Ⓔ   153. Ⓐ Ⓑ Ⓒ Ⓓ Ⓔ   175. Ⓐ Ⓑ Ⓒ Ⓓ Ⓔ   197. Ⓐ Ⓑ Ⓒ Ⓓ Ⓔ
132. Ⓐ Ⓑ Ⓒ Ⓓ Ⓔ   154. Ⓐ Ⓑ Ⓒ Ⓓ Ⓔ   176. Ⓐ Ⓑ Ⓒ Ⓓ Ⓔ   198. Ⓐ Ⓑ Ⓒ Ⓓ Ⓔ
133. Ⓐ Ⓑ Ⓒ Ⓓ Ⓔ   155. Ⓐ Ⓑ Ⓒ Ⓓ Ⓔ   177. Ⓐ Ⓑ Ⓒ Ⓓ Ⓔ   199. Ⓐ Ⓑ Ⓒ Ⓓ Ⓔ
134. Ⓐ Ⓑ Ⓒ Ⓓ Ⓔ   156. Ⓐ Ⓑ Ⓒ Ⓓ Ⓔ   178. Ⓐ Ⓑ Ⓒ Ⓓ Ⓔ   200. Ⓐ Ⓑ Ⓒ Ⓓ Ⓔ

Practice Test 1

DIRECTIONS: Choose the best answer for each question and mark the letter of your selection on the corresponding answer sheet

1. The posterior lobe of the pituitary gland in humans releases

   (A) TSH and FSH
   (B) ACTH and LH
   (C) oxytocin and vasopression
   (D) FSH and LH
   (E) prolactin and growth hormone

2. Microevolution is

   (A) a change in the genetic structure of a population
   (B) the discovery of small fossils
   (C) the genetic change that occurs in microorganisms such as bacteria
   (D) natural selection on islands
   (E) the survival of the smallest individual

3. Penicillin inhibits bacterial proliferation by

   (A) blocking ribosomal function
   (B) blocking the glycolytic pathway
   (C) stopping the electron transport chain
   (D) blocking cell wall synthesis
   (E) interrupting the active sites of vital enzymes

4. Electron transport and oxidative phosphorylation are thought to be coupled by a chemiosmotic mechanism. This hypothesis involves:

   (A) the use of energy derived from electron transport to maintain an equilibrium of hydrogen and hydroxyl ions on either side of a membrane
   (B) the use of energy derived from electron transport to form both an electrochemical and pH gradient across a membrane by inducing proton translocation
   (C) a transient wave of depolarization, opening voltage-gated ion channels in the plasma membrane
   (D) electron transport and phosphorylation in the thylakoid membrane of a chloroplast
   (E) phosphorylation and isomerization of glucose to a substrate for triose phosphates

5. Protein synthesis involves the following structures and/or components EXCEPT

   (A) rough endoplasmic reticulum (RER)
   (B) ribosomes
   (C) messenger RNA (mRNA)
   (D) transfer RNA (tRNA)
   (E) lysosomes

6. The most recent theories of the origin of life include all of the following elements in the primitive atmosphere EXCEPT

   (A) free oxygen
   (B) hydrogen
   (C) methane
   (D) ammonia
   (E) carbon dioxide

7. Albinism is a recessive trait. In a certain community of 200 people, 18 persons are albinos. How many people are normal homozygotes?

   (A) 182
   (B) 164
   (C) 100
   (D) 98
   (E) 84

8. All of the following are characteristics of chordates EXCEPT

    (A) a vertebral column
    (B) a dorsal hollow nerve cord
    (C) gill slits
    (D) a notochord
    (E) post-anal tail

9. A harmless animal that imitates the appearance of another species that is dangerous to the predator is an example of

    (A) Müllerian mimicry
    (B) Batesian mimicry
    (C) cryptic appearance
    (D) mutualism
    (E) altruism

10. In the gastrointestinal tract sucrose is broken down by

    (A) dehydrolysis
    (B) hydrophobic cleavage
    (C) hydrolysis
    (D) dehydration linkage
    (E) dephosphorylation

Question 11 refers to the following.

Huge areas in both the temperate and tropical regions of the world are covered by these biomes. They are typically areas with relatively low total annual rainfall or uneven seasonal occurrences of rainfall. This type of climate is unfavorable for forests but suitable for growth of grasses. The temperate and tropical versions of this biome are remarkably similar in appearance, although the particular species they contain may be very different. In both cases, there are usually vast numbers of large herbivores, which often include the ungulates (hoofed animals). Burrowing rodents or rodent-like animals are also common.

11. This describes which biome?

    (A) Taiga
    (B) Tundra
    (C) Chaparral

    (D) Grasslands
    (E) Desert

12. Chloroplasts, the cell organelles where photosynthesis occurs, contain all of the following EXCEPT

    (A) thylakoids
    (B) stroma
    (C) cristae
    (D) starch granules
    (E) lamellae

13. The eukaryotic cell would quickly self-destruct if powerful hydrolytic enzymes were released from

    (A) peroxisomes
    (B) endoplasmic reticulum
    (C) Golgi bodies
    (D) lysosomes
    (E) plastids

14. The effect of _____ on population growth increases as population size increases.

    (A) density-dependent factors
    (B) density-independent factors
    (C) dispersal
    (D) fitness
    (E) spacing

15. Epiphytic plants grow on the limbs of trees, using the tree only for support. This relationship is known as

    (A) predation
    (B) commensalism
    (C) mutualism
    (D) competition
    (E) amensalism

16. Which one of the following taxonomic groups includes all of the others?

    (A) Family
    (B) Genus
    (C) Class
    (D) Species
    (E) Order

17. A hereditary modification that increases an organism's chance of survival is called

    (A) a preadaptation
    (B) an adaptation
    (C) speciation
    (D) directional selection
    (E) stabilizing selection

18. Hormones from the pituitary affect secondary hormone release from all of the following EXCEPT

    (A) thyroid
    (B) adrenal cortex
    (C) ovaries
    (D) testes
    (E) adrenal medulla

19. The sequence of differentiative events that leads to the formation of mature sperm cells is

    (A) primary spermatocytes → secondary spermatocytes → spermatids → spermatogonia → sperm
    (B) spermatids → spermatogonia → primary spermatocytes → secondary spermatocytes → sperm
    (C) spermatogonia → spermatids → secondary spermatocytes → primary spermatocytes → sperm
    (D) spermatogonia → primary spermatocytes → secondary spermatocytes → spermatids → sperm
    (E) secondary spermatocytes → primary spermatocytes → spermatogonia → spermatids → sperm

20. A plant with no meristematic tissue will be unable to

    (A) photosynthesize
    (B) transport water
    (C) transport nutrients
    (D) produce fruits
    (E) respire

21. A plausible directional flow for membrane components in a eukaryotic cell is

    (A) Golgi apparatus → rough ER → smooth ER → nuclear envelope
    (B) nuclear envelope → rough ER → smooth ER → Golgi apparatus → secretory vesicles → plasma membrane
    (C) nuclear envelope → mitochondria → rough ER → smooth ER → secretory vesicles → plasma membrane
    (D) plasma membrane → secretory vesicles → rough ER → smooth ER → nuclear envelope
    (E) plasma membrane → smooth ER → rough ER → secretory vesicles → nuclear envelope

22. Sometimes it is found that viruses can transfer genetic material from one bacterial strain to another. This process is called

    (A) transduction
    (B) recombination
    (C) conjugation
    (D) transmission
    (E) mutation

23. "Stop" codons that specify the end of a protein are

    I.   UAA
    II. UAG
    III. UUA
    IV. UGA

    (A) I only
    (B) I, II, and III only
    (C) I and IV only
    (D) I, II, and IV only
    (E) I, II, III, and IV

24. Initiator tRNA with an anticodon recognizes and pairs with the initiation codon

    (A) 5′-CAU-3′
    (B) 5′-AUG-3′
    (C) 5′-CUA-3′
    (D) 5′-UUA-3′
    (E) 5′-TAC-3′

25. Which one of the following is NOT a density-dependent limitation on population growth?

    (A) Intraspecific competition
    (B) Interspecific competition
    (C) Physiological mechanism
    (D) Predator-prey system
    (E) Environmental changes

26. Which of the following factors affect(s) enzymatic activity?

    I.   Temperature
    II. Hydrogen ion concentration
    III. Enzyme poisoning
    IV. Water concentration

    (A) I
    (B) II and III
    (C) I, II, and III
    (D) I, III, and IV
    (E) I, II, III, and IV

27. An exocrine gastric product which combines with vitamin B12 so that it can be absorbed later in the small intestine is

    (A) pepsin
    (B) hydrochloric acid
    (C) mucus
    (D) an intrinsic factor
    (E) trypsin

28. The basic contractile unit in striated muscle is

    (A) the muscle fiber
    (B) the myofibril
    (C) the sarcomere
    (D) the myofilament
    (E) the thin filaments

29. After a Douglas fir forest is logged, beekeepers frequently keep their beehives on the logged land. The most reasonable explanation for this is

    (A) the climax community includes a major competitor of bees
    (B) the pioneer species and the bees have a mutualistic relationship

    (C) the bees are best kept away from people
    (D) the bees are needed to produce honey
    (E) the cleared forest is more accessible than the forest

30. Which of the following groups would be considered a society?

    (A) A swarm of flies attracted to a rotting fruit
    (B) A collection of different insects in a jar
    (C) A group of male crickets attracted to the same female
    (D) A collection of birds in a zoo
    (E) A pack of dogs chasing their prey in relays

31. The connective tissue sac enclosing the heart is called the

    (A) endothelium
    (B) myocardium
    (C) pericardium
    (D) vena cava
    (E) endocardium

32. Spores have all of the following characteristics EXCEPT that

    (A) they are haploid
    (B) they are usually unicellular
    (C) they are formed by meiosis
    (D) they germinate and develop into a gametophyte
    (E) they fuse with other cells

33. The taste buds of the tongue are what kind of receptors?

    (A) Proprioceptors
    (B) Tastereceptors
    (C) Osmoreceptors
    (D) Mechanoreceptors
    (E) Chemoreceptors

34. All of the following alter the Hardy-Weinberg equilibrium EXCEPT

    (A) mutations
    (B) genetic drift
    (C) natural selection

(D) sexual recombination

(E) sexual selection

35. Striated (skeletal) muscle fibers exhibit

(A) few mitochondria

(B) alternating A bands and I bands in a transverse pattern

(C) only one nucleus

(D) a spiral arrangement of muscle filaments

(E) a branching pattern of muscle filaments

36. Which one of the following group pairings is most closely related?

(A) Ciliates and sponges

(B) Flatworms and roundworms

(C) Sea anemones and flagellates

(D) Segmented worms and insects

(E) Clams and squid

37. After implantation of the embryo, the corpus luteum can secrete progesterone at high levels without being shut off because

(A) it is not under regulatory control

(B) the pituitary secretes FSH

(C) the levels of estrogen have decreased

(D) the hypothalamus is actively secreting

(E) the placenta secretes chorionic gonadotropin which is like luteinizing hormone (LH)

38. The formula presented below is the formula of the _____.

$$6Ru - P + 6CO_2 + 12NADPH_2 + 18\,ATP \rightarrow$$
$$C_6H_{12}O_6 + 18\,ADP + 18Pi + 6Ru - P + 6H_2O$$

(A) photosynthetic cycle

(B) respiration cycle

(C) fermentation

(D) Calvin cycle

(E) Krebs cycle

39. All of the following are secreted by the pancreas EXCEPT

(A) chymotrypsin

(B) trypsin

(C) pepsin

(D) carboxypeptidase

(E) lipase

40. In myelinated axons and dendrites, the action potential leaps from one node of Ranvier to the next by a process called

(A) hydrophobic induction

(B) saltatory conduction

(C) electron transport

(D) electrophillic conduction

(E) constriction and dilation

41. Hydrogen ions are NOT free to lower the blood's pH because they are

(A) removed through the action of carbonic anhydrase

(B) bound to water

(C) bound to hemoglobin

(D) removed by diffusion

(E) bound to carbon dioxide

42. The release of parathyroid hormone is stimulated by

(A) decreased plasma Ca^{++}

(B) increased venous pressure

(C) increased plasma Na^+

(D) increased TSH from the pituitary

(E) decreased muscle activity

43. Apoplastic movement of water into the xylem is prevented by

(A) the cortex

(B) the Casparian strip

(C) the endodermis

(D) the plasmodesmata

(E) the stele

44. All flagella and motile cilia have the following pattern of microtubules:

(A) 7 outer and 2 inner

(B) 9 outer and 0 inner

(C) 8 outer and 2 inner

(D) 9 outer and 2 inner

(E) 11 outer and 3 inner

45. In protein-driven chemiosmotic phosphorylation, ATP energy is generated as protons pass from a region of high concentration to one of low concentration, through complex structures of the thylakoid membranes called the

    (A) FAD enzyme complex
    (B) CF-1 particles
    (C) Fe S1 proteins
    (D) cytochromes
    (E) F-1 particles

46. Release of simple substances into the ecosystem for use by producers depends on activity of the

    (A) carnivores
    (B) decomposers
    (C) herbivores
    (D) primary consumers
    (E) secondary consumers

47. Which of the following does NOT perform excretory functions?

    (A) Kidneys
    (B) Lungs
    (C) Skin
    (D) Liver
    (E) Lymph nodes

48. A man was involved in an accident and suffered extensive damage to the cerebellum. Which of the following functions would he be unable to perform?

    (A) Recalling facts prior to his accident
    (B) Driving his car
    (C) Reading for long periods
    (D) Digesting his food
    (E) Distinguishing between hot and cold objects

49. According to the trophic levels of ecosystems, herbivores are

    (A) primary consumers
    (B) primary producers
    (C) secondary consumers
    (D) tertiary consumers
    (E) quaternary producers

50. A male European robin in breeding condition will attack a tuft of red feathers placed in his territory. Since red feathers are usually on the breast of his competitor, it is to his reproductive advantage to behave aggressively at the sight of them. This is due to

    (A) instinctive behavior
    (B) operants
    (C) releasers
    (D) fixed action pattern
    (E) appetitive behavior

51. The best technique for separating amino acids in hydrolysate is

    (A) centrifugation
    (B) chromatography
    (C) spectrophotometry
    (D) hydrolysis
    (E) ligation

52. Which of the following is NOT an example of instinctive behavior?

    (A) the pattern of breeding behavior in male sickleback fish
    (B) a dog's salivating at the sight of food
    (C) a human's heartbeat accelerating at the sight of a charging mad bull
    (D) the flight of birds to warmer places during winter
    (E) a moth's attraction to light

53. Motility in the alimentary tract is important to all of the following functions EXCEPT

    (A) translocation of food
    (B) increased surface area
    (C) expulsion of food
    (D) continuous exposure of material to secretory and absorbing surfaces
    (E) mechanical grinding and mixing

54. Meiotic drive is

    (A) preferential segregation of genes during meiosis in a heterozygous individual
    (B) preferential segregation of genes during meiosis in a homozygous individual

(C) a force that prevents meiosis from occurring
(D) a force that disrupts the normal segregation processes in meiosis
(E) a force that causes somatic cells to lyse

55. The gram stain, which is used to differentiate bacterial cells, is based on

(A) the protein content in the respective bacterial cell wall
(B) the carbohydrate content in the respective bacterial cell wall
(C) the lipid content in the respective bacterial cell wall
(D) the diffusion rate of staining fluid through the bacterial cell wall
(E) the lack of a cell wall in gram negative bacteria

56. Which of the following consists mostly of repetitive DNA?

(A) Euchromatin
(B) Constitutive heterochromatin
(C) Nucleosomes
(D) Z-DNA
(E) Faculative heterochromatin

57. Which one of the following terms best describes plankton?

(A) Small carnivores
(B) Small autotrophs
(C) Small heterotrophs
(D) Small omnivores
(E) Small pelagic organisms

58. The order of the contraction of the heart is

(A) A-V node → atria → bundle of His → ventricles
(B) S-A node → A-V node → bundle of His → ventricles
(C) atria → S-A node → A-V node → ventricles → bundle of His
(D) S-A node → atria → A-V node → bundle of His → ventricles
(E) atria → A-V node → S-A node → bundle of His → ventricles

59. The order of the clotting process is

(A) prothrombin → thrombin → fibrinogen → fibrin → clot
(B) fibrinogen → fibrin → prothrombin → thrombin → clot
(C) fibrin → fibrinogen → thrombin → prothrombin → clot
(D) thrombin → prothrombin → fibrin → fibrinogen → clot
(E) thrombin → prothrombin → fibrinogen → fibrin → clot

60. The myelin sheath of many axons is produced by the

(A) node of Ranvier
(B) nerve cell body
(C) Schwann cell
(D) astrocytes
(E) axon hillock

61. All of the following occur as muscles contract EXCEPT

(A) Z bands come closer
(B) H zones stay the same
(C) A bands stay the same
(D) I bands decrease
(E) Thick and thin filaments slide past each other

62. Under which of the following conditions can competitive inhibition of enzymes be reversed?

(A) The inhibitor attaches to the enzyme by weak bonds.
(B) The inhibitor attaches to a site on the enzyme away from the active site.
(C) The inhibitor changes the shape of the enzyme.
(D) The enzyme changes form.
(E) The concentration of the substrate is decreased.

63. A classification scheme that places pine trees, tomato plants, and pepper plants in one group and ferns and liverworts in another group is based on

I.   vascular versus nonvascular
II. flowering versus nonflowering
III. seed-forming versus non-seed-forming

(A) I only
(B) II only
(C) III only
(D) I and II
(E) I and III

64. The genes which serve as a binding site for RNA polymerase in DNA transcription are called

(A) operator genes
(B) structural genes
(C) promoter genes
(D) regulatory genes
(E) inhibitor genes

Use the following figure and information summary to answer questions 65 and 66.

Malaria in humans is caused by several different species of *Plasmodium*, a protist. The life cycle of the *Plasmodium* requires an *Anopheles* mosquito and a person, as shown below. The sporozoites enter the liver and divide, forming many merozoites, which enter the red blood cells. In the red blood cells, the merozoites multiply and then break out of the red blood cells into the plasma. Some of these merozoites re-enter the red blood cells and some form gametocytes, which are returned to a new mosquito after the person is again bitten. In the mosquito's gut, gametes are produced, fertilization forming a zygote occurs, and the zygote develops into an oocyte. The oocytes produce many sporozoites, which move up the gut to the salivary glands.

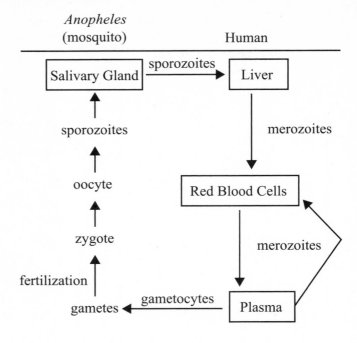

65. Some stages of the plasmodium have an apical complex that contains organelles that are specialized for penetrating cells and tissues of the host. In which of the following is the apical complex most probably present?

(A) Sporozoites and mesozoites
(B) Gametocytes and oocytes
(C) Gametocytes and sporozoites
(D) Sporozoites, mesozoites, and gametocytes
(E) Gametes, zygotes and oocytes

66. The diet of the *Plasmodium* is hemoglobin and its waste products, called hemozoin, are released when the host cell bursts. This results in a recurrence of chills and fever. The chills and fever are due to the

(A) release of the merozoites from the red blood cells to the plasma and their subsequent uptake into the red blood cells
(B) abundance of mosquitoes
(C) synchronized development of the sporozoite and merozoites
(D) response of the white blood cells in the plasma to the foreign hemozoin
(E) response time of the temperature centers in the hypothalamus

67. Which of the following correctly lists the compounds in order of decreasing free energy (under intracellular conditions)?
    a. phosphoenol pyruvate
    b. fructose-1,6-diphosphate
    c. glucose-6 phosphate
    d. pyruvate
    e. 1,3 diphosphoglycerate

    (A) c-b-e-a-d
    (B) c-b-a-e-d
    (C) c-b-e-d-a
    (D) b-c-e-a-d
    (E) b-e-c-a-d

68. Due to the dominant fauna in this period, the Mesozoic Era is often referred to as the

    (A) "Age of Reptiles"
    (B) "Age of Fishes"
    (C) "Age of Amphibians"
    (D) "Age of Birds"
    (E) "Age of Mammals"

69. Food acquisition by extracellular digestion and subsequent absorption is used by

    (A) fungi
    (B) mosses
    (C) ferns
    (D) slime molds
    (E) eubacteria

70. The placenta originates from

    I.   embryonic cells
    II.   maternal cells
    III. paternal cells

    (A) I
    (B) II
    (C) III
    (D) I and II
    (E) II and III

71. The prostaglandins, which may eventually prove to be effective birth control substances, are classified as

    (A) steroids
    (B) 20-carbon fatty acids
    (C) carbohydrates
    (D) lipoproteins
    (E) proteins

72. Which of the following DOES NOT involve the use or presence of fungi?

    (A) Production of the antibiotic penicillin
    (B) Production of soy sauce
    (C) Ringworm
    (D) Production of yogurt
    (E) Preparation of some cheeses such as Camembert and Roquefort

73. Which of the following applies to asexual reproduction in Ascomycota (cup fungi)?

    (A) Formation of zygosporangia
    (B) Formation of sporangia
    (C) Production of ascospores
    (D) Production of conidia
    (E) Production of basidospores

74. The primitive gut formed during gastrulation is called the

    (A) blastopore
    (B) gastrocoel
    (C) blastocoel
    (D) archenteron
    (E) ventral pore

75. Evolutionary change may occur in all of the following situations EXCEPT

    (A) Some bacteria survive after exposure to antibiotics because they are antibiotic-resistant.
    (B) A small number of individuals survive a hurricane that kills most of the people on an island.
    (C) A species of birds with a variety of genetically different forms migrates to a new island.

215

(D) dogs are selected for breeding because of certain behavioral characteristics that they did not acquire during training.

(E) horses raised on a nutritionally well-balanced diet are selected for breeding.

76. What evolutionary adaptation(s) allowed vertebrates, for the first time, to explore land for food?

I. Lobed fins
II. Amniote egg
III. Internal nostrils and primitive lungs
IV. Internal fertilization

(A) I only
(B) II only
(C) III only
(D) I and III
(E) II and IV

77. Usually a recessive sex-linked trait

(A) is expressed more often in males than in females
(B) is expressed more often in females than in males
(C) is expressed to the same extent in both males and females
(D) is only expressed in males
(E) is only expressed in females

78. The "10 percent rule" in ecology

(A) refers to the precentage of similar species that can coexist in one ecosystem
(B) refers to the average death total of all mammals before maturity
(C) is the percent of animals not affected by DDT
(D) refers to the level of energy production in a given trophic level that is used for production by the next higher level
(E) refers to the average birth rate in a climax community

79. In humans, the large bone extending from the hip to the knee is called the

(A) tibia
(B) fibula
(C) patella
(D) humerus
(E) femur

80. Prokaryotic cells differ from eukaryotic cells in that the former lack

(A) ribosomes
(B) a plasma membrane
(C) endoplasmic reticulum
(D) a cell wall
(E) DNA

81. On a small island off the coast of China, a large number of people are polydactyl (having more than five fingers or toes). Which factor most likely contributed to this phenomenon?

(A) Overcrowding
(B) Overproduction
(C) Variation
(D) Natural selection
(E) Isolation

82. Two unrelated organisms that become similar in appearance and ways of life as they adapt to similar environmental situations exhibit

(A) adaptive radiation
(B) convergent evolution
(C) divergent evolution
(D) homology
(E) parallel evolution

83. Choose the statement that best describes the climax stage of an ecological succession.

(A) It is usually populated only by plants.
(B) It is usually populated only by animals.
(C) It represents the initial phases of evolution.
(D) It changes rapidly from season to season.
(E) It remains until there are severe changes in the environment.

84. White and pink flamingoes are found in the wild and in zoos. The pink color in the feathers of flamingoes is due to a dietary pigment. If two pink flamingoes are bred, the offspring will be

    (A) all white
    (B) all pink
    (C) pink in half of the offspring
    (D) pink and white
    (E) pink in a quarter of the offspring

85. Eastern and western meadowlarks are two different species of birds. However, they look very similar to each other and breed in the spring in large, overlapping breeding areas. Also, they both eat seeds and obtain food on the ground. They are most easily distinguished by their calls, particularly the breeding songs of the males. This is an example of

    (A) physical mating barriers
    (B) gene flow
    (C) temporal isolation
    (D) habitat isolation
    (E) behavioral isolation

86. Translocation is a type of chromosomal mutation where

    (A) a segment of the chromosome is missing
    (B) a portion of the chromosome is represented twice
    (C) a segment of one chromosome is transferred to another nonhomologous chromosome
    (D) a segment is removed and reinserted
    (E) a segment is removed and destroyed

87. The denaturing of a protein by heat or radiation is caused by the destruction of which structures?

    (A) Secondary and tertiary
    (B) Secondary and quarternary
    (C) Secondary, tertiary, and quarternary
    (D) Tertiary and quarternary
    (E) Quarternary

88. Unlike collenchyma and sclerenchyma tissues, parenchyma tissue DOES NOT function in

    (A) support
    (B) gas exchange
    (C) nutrient exchange
    (D) both gas exchange and nutrient exchange
    (E) both support and gas exchange

89. A cell is placed in a solution of dye. After a while, the intracellular concentration of dye becomes much greater than the extracellular concentration. Upon addition of a metabolic inhibitor to the solution, the dye equilibrates across the cell membrane until the intra- and extra-cellular concentrations are equal. A possible role for this metabolic inhibitor might be to

    (A) inhibit protein synthesis
    (B) delay chromosomal replication
    (C) accelerate aerobic respiration
    (D) inhibit ATP production
    (E) accelerate meiotic processes

90. The oleander (Nerium) which lives in a very dry habitat has stomata that are located in

    (A) shallow hair-lined depressions in the upper epidermis.
    (B) deep hair-lined depressions in the lower epidermis.
    (C) very thin upper epidermis.
    (D) deep hair-lined depressions in the upper epidermis.
    (E) shallow hair-lined depressions in the lower epidermis.

91. There are two extreme life-history strategies that are found in many populations: r-selected and K-elected. In general, r-selected populations have a large number of small offspring with a high mortality rate and K-selected populations have a small number of large offspring with a low mortality rate. Which of the following combinations would be characteristic of an r-selected population?

    (A) Internal fertilization and internal development
    (B) Internal fertilization and external development

(C) External fertilization and internal development

(D) External fertilization and external development

(E) No combination has a greater need than another

92. Aggressive behavior in animals

(A) consists mostly of encounters between members of different species (excluding predation)

(B) occurs most frequently in contests over food

(C) usually consists of nonviolent displays within species which avoid serious injury

(D) occurs equally in both sexes

(E) tends to occur at only one time of the year

93. The climax organism growing above the tree line on a mountain would be the same as the climax organism found in the

(A) taiga

(B) tundra

(C) tropical forest

(D) desert

(E) temperate regions

Questions 94–96 refer to the typical flower of an angiosperm.

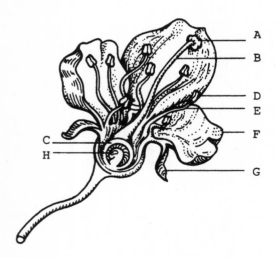

94. The structure labeled D is a/an

(A) stigma

(B) pedicel

(C) calyx

(D) filament

(E) anther

95. The function of this flower structure is to secrete a moist, sticky substance to which pollen grains can adhere.

(A) A

(B) B

(C) C

(D) D

(E) F

96. Which one of these structures may be absent if the flower is pollinated by the wind?

(A) A

(B) C

(C) D

(D) F

(E) H

97. Which of the following correctly describes polyoma virus?

(A) It is a DNA tumor virus.

(B) When a permissive cell is infected, a transformation event may occur in one out of every 104 to 105 cases.

(C) An RNA-DNA hybrid is formed from viral RNA and host deoxyribonucleotides.

(D) It contains a reverse transcriptase.

(E) It is an RNA virus.

98. In eukaryotes, a region of nucleic acid that specifies a polypeptide sequence is known as

I. an intron

II. a cistron

III. a structural gene

(A) I

(B) II

(C) III

(D) II and III

(E) I and III

99. All of the following are involved in gene expression in eukaryotes EXCEPT

    (A) the synthesis of primary RNA transcripts
    (B) the degradation of DNA in the nucleus
    (C) the translation of messenger RNA
    (D) the degradation of messenger RNA molecules in the cytoplasm
    (E) the processing of primary RNA transcripts

100. According to the endosymbiont theory

    (A) certain prokaryotic cells gradually changed to form eukaryotic cells
    (B) eukaryotic cells arose when certain prokaryotic cells entered and "took up residence" inside other prokaryotic cells
    (C) the earliest cells were eukaryotic
    (D) certain prokaryotic cells gradually degenerated to form viruses
    (E) certain eukaryotic cells gradually degenerated to form prokaryotic cells

101. The stage during development in which there is a hollow ball of cells is called

    (A) the blastula
    (B) the morula
    (C) the isolecithal stage
    (D) the gastrula
    (E) ovulation

Questions 102–104 refer to the following data computed from a field study.

Year	No. of Acres	No. of Pheasants	No. of Foxes
1980	12	200	6
1981	18	180	14
1982	10	80	18
1983	16	100	12
1984	20	140	8

102. A possible symbiotic process relating population change in the two species displayed is

    (A) amensalism
    (B) commensalism
    (C) mutualism
    (D) parasitism
    (E) predation

103. The greatest change in population density from either species occurs by the year

    (A) 1980
    (B) 1981
    (C) 1982
    (D) 1983
    (E) 1984

104. The population density of pheasants decreases from

    (A) 1980–1984
    (B) 1981–1984
    (C) 1982–1983 only
    (D) 1980–1983 only
    (E) The density never decreases

105. The percentage of individuals that when carrying a given gene in proper combination for its expression actually express the gene's phenotype is known as

    (A) interference
    (B) expressivity
    (C) penetrance
    (D) coincidence
    (E) regulation

Use the following cladogram showing a simplified phylogeny of animals to answer questions 106 and 107. The arrows represent the derived characteristics, and the ovals the hypothetical common ancestors.

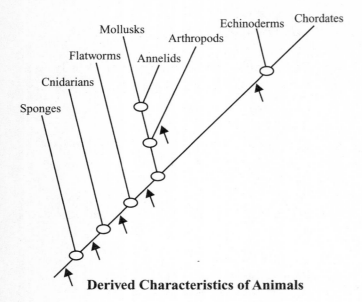

Derived Characteristics of Animals

106. Which of the following represent a monophyletic group?

   (A) Mollusks and arthropods
   (B) Chordates and cnidarians
   (C) Flatworms and annelids
   (D) Mollusks, annelids and arthropods
   (E) Sponges, cnidarians and flatworms

107. The derived characteristics of animals are

   (A) bilateral symmetry
   (B) true tissues
   (C) extracellular digestion
   (D) multicellular organization
   (E) cephalization

108. Damage to the right temporal lobe of the cerebral hemisphere results in

   (A) poor performance on IQ tests
   (B) poor performance on perceptual tests
   (C) no sensation on the right side of the body

   (D) lack of voluntary movement in the right arm, hand, leg, or foot
   (E) lack of movement in most body parts

109. The three periods within the Mesozoic Era are the

   (A) Permian, Carboniferous, and Devonian
   (B) Quaternary, Tertiary, and Cretaceous
   (C) Triassic, Jurassic, and Cretaceous
   (D) Silurian, Ordovician, and Cambrian
   (E) Tertiary, Jurassic, and Devonian

110. The biological membrane consists of a continuous double layer of lipid molecules with various membrane proteins embedded in it. Which one of the following lipid molecules is found only on the outer layer?

   (A) Glycolipid
   (B) Cholesterol
   (C) Phosphatidyl ethanolamine
   (D) Phosphatidyl serine
   (E) Sphingomyelin

111. The enzyme which adds the final nucleotide to seal the Okazaki fragments is

   (A) polymerase I
   (B) ligase
   (C) gyrase
   (D) polymerase II
   (E) polymerase III

112. The energy molecule used to attach the 50 S ribosomal subunit to the 30 S subunit is

   (A) GTP
   (B) ATP
   (C) NMP
   (D) NTP
   (E) ADP

Questions 113–115 refer to the growth of algae in a lake. The graph below shows the number of algae present throughout the year, as well as changes in nutrients, light, and temperature.

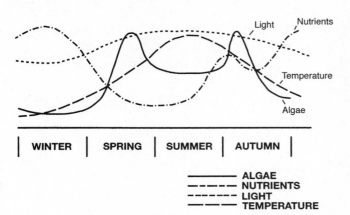

113. The rate of growth of algae is greatest in the

  (A) winter
  (B) spring and summer
  (C) spring and autumn
  (D) winter and summer
  (E) summer and autumn

114. During the summer, growth of algae is limited by

  (A) light
  (B) temperature
  (C) light and temperature
  (D) nutrients
  (E) light, temperature, and nutrients

115. Peaks in growth of algae depend upon

  (A) light alone
  (B) temperature alone
  (C) nutrients alone
  (D) light and temperature together
  (E) light, temperature, and nutrients

Questions 116–117 refer to the following diagram.

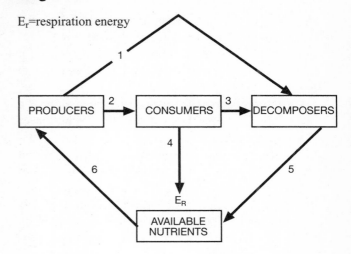

116. Which arrow(s) represent(s) only nutrient movement?

  (A) 1
  (B) 1 and 4
  (C) 2 and 3
  (D) 4
  (E) 5 and 6

117. Which one of the arrows represents only energy movement?

  (A) 1
  (B) 2
  (C) 3
  (D) 4
  (E) 5

118. Which one of the following is true about all temperate grassland areas?

  (A) The species composition is the same in similar biomes.
  (B) The abiotic requirements of the species are similar.
  (C) The pioneer species are the same.
  (D) The altitude is similar.
  (E) The amount of fresh water present is the same.

119. Inversion, a class of structural variation, is a type of chromosomal aberration in which

    (A) an extra chromosomal segment is included in the haploid pairing of chromosomes
    (B) a segment of chromosome is removed from one location and is placed at another site on the chromosome
    (C) chromosomal segments are placed along the chromosome thus causing the formation of new alleles
    (D) a segment of a chromosome is turned around 180° and is reinserted into the chromosome
    (E) there is a variation in the chromosome numbers, which ranges from the addition or loss of one or more chromosomes to the addition or loss of one or more pairs of haploid sets of chromosomes

120. Which one of the following represents a population?

    (A) The plants in the everglades.
    (B) The salamander and frogs in a marsh.
    (C) The decomposers in a marsh.
    (D) The red-winged blackbirds in a marsh.
    (E) The red worms and sediment in a marsh.

121. Which of the following is a true statement with regard to the differences between monocotyledon and dicotyledon plants?

    (A) The pattern of veins in the leaf is netlike in the dicot and parallel in the monocot.
    (B) In the seeds, one cotyledon is present in the dicot and two cotyledons are present in the monocot.
    (C) The flowers of dicots have three floral parts, or multiples of three; the flowers of monocots have four or five floral parts, or multiples of four or five.
    (D) The vascular system of dicots occurs in scattered bundles, while the vascular system of monocots is generally a neat ring of vascular bundles arranged in a circle around the stem.
    (E) In dicots, there is no vascular cambium; in monocots vascular cambium is present.

122. Suppose you made an incision around the entire circumference of a tree trunk (below the branches) such that the phloem was cut all around. The direct result would be that

    (A) roots will die because of a lack of nutrients
    (B) leaves will not be able to carry out photosynthesis because of a lack of water
    (C) buds will not grow above the cut, but roots will continue to grow
    (D) the leaves will die because of a lack of nutrients
    (E) the tree will not be affected

123. In the photolysis of water

    I. Two electrons enter the electron transport system.
    II. Photosystems are utilized.
    III. Two hydrogen ions are released into the atmosphere.
    IV. NAD is the final electron acceptor.

    (A) I, II, and III are correct
    (B) I and II are correct
    (C) IV is correct
    (D) II and IV are correct
    (E) III and IV are correct

124. In chloroplasts it has been found that light-independent reactions actually occur in

    (A) the thylakoid
    (B) the grana
    (C) the stroma
    (D) the lamellae
    (E) the quantasomes

125. Which of the following derived characteristics in animals allows the internal organs of animals to move independently of the outer body wall?

    (A) True tissue
    (B) Bilateral symmetry
    (C) Vascular tissues
    (D) Coelom
    (E) Archenteron

126. All of the following pigments act as accessory pigments in photosynthesis EXCEPT

   (A) carotenes
   (B) xanthophylls
   (C) flavenoids
   (D) phycobilins
   (E) chlorophyll b

127. From the fossils of the *Homo habilis*, discovered by Louis Leakey, we can say that *Homo habilis* did NOT

   (A) have a greater cranial capacity than *Australopithecus*
   (B) seem closer to the human form
   (C) belong to the genus *Homo*
   (D) possess strong jaws and teeth
   (E) appear closer to the human line than *Australopithecus*

128. Which of the following acts on the glandular epithelium of the uterus (the endometrium) and converts it to an actively secreting tissue?

   (A) Acetylcholine
   (B) Cytoldnin
   (C) Progesterone
   (D) Ecdysone
   (E) Actin

129. The dominant tropical coal swamp plants went extinct quite quickly in the Carboniferous period. The massive coal deposits in the northeastern United States and in Britain suggests that

   (A) angiosperms competed with the coal swamp plants driving them to extinction
   (B) gymnosperms competed with the coal marsh plants driving them to extinction
   (C) the large tar pits buried the coal marsh plants
   (D) during the Carboniferous period the northeastern U.S. and Britain were part of one continent
   (E) during the Carboniferous period there was extensive extinction of plant and animals

130. Which of the following has the greatest potential to produce a new species of plants?

   (A) Increased ploidy
   (B) Inbreeding
   (C) Increased fitness
   (D) Polymorphism
   (E) Mutations

DIRECTIONS: For each group of questions below match the numbered word, phrase, or sentence to the most closely related lettered heading and mark the letter of your selection on the corresponding answer sheet. Only one letter heading may be chosen for each question.

Questions 131–135

   (A) Cotyledons
   (B) Rhizoids
   (C) Sporophyte
   (D) Xylem
   (E) Palisade layer

131. Simple filaments performing water absorption

132. Chiefly responsible for photosynthesis

133. Reduced in size, cannot exist independently, especially in primitive plants

134. Functions mainly in nutrient absorption from endosperm in monocots

135. Transports nutrients up to leaves

Questions 136–138

   (A) Holoenzyme
   (B) Apoenzyme
   (C) Metalloenzyme
   (D) Coenzyme
   (E) Prosthetic group

136. An organic, nonprotein cofactor

137. A coenzyme tightly bound to the apoenzyme

138. A cofactor joined to an apoenzyme

Questions 139–142

    (A) Medulla
    (B) Midbrain
    (C) Cerebellum
    (D) Cerebrum
    (E) Thalamus

139. Immediately connected to the spinal cord

140. Relay center for sensory impulses

141. Regulates and coordinates muscle contraction

142. Contains gray matter

Questions 143–146

    (A) DNA
    (B) Ribosomal RNA
    (C) Transfer RNA
    (D) Messenger RNA
    (E) RNA polymerase

143. Involved in semiconservative replication

144. Mediates transcription of DNA

145. Is the anticodon which carries amino acids

146. Transports codons to the ribosomes

Questions 147–151

    (A) Gastrin
    (B) Duodenum
    (C) Lipase
    (D) Colon
    (E) Oral cavity

147. Digestion begins at this site

148. Produced by food distending the stomach walls

149. Structure where villi begin to appear

150. Is found in humans, where very large amounts of bacteria exist

151. Is released by pancreas

Questions 152–155

    (A) Capillaries
    (B) Birds
    (C) Sino-atrial node
    (D) Atrio-ventricular node
    (E) Reptiles

152. Initiates heartbeat in mammals

153. Possess a characteristic four-chambered heart

154. Most oxygen exchange takes place at this site

155. Particularly active during fever, causes increased heart rate

Questions 156–160

    (A) Pheromones
    (B) Display
    (C) Courtship
    (D) Müllerian mimicry
    (E) Adaptive behavior

156. Female moths attract males over very large areas using air currents.

157. Black-headed gulls remove conspicuous objects and broken eggshells from nests.

158. Two black-headed gulls meet; one gives the agonistic "upright" posture.

159. Two species of inedible insects evolve to resemble each other.

160. An Eyrebird sings his favorite song.

Questions 161–165

    (A) Cortex
    (B) Adventitious roots
    (C) Growth zone
    (D) Lenticel
    (E) Cutin

161. These are structures that grow from stem or leaves but not from the primary root.

162. Meristematic cells composed of small, thin-walled cells with large nuclei

163. Composed of parenchyma cells in both stem and root

164. Helps stems to breathe

165. Accounts for retarding amount of water loss due to evaporation

Questions 166–169

    (A) Echinoderms
    (B) Coelenterates
    (C) Arthropods
    (D) Annelids
    (E) Flatworms

166. Two part double ventral nerve cord which is segmented and which originates in the esophagus

167. Netlike system composed of separate neurons that cross each other but do not touch.

168. Nervous system consists of specialized neuroectoderm which forms shallow grooves along the surface of the arms.

169. Advanced planarians which have a longitudinal, bilaterally symmetrical body plan comprised of two nerve cords and whose neurons are concentrated in the head region.

Questions 170–171

    (A) Adenyl cyclase
    (B) Cyclic AMP (cAMP)
    (C) Epinephrine
    (D) Steroid hormone
    (E) Adrenocorticotrophic hormone (ACTH)

170. Mediates the activation of an enzyme which causes an increase in blood glucose level through the enzymatic breakdown of glycogen

171. Is known as the second messenger

Questions 172–174

    (A) Bottleneck effect
    (B) Gene flow
    (C) Founder effect
    (D) Nonrandom mating
    (E) Natural selection

172. Loss of alleles from the population due to a hurricane

173. It is usually considered adaptive

174. Change in the genotype frequency but not the allele frequency

Questions 175–179

    (A) Gap1 stage
    (B) Prophase
    (C) Cytokinesis
    (D) S stage
    (E) Anaphase

175. At this stage DNA replication and the formation of new chromosomes are observed. It is the stage at which the cell incorporates radioactive thymine.

176. The chromosomes were observed to condense at this stage and the nucleolus to have disappeared.

177. Chromosomes move rapidly apart and the fibers connected to their centromeres shorten.

178. The constricting ring has closed down on the fibrous remains of the spindle.

179. This is generally a very active period. It is the time when the cell synthesizes the enzymes and structural proteins necessary for cell growth.

Questions 180–184

(A) Homoplasy
(B) Analogy
(C) Serial Homology
(D) Allometry
(E) Homeosis

180. An explanation for the specialized chain of vertebra in vertebrates

181. A possible explanation for the difference in shape of the skull of a chimp and a human

182. A possible explanation for differences in the number of petals on different species of flowers

183. The condition that is used to explain the relationship between a salmon's tail fin and a seal's flipper

184. A similarity in function or behavior among organisms.

Questions 185–189

(A) Erythrocytes
(B) Neutrophils
(C) Eosinophils
(D) Lymphocytes
(E) Basophils

185. Phagocytic cells concerned primarily with local infections

186. Protein substance in blood that inactivates foreign protein

187. Leukocyte that probably produces heparin (an anticoagulant), histamine, and serotonin

188. Cells that transport O_2 and CO_2

189. Phagocytic cells concerned, primarily, with generalized infections

DIRECTIONS: The following groups of questions are based on laboratory or experimental situations. Choose the best answer for each question and mark the letter of your selection on the corresponding answer sheet.

Questions 190 and 191

Bacteria were grown in a medium containing heavy isotopes, the medium was then infected with phage and the bacteria were transferred to a medium containing light isotopes. Constituents produced before and after infection could be separated by density-gradient centrifugation. New high density RNA and proteins were isolated. Light ribosomes were absent.

190. These experiments showed that

(A) ribosomes were synthesized after infection
(B) ribosomes were not synthesized after infection
(C) light ribosomes were synthesized
(D) RNA was not synthesized after infection
(E) most of the radioactively labeled RNA were "light" ribosome

191. The new protein was labeled with heavy isotopes. This meant that

(A) ribosomes are specialized structures which synthesize proteins
(B) tRNA did not transcribe for protein synthesis
(C) no protein synthesis was recorded
(D) new proteins were synthesized in preexisting ribosomes
(E) tRNA is required to produce proteins

Questions 192–195

In *Drosophila*, the loci for the alleles cut wings (ct), yellow body (y), and vermilion eyes (v) are linked. Each of these alleles are recessive to the corresponding wild type alleles (+) for normal wings, normal body color, and normal eye color. The results below were produced for a mating between a heterozygous fly and one that was homozygous recessive.

F$_1$ Phenotypes

ct	+	+	=10
+	y	+	= 2
ct	y	v	= 4
ct	y	+	=35
+	+	+	= 4
ct	+	v	= 1
+	y	v	= 7
+	+	v	=37

192. Which of the following is a parental gamete?

  (A) ct   y   +
  (B) ct   +   +
  (C) +   y   v
  (D) ct   y   y
  (E) +   y   +

193. The alleles when mapped should align in this way

  (A)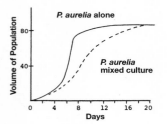

  (B)

  (C)

  (D)

  (E)

194. The coefficient of coincidence would be

  (A) 1.36
  (B) 2.36
  (C) 0.36

  (D) 0.136
  (E) 0.0036

195. The interference would be

  (A) –0.36
  (B) 1.00
  (C) 0.15
  (D) 1.51
  (E) –0.2

Questions 196–197 concern the curves that show the effect of competition between two species of *Paramecium*.

The solid curves show the growth of population volume of each species alone in a controlled environment with a fixed supply of food. The dotted curve shows the change in population volume of the same species when in competition with the other under similar conditions.

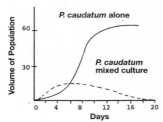

196. It was found that when grown in the mixed culture

  (A) *P. aurelia* grew normally
  (B) *P. caudatum* was inhibited more than *P. aurelia*
  (C) there was no noticeable effect on any of the species
  (D) *P. aurelia* growth was inhibited more than *P. caudatum*
  (E) *P. caudatum* growth was stimulated

197. The effect when the two cultures were mixed was most likely due to

  (A) mutualism
  (B) cooperation
  (C) parasitism
  (D) competition
  (E) eutrophication

Questions 198–200

Genetic damage in certain organisms was traced to mutagenic screening systems applied. The chart below shows a reflection of the types of genetic damage detected.

Screening System		Type of Damage Detected								
		Chromosome Aberrations					Mutations			
Category	Organisms	Deletions	Duplications	Dominant Lethality	Translocations	Nondis-junction	Forward or reverse or both	Multiple Specific Locus	Induced Recombination
Bacterial	E. Coli						●		
Fungal	Yeasts	●	●		●	●	●	●	
Plant	Vica fibre	●		●	●	●			
Insect	Bombyx	●	●	●	●	●	●	●	
Mammals	Chinese Hamster	●		●	●	●			
	Mouse	●		●	●	●			
	Humans	●	●	●	●	●	●		

198. Which of the following statements is true regarding the data obtained from these tests?

(A) *E. coli* suffered no particular mutation.

(B) Fungal organisms especially yeasts produced spores under these circumstances.

(C) Insects were more adapted to screening systems and therefore suffered little or no damage.

(D) Vica fibre and mouse cells had similar damages.

(E) Only vica fibre and humans did not suffer forward or reverse mutations.

199. From this data we can infer that

(A) screening systems can detect cancer

(B) screening systems are more helpful than hazardous to somatic cells

(C) the nature of the screening tests and the outcome are helpful in understanding what causes chromosome aberrations and mutations in some organisms

(D) screening systems are more useful for insects than bacteria

(E) screening systems do not detect induced mutations in humans

200. The most frequent type of chromosomal aberrations were

I. Dominant lethality
II. Translocations and forward or reverse mutations
III. Deletions and duplications
IV. Nondisjunction
V. Forward or reverse or both mutations

(A) II only

(B) II and IV

(C) II and III

(D) V only

(E) III and IV

Practice Test 1 Answer Key

1. (C)	35. (B)	69. (A)	102. (E)	135. (D)	168. (A)
2. (A)	36. (E)	70. (D)	103. (B)	136. (D)	169. (E)
3. (D)	37. (E)	71. (B)	104. (D)	137. (E)	170. (C)
4. (B)	38. (D)	72. (D)	105. (C)	138. (A)	171. (B)
5. (E)	39. (C)	73. (D)	106. (D)	139. (A)	172. (A)
6. (A)	40. (B)	74. (D)	107. (D)	140. (E)	173. (E)
7. (D)	41. (C)	75. (E)	108. (A)	141. (C)	174. (D)
8. (A)	42. (A)	76. (D)	109. (C)	142. (D)	175. (D)
9. (B)	43. (B)	77. (A)	110. (A)	143. (A)	176. (B)
10. (C)	44. (D)	78. (D)	111. (B)	144. (E)	177. (E)
11. (D)	45. (B)	79. (E)	112. (A)	145. (C)	178. (C)
12. (C)	46. (B)	80. (C)	113. (C)	146. (D)	179. (A)
13. (D)	47. (E)	81. (E)	114. (D)	147. (E)	180. (C)
14. (A)	48. (B)	82. (B)	115. (E)	148. (A)	181. (D)
15. (B)	49. (A)	83. (E)	116. (E)	149. (B)	182. (E)
16. (C)	50. (C)	84. (A)	117. (D)	150. (D)	183. (A)
17. (B)	51. (B)	85. (E)	118. (B)	151. (C)	184. (B)
18. (E)	52. (D)	86. (C)	119. (D)	152. (C)	185. (B)
19. (D)	53. (B)	87. (A)	120. (D)	153. (B)	186. (C)
20. (D)	54. (D)	88. (A)	121. (A)	154. (A)	187. (E)
21. (B)	55. (C)	89. (D)	122. (A)	155. (C)	188. (A)
22. (A)	56. (B)	90. (B)	123. (B)	156. (A)	189. (D)
23. (D)	57. (E)	91. (D)	124. (C)	157. (E)	190. (B)
24. (B)	58. (D)	92. (C)	125. (D)	158. (B)	191. (D)
25. (E)	59. (A)	93. (B)	126. (C)	159. (D)	192. (A)
26. (C)	60. (C)	94. (E)	127. (D)	160. (C)	193. (C)
27. (D)	61. (B)	95. (A)	128. (C)	161. (B)	194. (A)
28. (C)	62. (A)	96. (D)	129. (D)	162. (C)	195. (A)
29. (B)	63. (C)	97. (A)	130. (A)	163. (A)	196. (B)
30. (E)	64. (C)	98. (D)	131. (B)	164. (D)	197. (D)
31. (C)	65. (A)	99. (B)	132. (E)	165. (E)	198. (D)
32. (E)	66. (D)	100. (B)	133. (C)	166. (D)	199. (C)
33. (E)	67. (A)	101. (A)	134. (A)	167. (B)	200. (E)
34. (D)	68. (A)				

Practice Test 1
Detailed Explanations of Answers

1. (C)

The pituitary gland, also known as the hypophysis, lies in a pocket in the skull known as the sella turcica. It is connected to the brain by the infundibular stalk. The pituitary is a compound organ made up of anterior, posterior, and intermediate lobes. The anterior lobe is made up of glandular tissue which produces at least six different protein hormones: TSH, ACTH, FSH, LH, prolactin, and growth hormone. The posterior lobe, which is true neural tissue, releases two hormones: oxytocin and vasopressin.

2. (A)

Microevolution is the change in the genetic structure of a population as measured by changes in frequencies of alleles and genotypes. The discovery of small fossils may contribute information about microevolution but it is not microevolution. The genetic change in microorganisms is microevolution but microevolution is not restricted to changes in microorganisms. Natural selection is an agent of microevolution.

3. (D)

Penicillin is effective only against actively growing bacteria because it blocks the synthesis of cell walls. Penicillin prevents the incorporation of N-acetyl muramic acid into the structure that constitutes the bacteria's cell wall. If cell wall formation is complete, penicillin has no effect. Thus, only actively growing cells are killed by this antibiotic.

4. (B)

The hypothesis of chemiosmotic coupling proposes that the energy derived from electron transport causes translocation of protons across the inner mitochondrial membrane, thus creating a pH and electrochemical gradient across that membrane. The free energy which results from the flow of protons back across the membrane is used to drive the formation of ATP from ADP and phosphate. A similar chemiosmotic mechanism is believed to be involved in photosynthetic ATP synthesis.

5. (E)

All are clearly involved, except lysosomes. The RER is the site to which bound ribosomes are attached; polypeptides produced on bound, ribosomes are characteristically destined to be exported from the cell and/or localized in the cell membrane. There is a contrast, in function, between free and bound ribosomes: polypeptides (proteins) produced by free ribosomes are released into the cytoplasm of the cell.

6. (A)

The primitive atmosphere had essentially no free oxygen. All oxygen was in the form of water and oxides.

7. (D)

First, we must find the percentage of albinos in the population:

$$18/200 \times 100 = 9\%.$$

We know that albinism is the homozygous recessive trait (aa), so to find the frequency of the recessive allele (a) we take the square root of 9% $(0.09) = \sqrt{.09} = 0.30$. We also know that the frequency of the recessive allele (a) added to the frequency of the dominant allele (A) must equal 1. Thus, the frequency of the non-albino allele = $1 - 0.30 = 0.70$. Squaring the frequency of the non-albino allele will give us the frequency of homozygous non-albinos (AA).

$$(0.70)^2 = 0.49 \text{ or } 49\%.$$

Then we multiply 0.49×200 and see that there are 98 homozygous dominant nonalbinos.

8. (A)

A vertebral column is found in vertebrates. Vertebrates are one group of chordates and chordates during development have the four remaining characteristics listed. However, not all chordates have a vetebral column.

9. (B)

Cryptic appearance, Müllerian mimicry, and Batesian mimicry are ways in which organisms avoid becoming the victims of predators. The cryptic appearance of some organisms enables them to blend into their background, becoming invisible to their potential attackers. Müllerian mimicry involves the evolution of two or more inedible or unpleasant-tasting species to resemble one another. Batesian mimicry, however, involves the resemblance of an unprotected, harmless species to a dangerous species. This makes it difficult for the predator to distinguish between the two forms. Once the predator has tasted the noxious species, it tends to stay away from both species.

An Example of Batesian Mimicry

(A) The Monarch butterfly, a distasteful species. (B) The Viceroy, not so distasteful a species that mimics the Monarch. Species in the group to which the Viceroy belongs ordinarily have a quite different appearance.

10. (C)

Hydrolysis is the process applied to break down sucrose into glucose and fructose, the two monomers of which it is made. It is just the opposite of a dehydration linkage, which links glucose and fructose together to form sucrose.

In hydrolytic cleavage a water molecule is added to the linkage breaking the sucrose into its component parts.

11. (D)

The biome described has relatively low rainfall, is suitable for growth of grasses but not trees, and is inhabited by many large herbivores. The taiga is characterized by largely evergreen forest vegetation, which eliminates this choice since the biome described cannot support trees. The tundra has low temperatures unsuitable for trees or grasses, thereby eliminating this choice. The chaparral is an incorrect choice because it has abundant rain in the winter and its land is favorable for the growth of evergreen trees. The desert does receive little rainfall but vegetation is sparse, making it difficult for large numbers of herbivores to live in this biome. The grassland is the correct answer. The grassland biome is characterized by a low amount of rainfall (25–75 cm/year) and it provides natural pastures for grazing animals and vegetation to support the large number of herbivores that exist there.

12. (C)

Cristae refers to the inner membrane of the mitochondria, while all the others, thylakoids, stroma, starch granules, and lamellae are of the chloroplasts.

13. (D)

Lysosomes are membrane bound sacs that are roughly spherical. They are bags of powerful hydrolytic enzymes which are packaged and synthesized by the Golgi apparatus. If these enzymes were freely floating around in the cytoplasm, the cell would quickly self-destruct.

14. (A)

Only density-dependent factors always affect the growth of populations as population size increases. Density-independent factors have an effect on both small and large populations. Dispersal affects the genetic diversity and the distribution but not specifically the growth. Fitness contributes to the growth of populations, but it is independent of the size of the population. Spacing affects the distribution of the populations.

15. (B)

The tree provides only support and thus is not affected, whereas the epiphytic plant benefits from the support. No other interaction results in a benefit to one species with no effect on the second. Predation and mutualism result in a benefit to one species, but there is either a harm or a benefit to the second species. Competition results in a negative effect to both species. Amensalism results in harm to one species with no effect on the second species.

16. (C)

The correct sequence of taxonomic categories from largest to smallest is kingdom, phylum, class, order, family, genus, and species.

17. (B)

By definition, an adaptation is the change that results from selection. It is usually positive because it is selected for under the given conditions. A preadaptation is a trait that is present before it is used. Origin of a new species, or speciation, is usually a result of many adaptations.

18. (E)

The pituitary is the master gland which regulates all of these structures except the adrenal medulla. The adrenal medulla arises from the same source as the nervous system. Therefore, it is regulated by the stimulation of sympathetic preganglionic nerve fibers. Once stimulated, the adrenal medulla can secrete adrenalin (epinephrine) and noradrenalin (norepinephrine), which are responsible for the fight-or-flight response.

19. (D)

Mature sperm cells are the result of two stages of meiosis. One spermatogonium from the testes produces four mature sperm cells.

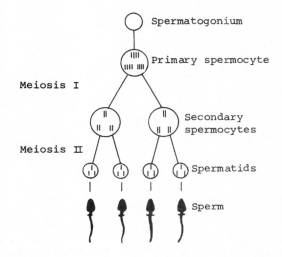

20. (D)

Fruit production requires active cell division. In plants, only regions of meristematic tissue are capable of active cell division. The removal of all of the meristematic tissue from a plant leaves it incapable of cell division and therefore incapable of producing fruits.

21. (B)

This is the endomembrane system. There are direct structural connections from the nuclear envelope, rough ER to the smooth ER and subsequent functional relationships from the Golgi apparatus to the secretory vesicles to the plasma membrane.

22. (A)

When a virus infects a bacterium, its DNA is integrated into the bacterial genome which corresponds to the lysogenic phase of its life cycle. When the viral DNA emerges from the bacterial genome, it usually takes up some of the bacterial DNA through an imprecise excision process. The virus then becomes a transducing phage which acts as a vector to promote the exchange of bacterial genes.

23. (D)

Of the 64 codons, three are "stop" codons that specify the end of a protein. These three are UAA, UAG, and UGA. The remaining 61 codons specify the 20 amino acids.

24. (B)

In eukaryotes a special initiator tRNA with a 5'-CAU-3' anticodon recognizes and pairs with the initiation codon 5'-AUG-3'.

25. (E)

Competition is the chief density-dependent limiting factor, because a limited supply of resources can only support a population of certain size. Organisms of the same species having the same living requirements experience competition. Organisms of different species can also experience competition, but its intensity depends on the similarity of their living conditions. Physiological mechanism is also a density-dependent limiting factor since a density-reduced resistance to pathogens and parasites has been evidenced. Environmental changes such as flooding, fire, and abnormal weather changes affect the population size regardless of population density.

26. (C)

Temperature extremes affect enzymatic activities. High temperatures denature the enzyme while low temperatures cause the rate of enzymatic action to slow down. The concentration of hydrogen ions, or pH, also has a great effect on enzymatic activity. Finally, enzyme-poisoning may also disrupt the activity of enzymes.

27. (D)

Intrinsic factor is a glycoprotein that combines with vitamin B12 to form a complex necessary for the absorption of vitamin B12 in the ileum.

28. (C)

The sarcomere is the area between Z lines of a myofibril. It is the basic, repeating contractile unit containing the A band in which alternating actin and myosin filaments slide together during contractions.

29. (B)

The forest is the climax community and once it is logged, secondary succession begins. The pioneer species are frequently species such as fireweed that are used by bees. In this question, it was not necessary to know the identity of pioneer species but to recognize that the pioneer species would not be trees, but rapidly growing small flowering plants, and that bees are pollinators that have a mutualistic relationship with flowering plants. Whether or not there was a competitor of bees in the climax forest (A) is not important because bees pollinate flowering plants, not conifers. Some people are very allergic to bee stings and like to have bees kept away from them (C), but this is not why the bees are brought to the logged land. It is true that bees are needed to make honey (D) but this is not related to why they are brought to the logged land. It is true that the cleared forest is more accessible (E), but there must be a reason to bring the bees.

30. (E)

A society is made up of a group of animals that belong to the same species and are organized in a cooperative manner. These animals are usually bound together by reciprocal communication, which leads to this type of cooperative behavior.

31. (C)

The pericardium is the membranous sac that encloses the heart. Endothelium is a term that refers to endothelial cells. Endocardium and myocardium refer to the heart muscle. The vena cava is the largest vein in the body.

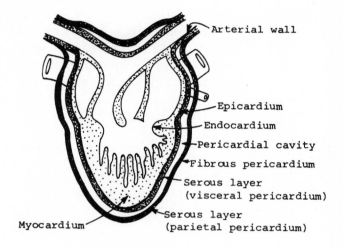

The Surrounding Heart Layers

32. (E)

Spores do not fuse with other cells; they develop by mitosis into gametophytes (D). Spores are usually unicellular (B) and are produced by meiosis (C), and are therefore haploid (A).

33. (E)

The taste buds contain very specialized chemoreceptors which can distinguish certain chemicals. In this way, one is able to distinguish sweet, salty, sour, and bitter tastes.

34. (D)

According to the Hardy-Weinberg Theorem and the associated equilibrium, the frequencies of alleles and genotypes in a population are constant unless acted upon by factors other than sexual recombination. These factors are: mutations, genetic drift, natural selection, sexual selection as well as nonrandom mating and migration.

35. (B)

Striated (skeletal) muscle fibers are multinucleated cylindrical cells arranged in parallel bundles. The most striking feature of such fibers is the transverse pattern of alternating dark A and light I bands. Both skeletal and cardiac muscle exhibit such patterns; smooth muscle does not.

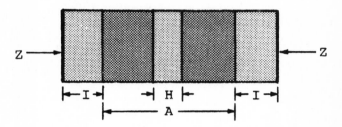

A Sarcomere of a Myofibril Showing Bands

36. (E)

This is a question about systematics. Four of the pairs of organisms represent different phyla:

cilates (Ciliata)—sponges (Porifera)

flatworms (Platyhelminthes)—roundworms (Aschelminthes)

sea anemones (Cnidaria)—flagellates (Mastigophora)

segmented worms (Annelida)—insects (Arthropoda)

Only the clams and squid are members of the same phylum (Mollusca) and, therefore, are most closley related.

37. (E)

The chorion (the embryo's part of the placenta) secretes a hormone called chorionic gonadotrophin, which is functionally similar to LH (luteinizing hormone). Its function is to take the place of LH in preserving the corpus luteum because of the inhibition of LH production by a high progesterone level. The corpus luteum is then able to secrete progesterone at high levels without being shut off.

38. (D)

$$6Ru - P + 6CO_2 + 12NADPH_2 + 18 ATP \rightarrow$$

$$C6H_{12}O_6 + 18 ADP + 18Pi + 6Ru - P + 6H_2O$$

The formula presented is the formula of the Calvin Cycle, also known as the light-independent reactions. The end products are carbohydrates, ADP, Ru − P, and water.

39. (C)

Pepsin is an enzyme of the stomach. It starts protein digestion in the stomach by splitting the long polypeptides into shorter fragments. These peptides are further digested in the intestine. The pancreatic secretions contain enzymes that finish the digestion of food in the duodenum, the first part of the small intestine.

40. (B)

When a myelinated neuron is activated, the action potential moves along the axon by jumping from one node to another. This type of transmission is called saltatory transmission and is much faster than point-to-point propagation.

41. (C)

Hydrogen ions combine with the ionized form of hemoglobin (Hb−) to form acid hemoglobin (HHb). Acid hemoglobin binds the H^+ ions relatively strongly so that they can be effectively removed from the blood. This keeps the pH constant if there is an excess of hydrogen ions.

42. (A)

Parathyroid hormone acts to control plasma Ca^{++} by releasing Ca^{++} from bone increasing intestinal absorption of Ca^{++}, and decreasing renal excretion of Ca^{++}. Parathyroid hormone secretion is stimulated by reduced plasma Ca^{++}.

43. (B)

Apoplastic movement is through the extracellular fluid between the cell walls. This movement occurs through the cortex (A) to the endodermis (C). A waxy material on the walls of the endodermal cells called the Casparian strip (B) prevents water movement into the stele (E) where the xylem is located. Plasmodesmata are channels that connect cells.

44. (D)

The core of all motile cilia and flagella consist of two central and 9 peripheral pairs of microtubules composed of the proteins alpha and beta tubulin. Nonmotile cilia lack the two central pairs.

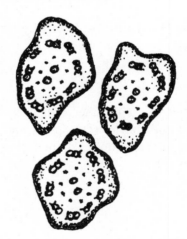

Cross-Sections of Motile Cilia

45. (B)

In protein-driven chemiosmotic phosphorylation, ATP energy is generated as protons pass from a region of high concentration to one of low concentration through CF-1 particles, complex structures of the thylakoid membranes.

46. (B)

Decomposers are the bacteria, fungi, and other microorganisms that digest organisms' carcasses and waste. Their actions return elements from an ecosystem's biotic component back to the abiotic sector of the environment.

47. (E)

Excretion is the removal of metabolic wastes that can become toxic if allowed to accumulate. The kidneys and skin remove salts, urea, and other organic compounds. The lungs remove water and carbon dioxide that are waste products of the circulatory system. The liver breaks down bile pigments, red blood cells, and some proteins and drugs. Lymph nodes, on the other hand, are filters for invading bacteria and indigestible particles such as dust and soot. Thus, the lymphatic system is not excretory in the strict sense of the word since it rids the body of foreign, invading substances rather than waste products of its own metabolism.

48. (B)

The cerebellum is found towards the back of the brain above the medulla. It controls balance, equilibrium, and coordination. Damage to it would lead to loss of these functions. Since driving a motor car requires a great deal of coordination, one would expect that this ability would be impaired when the cerebellum is extensively damaged.

49. (A)

The ecosystem is made up of producers, consumers, and decomposers. Consumers are divided into different trophic levels in the food chain. Herbivores are primary consumers, while carnivores, that feed on herbivores, are secondary consumers.

50. (C)

Releasers are sign stimuli that are usually emitted by a member of the same species. These stimuli are particularly effective in triggering behavioral response. The response of the male European robin to the tuft of red feathers is an instinctive pattern because tufts of red feathers are normally on the breast of competitors. This is an illustration of the action of releasers.

51. (B)

By far the most useful and precise method for separating amino acids in hydrolysate is chromatography. It is the quantitative separation of the amino acids on a column of an ion-exchange resin, followed by the complete elutriation of each amino acid. When this method is used, the position of each amino acid from the protein hydrolysate is known from its position in the chromatography column or its time of elution.

52. (D)

An instinctive behavior is one that depends on an inhibitor or a block. Innate behavior is based on fixed patterns. Not all innate behavior is instinctive. The flight of birds to warm places during winter is an innate behavior that is not instinctive.

53. (B)

The motility is the ability of the alimentary canal to contract and propel food. While doing so it moves the food, is responsible for expulsion of food, grinds and mixes the food and exposes the food to the secretory and absorbing surfaces. Motility does not increase the surface area.

54. (D)

Meiotic drive is a factor that may alter allelic frequencies in a gene pool. Meiotic drive is the term for preferential segregation of genes that may occur in meiosis. For example, if a particular chromosome is continually segregated to the polar body in female gametogenesis, its genes would tend to be excluded from the gene pool since the polar bodies are nonfunctional and will disintegrate. There is significant evidence that, due to physical differences between certain homologous chromosomes, preferential selection of one over the other often occurs at other than random proportions.

55. (C)

Gram-staining is one of the most important differential staining techniques used today to determine differences between bacterial cells. Bacteria may either be gram-positive, staining a violet color, or gram-negative, staining a red color. The difference in staining is based on the varying lipid contents of the cell walls of the bacteria.

56. (B)

Constituitive heterochromatin consists mostly of repetitive DNA. Euchromatin is devoid of repetitive sequences. Faculative heterochromatin varies in state in different cell types or at different states of development and sometimes between homologous chromosomes. Nucleosomes and z-DNA refer to the structure of DNA not the sequencing.

57. (E)

Plankton, the small pelagic organisms, includes both phytoplankton and zooplankton. Plankton includes carnivores and omnivores. Phytoplankton are autotrophs and zooplankton are heterotrophs. This is an example of the importance of understanding the exact meaning of a word.

58. (D)

The order of the contraction of the heart is

S-A node → atria →A-V node →
Bundle of His → ventricles.

The origin of the heartbeat is the S-A node (sinoatrial node). The S-A node causes the atrium to contract. The impulse is then transmitted to A-V node (atrioventricular node) and then to the bundle of His, which then initiates ventricular contraction.

59. (A)

There are two basic proteins in the clotting process of blood: prothrombin and fibrinogen. When the vessel is damaged, the damaged cells release thromboplastins. These enzymes break down prothrombin into thrombin.

Thrombin in turn breaks apart fibrinogen, thus causing fibrin to be formed. Fibrin fibers, along with damaged platelets, form the clot.

60. (C)

Schwann cells are the myelin-forming cells of the peripheral nervous system. Each Schwann cell forms a single myelin internodal segment around a portion of an axon. Schwann cells may also surround unmyelinated axons, without producing myelin.

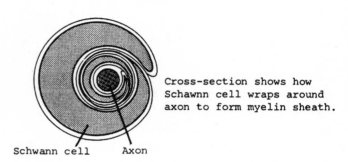

Cross-section shows how Schawnn cell wraps around axon to form myelin sheath.

Schwann cell   Axon

61. (B)

The sliding filament model of muscle contraction states that thick and thin filaments slide past each other while their lengths remain the same. The H zone is composed of an area of thick filament alone; the overlap with thin filaments is not part of this area. As muscles contract, the Z bands come closer and the H zone decreases.

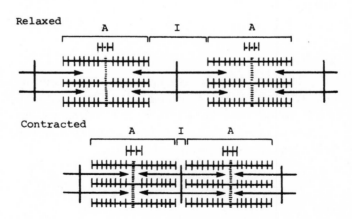

The Sliding Filament Theory of Muscle Contraction

The thin filaments slide over the thick filaments during contraction. Filaments do not shorten during contraction. The sarcomere itself shortens since adjacent Z-lines are drawn together as actin filaments slide over myosin filaments. Actin filaments on either side of a Z-line have opposite polarity, so that each sarcomere contracts during filament sliding.

62. (A)

If there are weak bonds between the enzyme and the inhibitor, the bonds can be broken and the substrate can compete for the binding sites, particularly if more (not less (E)) substrate is added. Noncompetitive inhibitors bind at sites away from the active site (B) and change the shape of the enzyme (C). Some allosteric enzymes have an active-inactive form (D) but the inactive form is not due to competitive inhibition.

63. (C)

Pines, tomato plants, pepper plants, and ferns are vascular plants. Of the plants that are mentioned, only tomato and pepper plants are flowering plants. However, pine trees, tomato plants, and pepper plants all produce seeds. Ferns and liverworts produce spores.

64. (C)

A controllable unit of transcription is called an operon. An operon consists of a binding site for RNA polymerase (promoter), a binding site for a specific repressor (operator), and one or more structural genes. RNA polymerase is the enzyme responsible for transcription of DNA.

The operator gene is located between the promoter and structural genes on the chromosome. When a repressor binds to the operator, the repressor prevents the movement of RNA polymerase along the DNA molecule, thereby inhibiting transcription of the structural genes. When the repressor is not bound to the operator, transcription is free to occur.

65. (A)

The sporocytes must enter the liver and the merozoites must enter the red blood cells. The gametocytes are in the plasma and taken up into the gut of the mosquito when the mosquito bites the person. The gametes, zygote and oocytes are all produced and stay in the digestive ract of the mosquito.

66. (D)

The plasmodium feeds on hemoglobin and thus feeds when in the red blood cells. The wastes are released into the plasma where the waste are recognized as foreign matter and attacked by the white blood cells. This causes fever and associated chills. The release of the merozoites from the red blood cells to the plasma and their subsequent uptake into the red blood cells (A) and synchronized development of the sporozoite and merozoites may contribute to the recurrence of the chills and fever but not the chills and fever themselves. The mosquitoes (C) only introduce the plasmodium to the humans. The hypothalamus (E) is important in temperature regulation. It does not cause chills and fevers.

67. (A)

These compounds are all glycolytic intermediates; clearly, the metabolic pathway is thermodynamically favorable under intracellular conditions and so it is only required to write down the order of compounds as they are formed in glycolysis.

68. (A)

The Mesozoic Era began some 230 million years ago and was characterized by a wide variety of reptiles. In fact, the Mesozoic Era is commonly referred to as the "Age of Reptiles"; common reptiles of this era were the primitive lizards: snakes, turtles, alligators, crocodiles, pterosaurs (flying reptiles), and later in the era, the dinosaurs. All of these, and also the mammals which came later, evolved from an important paleozoic group called the stem or root reptiles (cotylosaurs).

69. (A)

Fungi are saprobes that cannot produce their own carbohydrates. Many nonphotosynthetic plants, such as mushrooms, cannot produce their own carbohydrates from the air and water. Instead, they must obtain the carbohydrates they require from their surroundings. This is usually supplied by dead plant matter, which is mainly cellulose, and other polysaccharides, which must be first digested. Specialized secretary cells in these organisms release the enzymes for digestion into the surrounding organic matter. The products of this extracellular digestion are then directly absorbed through the cells' surfaces, sometimes by specialized absorbing structures.

70. (D)

The placenta is a region where a portion of the embryonic chorion and the maternal uterine wall join. It functions in the exchange of nutrients, wastes, and gases between the mother and the fetus.

71. (B)

The compounds known as prostaglandins have been identified as cyclic, oxygenated, 20-carbon fatty acids. The prostaglandins, which are secreted by the seminal vesicles and other tissues, appear to mediate hormonal action by influencing the formation of cyclic AMP.

72. (D)

The conversion of milk to yogurt is accomplished by bacteria. Penicillin is produced from the species of the fungus *Penicillium*. Similarly, species of the fungus *Penicillium* are used to produce the colors and flavors of some cheeses. Soy sauce is produced by fermenting soybeans with fungus. Ringworm is a fungus that colonizes on human skin.

73. (D)

The conidia are spores produced asexually in Ascomycetes. Sporangia are produced asexually in Zygomycota (zygote fungi). Zygosporangia, ascospores, and basidospores are sexually derived.

74. (D)

A gastrula is a two- or three-layered ball of cells; these layers are known as the germ layers. In chordates and echinoderms, the gastrula is formed by the inward folding of one side of the blastula, partially obliterating the original cavity, or blastocoel. A new cavity is formed which is open to the outside and is called the archenteron or primitive gut. The blastopore is the opening from the archenteron to the outside.

75. (E)

To answer this question you must know how evolution occurs, and you must figure out whether specific situations lend themselves to evolutionary change. In order for evolutionary change to occur, organisms must exist that are genetically different from each other. Acquired changes in the individual will not contribute to evolution because there is no genetic basis for these changes. Even if the acquired changes increase reproductive success, the offspring will not show an increase in the frequency of the acquired trait.

Thus, a nutritionally well-balanced diet may improve the horses that are given such a diet, but it will not change the next generation of horses. On the other hand, genetically different forms of birds provide the raw materials for evolution. Those organisms that are genetically better adapted to the environment will be the most likely to survive and reproduce. For example, antibiotic-resistant bacteria will be selected for after exposure to antibiotics. Behavioral characteristics that are not acquired would be inherited and could be selected either during natural selection or selective breeding.

Another source of evolutionary change is genetic drift. Given genetic variation in a small population, it is possible that some organisms may survive based upon luck. If the individuals surviving a hurricane happen to be genetically different from the rest of the population, then evolutionary change will occur. Genetic drift does not necessarily mean that the surviving organisms are the best adapted or most fit.

76. (D)

Lobed fins, internal nostrils, and primitive lungs are the invention of the Crossopterygians. These genetic mutations allowed these ancient fish to move and respire for short periods of time on land. It is thought that amphibians evolved from such lobed finned fish.

77. (A)

If a sex-linked trait is recessive, a male would have a greater chance of expressing the trait. In order for a girl to express that trait, she would have to have two copies of the recessive gene, because its expression would be masked by the presence of a normal X chromosome or one with a dominant allele. Thus, both her parents would have to carry the gene. A boy, however, need only have one copy of the gene in order to express its trait, because his Y chromosome does not carry any genes that would mask the recessive gene. So only his mother needs to carry the gene. The chances of this happening are much greater than the chance that two people carrying the gene will mate and have a female. Therefore, the trait is more commonly expressed in males.

78. (D)

As energy flows through the various food chains, it is being constantly channeled into three areas. Some of the energy goes into production, which is the creation of new tissues by growth and reproduction. Energy is also used for the manufacture of storage products such as fats and carbohydrates. The rest of the energy is lost to the ecosystem by respiration and decomposition. The loss of one energy due to respiration is very high and only a small fraction of energy is transferred successfully from one trophic level to the next.

Each trophic level depends on the preceding level for its energy source. Ecological efficiencies vary widely, but it has been shown that the average ecological efficiency of any one trophic level is about 10%.

79. (E)

The femur is the large, upper leg bone which extends from the hip (pelvis) to the knee. The tibia and fibula are two smaller bones which extend from the knee to the ankle. The patella is the scientific name for the knee cap, and the humerus is the upper arm bone.

80. (C)

Prokaryotic cells lack the internal membranous structure characteristic of eukaryotic cells, and generally belong to the simple life forms, such as bacteria. Prokaryotic cells lack endoplasmic reticulum, Golgi apparati, lysosomes, vacuoles, and a distinct nucleus. They are generally less organized and less advanced than eukaryotic cells.

81. (E)

Isolation and associated inbreeding of the inhabitants resulted in development of characteristics different than those of the inhabitants of mainland China. The sea that separates the island from the mainland is the reproductive barrier that isolated the population.

82. (B)

Divergent evolution and adaptive radiation are associated with increasing dissimilarity. The independent occurrence of similarities resulting from common ancestry is termed homology and the process is termed parallel evolution. Similar features (e.g., wings of insects and binds) that occur independently in distantly related organisms in response to adapting to similar environments illustrate convergence. From different evolutionary origins, unrelated organisms become similar. One example is the evolution of wings in insects and birds.

83. (E)

If no disruptive factors interfere, most successions eventually reach a stage that is much more stable than those that preceded it. The community of this stage is called the climax community. It has much less tendency than earlier successional communities to alter its environment in a manner injurious to itself. In fact, its more complex organization, larger organic structure, and more balanced metabolism enable it to control its own physical environment to such an extent that it can be self-perpetuating.

84. (A)

The pink color is dietary and not inherited, thus, all offspring will be white.

85. (E)

This is a form of behavioral isolation. The different mating songs are different behaviors. Both species breed in the spring and on overlapping breeding grounds, so it is not temporal or habitat isolation. There is no information about breeding compatibility and thus the possibility of physical mating barriers. Because they are two different species, there is no gene flow between them.

86. (C)

When a segment of one chromosome is transferred to another non-homologous chromosome, the mutation is known as a translocation. A deletion is a mutation in which a segment of the chromosome is missing. In duplication, a portion of the chromosome is represented twice. An inversion results when a segment is removed and reinserted in the same location, but in the reverse direction.

A B C D E F	Normal
A B C E F	Deletion of segment D
A B C C D E F	Duplication of segment C
A E D C B F	Inversion of segment B-E
G H I J K L A B C	Translocation from A-C to chromosome GHIJKL

Mutations Involving Chromosome Structure

87. (A)

The secondary and tertiary structures of polypeptides are destroyed by denaturation.

88. (A)

Parenchyma tissue, which occurs in roots, stems, and leaves, consists of small cells with a thin cell wall and a thin layer of cyptoplasm surrounding a large vacuole. The cells are loosely packed, resulting in abundant spaces in the tissue for gas and nutrient exchange. Most of the chloroplasts of leaves are found in these cells.

89. (D)

Active transport initially enabled the cell to maintain a higher concentration of dye inside the cell. However, if the inhibitor were to interfere with ATP production and, consequently, active transport, normal diffusion processes would take over, resulting eventually in equal intra- and extra-cellular concentrations of the dye.

90. (B)

In most plants the stomata are located mainly in the lower epidermis. The lower epidermis is the side of the leaf usually turned away from the sun's rays where the drying tendency is less harsh. It is also covered with short hairs which reduce direct air currents across the stomatal openings. The oleander (Nerium) which lives in a very dry habitat has stomata, which are located in deep hair-lined depressions in the lower epidermis.

91. (D)

External fertilization means that eggs are fertilized outside the female's body. It follows that if eggs are fertilized externally, then development of the embryo will take place externally also. Organisms in which fertilization is external must produce an enormous number of eggs. Since the egg cells (which are discharged from the body of the female) are unprotected, many of them are eaten or otherwise lost. These are r-selected species. External fertilization and external development of the young are the reproductive patterns present in most fish and amphibians.

92. (C)

It is necessary to distinguish aggressive behavior between members of the same species from violent predatory behavioral patterns, which are usually directed at members of different species. Aggressive behavior is usually exhibited between same-species males in defending their territory or establishing their status in a social order. The critical factor in aggressive behavior is that most of the fighting consists of display. Display consists of ritualized, highly exaggerated movement or sound that conveys the attack motivation of the contestants. In attempting to appear as formidable as possible, the animal often changes the shape of certain body parts in an attempt to make itself appear as large as possible. The adaptive significance of display is that the two contestants are rarely seriously hurt.

93. (B)

The environment above the tree line would resemble most closely the climate of the tundra.

Because of the tundra's intense coldness, trees are unable to grow and vegetation consists mostly of grasses.

94. (E)

The parts of a flower are as follows.

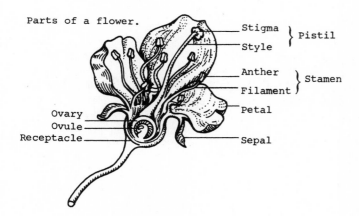

Parts of a flower.
Stigma } Pistil
Style
Anther } Stamen
Filament
Petal
Sepal
Ovary
Ovule
Receptacle

95. (A)

Structure A is known as the stigma; it is part of the flower's female reproductive organ called the pistil.

96. (D)

Flowers which are pollinated by the wind need not be attractive to birds and insects, and so their petals tend to be less showy, and may even be absent.

97. (A)

(A) is the only correct answer. A permissive cell permits viral replication—it is one in every 104 to 105 nonpermissive cells that may be transformed by an oncogenic virus. RNA-DNA hybrids and reverse transcriptases are characteristic of infection by RNA viruses.

98. (D)

In eukaryotes, the group of codons necessary to produce one polypeptide is called a cistron both in mRNA and DNA. In DNA, the cistron is also referred to as the "structural gene," because it determines protein structure.

99. (B)

Any step involved in the synthesis of protein is subjected to control. DNA carries the genetic information of how a protein is made. This information needs to be transcribed and translated. The first level of control rests on how or when a gene is transcribed. Newly formed transcripts should be processed with the addition of a poly A tail to the 3′ end and a methylated G nucleotide (called a cap) to the 5′ end of mRNA molecules in eukaryotes. Following transcription and processing, a transcript is selected to be transported across the nuclear membrane to the cytoplasm. Once in the cytoplasm, mRNA is either translated by ribosomes or degraded. The DNA is not degraded.

100. (B)

Many biologists believe that eukaryotic cells evolved when formerly free-living prokaryotes established a symbiotic relationship. Perhaps one prokaryotic cell ingested another and instead of digesting it began to coexist with it. Much of the evidence for the symbiont theory involves similarities between bacteria and two of the main organelles of eukaryotic cells: chloroplasts and mitochondria.

101. (A)

The developmental stage of cells where the cells resemble a hollow ball is known as the blastula. A morula is a developmental stage found in echinoderms. The isolecithal stage is not a stage of development, but a description of the arrangement of the yolk and albumin in an egg. The gastrula is a 3-layered cell stage formed from the blastula. Ovulation is the release of one or more eggs from the ovaries.

102. (E) 103. (B) 104. (D)

The fox is a possible predator of pheasants in an ecosystem food chain. Notice how a decrease in pheasant prey correlates with annual gains in fox success by numbers. The density of a population is determined by dividing the number of individuals in the population by the number of acres. The largest change in the population density of either organism occurs in the year 1981. The pheasant population density decreases from 16 to 10 to 8 to 6 from 1980 to 1983, respectively. The population density of pheasants increases slightly in 1984.

If the pheasants were eating something toxic to the fox but not to themselves, one might expect the fox population density to increase slightly in response to the large number of pheasants. However, because the pheasants would be poisonous, the fox population would then suddenly decrease. The decrease would be quite rapid. Under such conditions, the number of pheasants would in turn increase rapidly due to the decreased number of predators.

105. (C)

Penetrance of a mutation is defined as the percentage of individuals that show some degree of expression of a mutant genotype.

106. (D)

A monophyletic group is a terminal group, or a group that contains a common ancestor and all of its descendants. Separately, each group is a terminal group and thus a monophyletic group. However, of the combinations given, only the mollusks, annelids and arthropods are a monophyletic group.

107. (D)

The protozoa or animal-like protists gave rise to the animals. The protozoa are unicellular and animals are multicellular and thus the derived trait of animals is multicellular organization. True tissues are the derived characteristic of cnidarians. All animals have intracellular, not extracellular digestion. Bilateral symmetry and cephalization are derived characteristics of flatworms.

108. (A)

Damage to the right temporal lobe of the cerebral hemisphere could result in an individual's poor performance on an IQ test.

109. (C)

The following chart shows the official categorization of era, period, and epoch.

Era	Period	Epoch
Cenozoic	Quaternary	Recent
		Pleistocene
	Tertiary	Pliocene
		Miocene
		Oligocene
		Eocene
		Paleocene
Mesozoic	Cretaceous	
	Jurassic	
	Triassic	
Paleozoic	Permian	
	Carboniferous	
	Devonian	
	Silurian	
	Ordovician	
	Cambrian	

110. (A)

Membrane lipid molecules are amphipathic: they contain both hydrophilic and hydrophobic ends. The three major classes of membrane lipid molecules are phospholipids, cholesterol, and glycolipids. Different lipid compositions on the inner and outer monolayers contribute to the asymmetry of plasma membrane. For example, glycolipids are present only in the outer monolayer with their sugar groups exposed at the cell surface. The function of these oligosaccharide side chains is unknown. But their structural complexity and positions on cell surface may suggest a role in intercellular communication. Phosphatidyl ethanolamine, phosphatidyl serine, and sphingomyelin fall under the category of phospholipids.

111. (B)

The DNA enzyme which seals the Okazaki fragments in the lagging strand of a replication fork is ligase.

112. (A)

The energy molecule used to attach the 50S ribosomal subunit to the 30S subunit is GTP.

113. (C)

This question asks you to interpret the data shown on the graph. You must look at the unbroken line showing the number of algae present and find the steepest portions slanting upward. There are marked increases in growth in both the spring and autumn.

114. (D)

During the summer, temperature and light are high and therefore do not limit the growth of the algae. Nutrients are low and thus limit gowth of algae. Although there is a slight increase in nutrient levels in the latter part of the summer, there is a lag period between increase in nutrients and increase in algae. This lag is also apparent in spring.

115. (E)

To answer this question you must compare the changes in environmental factors with peaks in algal growth. In the winter, nutrient levels are high, but light and temperature levels are at their lowest. Therefore, light and temperature must be limiting factors. Maximum growth in the spring corresponds to high nutrient levels with increasing amounts of light and temperature. As the algae grow they use up nutrients in the environment, which limits their continued growth. In the fall, nutrient levels begin to increase again. Growth increases and is sustained while temperature and light levels are still high.

116. (E)

The decomposers release nutrients (5), which are taken up by organisms (6).

117. (D)

Respiratory energy (4) is heat that is unavailable to other organisms. Both nutrients and energy move from the producer to the decomposer (1), the producer to the consumer (2), and the consumer to the decomposer (3). Only nutrients are made available by the decomposer (5).

118. (B)

This question is about characteristics of biomes, not specifically about temperate grasslands. An important characteristic of biomes is the similar abiotic conditions. The types of species present, not the particular species composition (A), are similar. This is also true for pioneer species (C) in succession. The altitude (D) may or may not be similar. Altitude affects the parameters that define the abiotic conditions but by itself does not determine the biome. The amount of fresh water (E) within a terrestrial ecosystem can affect the particular characteristics of the terrestrial system, but it is not consistent among any one terrestrial ecosystem.

119. (D)

Inversion is a type of chromosomal aberration in which a segment of a chromosome is turned around 180° and reinserted into the chromosome. It is produced by two breaks in a chromosome. The presence of an inversion does not imply that any new genes are present.

120. (D)

A population comprises all of the individuals of a particular species in a particular area. The red-winged blackbirds in a marsh fit the definition of a population. The plants are a large taxonomic group that includes many species. The salamanders and frogs are two different species in the marsh. The decomposers are one group in a trophic level that includes many species. The red worms and sediment include only one species but also include an abiotic factor, the sediment.

121. (A)

Angiosperms are divided into monocots and dicots according to five major structural differences. Answer (A) is correct: the pattern of veins in the leaf is netlike in the dicot and parallel in the monocot. Answers (B) through (E) have the characteristics of dicots and monocots reversed.

122. (A)

Recall that the phloem serves to transport nutrients from leaves to the rest of the plant, and not to transport water.

123. (B)

In the photolysis of water, two hydrogen ions are released into the interior space of the thylakoid, and two electrons enter the electron transport system. Photosystems P680 and P700 are used.

124. (C)

The lamellae extend continuously between the thylakoids, through an amorphous region called the stroma. The thylakoids with their photosynthetic pigments are involved in the immediate, light-related events of photosynthesis—the light reaction. However, carbohydrate formation actually begins in the stroma. This part of photosynthesis does not involve light directly and is termed the "dark reaction" or the light-independent reaction.

125. (D)

The coelom is a fluid-filled cavity that is lined by mesoderm and lies between the digestive tract and the outer body wall. It is found in mollusks, annelids, arthropods, echinoderms, and chordates. True tissues and bilateral symmetry are important derived characteristics in animal phylogeny but they are not cavities.

Vascular tissues are found in plants. The archenteron is an embryonic cavity.

126. (C)

Flavenoids are pigments that give color to flowering plants. All of the other pigments are in chloroplasts and absorb radiant energy and transfer it to chlorophyll a.

127. (D)

Homo habilis is the name given by Louis Leakey to certain fossils found in Africa that seemed closer to the human line. He placed them in the genus *Homo*. *Homo habilis* has a much greater cranial capacity than *Australopithecus* and appears closer to the human line than *Australopithecus*. *Australopithecus*, however, was the one that possessed strong jaws and teeth.

128. (C)

Progesterone acts on the glandular epithelium of the uterus called the endometrium and converts it to an actively secreting tissue. The glands become coiled and filled with glycogen, the blood vessels become more numerous, and various enzymes accumulate in the glands and connective tissue of the endometrium. These changes are ideally suited to provide a favorable environment for implantation of a fertilized ovum. Progesterone also causes the mucus secreted by the cervix (a muscular ring of tissue at the mouth of the uterus which projects into the vagina) to become thick and sticky. This forms a "plug" which may constitute an important blockage against the entry of bacteria from the vagina. This is a protective measure for the fetus should conception occur.

129. (D)

The position of the continents has changed over billions of year and at the time of the death of the coal marsh plants, the northeastern USA and Britain were part of one landmass. Gymnosperms and angiosperms dominated in latter eras. The tar pits are more recent and occur in southwestern, not northeastern, USA. The carboniferous period is in the Paleozoic era. Mass extinction occurred at the end of the Paleozoic, not in the midst of this era.

130. (A)

Increasing the number of sets of chromosomes is a known way of producing new species. It is common in plants. Mutations lead to altered gene frequencies and microevolution, but will only result in a new species if the phenotype containing the gene is selected for and the individuals within the population are isolated. Inbreeding changes the genotype frequency, not the allele frequency, and does not produce new species. Polymorphism is the phenotypic expression resulting from genotypes for which there are several possible alleles. It contributes to variation and in many cases to stability, not speciation. Increased fitness is the increased contribution of alleles by an individual to the next generation. Speciation may result in increased fitness of individuals, but fitness will not lead to speciation.

131. (B)

Rhizoids are usually found in bryophytes such as mosses and liverworts. They are simple filaments of cells or cellular projections performing the function of water absorption. The rhizoids are, however, not very efficient absorbers and in relatively dry areas cannot withdraw adequate materials from the ground.

132. (E)

The cells of the palisade layer contain large numbers of chloroplasts that are responsible for photosynthesis.

133. (C)

The mature sporophyte is composed of a foot in the archegonium of the gametophyte and a leafless, spindle-like stalk or seta which rises above the gametophyte. The sporophyte is nutritionally dependent on the gametophyte, absorbing water and nutrients from the archegonium via the tissues at the foot.

134. (A)

The cotyledon is a primary leaf structure found in the early angiosperm embryo. In the monocots, the single cotyledon serves primarily to absorb the endosperm tissue, rather than store it, as dicots do. The portion of the embryo lying below the point of attachment of the cotyledons is called the hypocotyl and the part above is called the epicotyl. In monocots, the endosperm usually persists even after germination, and the cotyledon continues to absorb the nutrient material for the seedling until it can synthesize its own nutrients.

135. (D)

The xylem of the vascular system found in plants is involved in the conduction of absorbed water and minerals up the stem to the leaves.

136. (D)

The structure of enzymes differ significantly. Some are composed solely of protein (for example, pepsin). Others consist of two parts, a protein part (also called an apoenzyme), and a nonprotein part or cofactor. The cofactor may be either a metal ion or an organic molecule called a coenzyme. Coenzymes usually function as intermediate carriers of the functional groups, atoms, or electrons that are transferred in an overall enzyme transfer reaction.

137. (E)

A coenzyme that is very tightly bound to the apoenzyme is called a prosthetic group.

138. (A)

When an apoenzyme and a cofactor are joined together, they form what is called a holoenzyme. Often, both a metal ion and a coenzyme are required in a holoenzyme. Those apoenzymes needing a metal ion to function are also called metalloenzymes.

139. (A)

The most posterior part of the brain, connected immediately to the spinal cord, is the medulla. Here the central canal of the spinal cord (spinal lumen) enlarges to form a fluid-filled cavity called the fourth ventricle. The medulla has numerous nerve tracts (bundles of nerves) which bring impulses to and from the brain. The medulla also contains a number of clusters of nerve cell bodies, known as nerve centers. These reflex centers help control respiration, heart rate, the dilation and constriction of blood vessels, swallowing, and vomiting.

140. (E)

The thalamus of the forebrain serves as a relay center for sensory impulses. Fibers from the spinal cord and parts of the brain synapse here with other neurons going to the various sensory areas of the cerebrum. The thalamus seems to regulate and coordinate the external signs of emotions. By stimulating the thalamus with an electrode, a sham rage can be elicited in a cat—the hair stands on end, the claws protrude, and the back becomes humped. However, as soon as the stimulation ceases, the rage responses disappear.

141. (C)

Above the medulla is the cerebellum, which is made up of a central part and two hemispheres extending sideways. The size of the cerebellum in different animals is roughly correlated with the amount of their muscular activity. It regulates and coordinates muscle contraction and is relatively large in active animals such as birds. Removal or injury of the cerebellum is accompanied not by paralysis of the muscles but by impairment of muscle coordination. A bird, with its cerebellum surgically removed, is unable to fly and its wings seem to thrash about without coordination.

142. (D)

The cerebrum, consisting of two hemispheres, is the largest and most anterior part of the human brain. In human beings, the cerebral hemispheres grow back over the rest of the brain, hiding it from view. The outer portion of the cerebrum, the cortex, is made up of gray matter which comprises the nerve cell bodies. The gray matter folds greatly, producing many convolutions of the cerebral surface. These convolutions increase the surface area of the gray matter. The inner part of the brain is the white matter which is composed of masses of nerve fibers.

143. (A)

Semiconservative replication is the mechanism of DNA replication; the process by which the two complementary strands of the double helical DNA replicate to form new complementary strands. This model states that the replication of one DNA molecule yields two hybrids, each composed of one parental strand and one newly synthesized strand.

144. (E)

Inside the nucleus, transcription of DNA is mediated by an enzyme, RNA polymerase. DNA undergoes a localized unfolding in the vicinity of the gene to be transcribed. RNA polymerase then selects precursor ribonucleotides complementary to the DNA template which are then polymerized to give a strand of mRNA. The newly synthesized single-stranded mRNA subsequently peels away from the template allowing a new RNA polymerase molecule to attach or the DNA strand to reunite.

145. (C)

Transfer RNA is the smallest type of RNA and is involved in the carrying of amino acids to the ribosomes where translation of mRNA into proteins takes place. There are at least 20 different tRNAs (one specific for each amino acid). Unlike the other two RNAs (mRNA and rRNA), tRNA forms loops and double-stranded sections. In one of these loops is located the anticodon, which is what distinguishes the different types of tRNA from each other. The function of tRNA is to insert the amino acid specified by the codon on the mRNA into the growing polypeptide chain at the ribosomes, and it is through the complementation of anticodon and codon that the appropriate amino acid is incorporated.

146. (D)

Messenger RNA carries the genetic information coded for in DNA and is responsible for the translation of that information into a polypeptide chain. Each set of three bases of the mRNA comprises a codon which directs the incorporation of a specific amino acid into the polypeptide chain. Messenger RNA binds reversibly to the smaller subunit of the ribosome, where protein synthesis is initiated. It can dissociate from the ribosome without jeopardizing the integrity of the ribosome. Messenger RNA is heterogeneous in size due to the different lengths of polypeptide chains which a cell needs to synthesize; therefore, it is the most variable type of RNA.

147. (E)

The human and other animals' digestive system begins at the oral cavity. The teeth break up food by mechanical means, increasing the substrate's surface area available to the action of digestive enzymes. In addition to tasting, the tongue manipulates food and forms it into a semipherical ball (bolus) with the aid of saliva.

148. (A)

When food reaches the stomach, the distension of the stomach walls stimulates an increase in the rate of stomach movement and the production of gastrin.

149. (B)

The first part of the small intestine, which is mostly involved in absorption, is the duodenum. Small finger-like protrusions, called villi, line the entire small intestine. Villi greatly increase the intestinal surface area and it is through the villi that most of the nutrient absorption takes place.

150. (D)

In an area of the large intestine called the colon, massive numbers of bacteria exist. Their function is not fully understood, but some can synthesize vitamin K, which is of great importance to the human blood clotting mechanism.

151. (C)

Lipase, produced by the pancreas, digests the lipids in our foods, with the help of bile.

152. (C)

The initiation of the heartbeat and the beat itself are intrinsic properties and are not dependent upon stimulation from the central nervous system. The initiation of the heartbeat originates from a small strip of specialized muscle in the wall of the right atrium called the sino-atrial (S-A) node, which is also known as the pacemaker of the heart. It is the S-A node which generates the rhythmic self-excitatory impulse, causing a wave of contraction across the walls of the atria. This wave of contraction reaches a second mass of nodal muscle called the atrio-ventricular node, or A-V node. The A-V node then helps the rest of the heart to undergo contraction.

153. (B)

A four-chambered heart is characteristic of "warm-blooded" animals such as birds and mammals. Since these animals maintain a relatively high constant body temperature, they must have a fairly high metabolic rate. To accomplish this, much oxygen must be continually provided to the body's tissues. A four-chambered heart helps to maximize this oxygen transport by keeping the oxygenated blood completely separate from the deoxygenated blood.

154. (A)

It is in the capillaries that the most important function of circulation occurs; that is, the exchange of nutrients and waste materials between the blood and the tissues.

Capillaries are the blood vessels that connect the arteries with the veins. The capillaries have walls composed of endothelium only one cell thick. It is the thinness of the capillary walls which allows for the diffusion of oxygen and nutrients from the blood into the tissues and for carbon dioxide and nitrogeneous wastes to be removed from the tissues by the blood.

155. (C)

During a fever or exercise, excess heat is produced which raises the body temperature a few degrees. This extra heat causes the sino-atrial node to increase its rate of stimulation, thus increasing the number of beats per minute. It is thought that increased temperature increases the permeability of the muscle membrane of the S-A node to various ions (Na^+ and K^+), thus accelerating the stimulatory process. (It is the change in membrane permeability to Na^+ ions that causes an impulse to arise.) This temperature sensitivity of the S-A node explains the increased heart rate accompanying a fever.

156. (A)

Pheromones, specific chemicals involved in social communication, are released by an organism and act upon another organism at a distance from their point of release and in very specific ways. Pheromones can be classified in two groups: those that possess releaser effects, which elicit immediate behavioral responses, and those that have primer effects, which work by altering the physiology and subsequent behavior of the recipient. Releaser pheromones are used by a variety of animals for different purposes, such as the attraction of mates, individual recognition, or trail or territorial marking. For example, the females of many moth species secrete pheromones to attract males over large areas.

157. (E)

Behavior that contributes to the immediate survival and reproduction of the animal is adaptive. The egg-shell-and-object-removing habit of these gulls is such adaptive behavior. The gulls remove not only broken shells, but also any conspicuous objects placed in the nest during the breeding season. They seem to discard conspicuous objects as a defense mechanism against visual predators. When investigators placed conspicuous objects along with eggs in nests, these nests were robbed of eggs (by other gulls) more often than nests having only eggs. Thus, object-removing behavior is significant and adaptive in that it reduces the chances of a nest being robbed, thus enhancing the survival of offspring.

158. (B)

Displays, mostly visual, serve as signals to convey their potential behavior. Such displays as observed with the gulls convey an animal's readiness to mate, to attack, or to retreat.

159. (D)

Müllerian mimicry involves the evolution of two or more inedible or unpleasant species to resemble each other. Since each species serves as both model and mimic, greater protection is afforded to them because their repellent qualities are more frequently advertised.

160. (C)

Courtship is an important means of communication between some animals that involves a multitude of precopulatory behavior patterns which serve in many instances to advertise the presence of a sexually receptive individual and inhibit aggression by a prospective mate.

161. (B)

Additional roots that grow from the stem or leaf, or any structure other than the primary root or one of its branches, are termed adventitious roots. Adventitious roots of climbing plants such as the ivy and other vines attach the plant body to a wall or a tree. Adventitious roots will arise from the stems of many plants when the main root system is removed. This accounts for the ease of vegetative propagation of plants that are able to produce adventitious roots.

162. (C)

Meristematic cells comprise the primary growth zones of both roots and plants. The meristem consists of actively dividing cells from which all the other tissues of the root or stem are formed.

163. (A)

Inside the epidermis of both stem and root is the cortex. The cortex of the root is a wide area composed of many areas of large, thin-walled, nearly spherical parenchymal cells. The cortex of the stem is narrower than that of the root. The cortical cells differ from those of the root in that the former is photosynthetic while the latter is not. In addition, there is an outer layer of thick-walled collenchymal cells in the stem, which serves as a supportive tissue. Collenchymal cells are not present in the root.

164. (D)

Lenticels are masses of cells which rupture the epidermis and form swellings. Lenticels represent a continuation of the inner plant tissues with the external environment, and permit a direct diffusion of gases into and out of the stem or twig. Such direct passages are necessary because the cambium forms a complete sheath around the vascular bundles and effectively obstructs the ventilation of the vascular bundles.

165. (E)

Cutin is a waxy organic substance secreted by the epidermal cells of the stems and leaves, but not the roots. Its waterproof property retards water loss due to evaporation.

166. (D)

Annelids are segmented worms whose central nervous system is a pair of longitudinal segmented cords in which the cell bodies of the neurons form masses called ganglia. The ganglia originate in the esophagus of the head. Almost all the cell bodies are located in the ganglia.

167. (B)

It is in the coelenterates that the most primitive true nervous system appears. This is a netlike system that is composed of separate neurons which cross each other without touching. Coelenterates may have two kinds of neurons. These are slow-conducting neurons and fast-conducting neurons.

168. (A)

Echinoderms are radially symmetrical when mature. Their nervous system consists of a nerve ring around the mouth which gives rise to five nerve trunks. In some groups of echinoderms, the five nerve trunks consist of specialized neurectoderm which form shallow grooves along the surface of the arms.

169. (E)

Flatworms are advanced planarians which possess a longitudinal, bilaterally symmetrical body plan comprised of two nerve cords. The neurons are concentrated in the head region.

170. (C)

Epinephine, also known as adrenalin, is a hormone produced by the adrenal medulla. It causes specific changes in the body usually in response to stress. One of its functions is to cause an increase in blood glucose level through the enzymatic breakdown of glycogen by mediating the activation of an enzyme.

171. (B)

Cyclic AMP or cAMP is now known as the second messenger. It was discovered to be a participant in hormone activity. It is an intracellular second messenger and receives "messages" from hormones which bring their message to the target cell membrane. The response from within the cell is actually started by cAMP.

172. (A)   173. (E)   174. (D)

All of the possible answers can affect the Hardy-Weinberg Equilibrium. Founder and bottleneck effects are both examples of genetic drift enhanced after a large population is reduced in size. However, bottleneck effect is the reduction in size of a large population due to a catastrophe such as a hurricane, whereas founder effect is due to colonization of new areas. Gene flow or migration results in the addition or loss of alleles due to movement of organisms. Nonrandom mating alters the frequency of genotypes but not alleles. Natural selection is usually adaptive.

175. (D)   176. (B)   177. (E)   178. (C)   179. (A)

Each cell cycle is divided into four major parts: M, G1, S, and G2. The M stage represents mitosis and cell division and includes four stages: prophase, metaphase, anaphase, and telophase. Mitosis is followed by G1 where the cell busily synthesizes proteins, builds structures, and carries on all types of metabolic activities. The S stage is the time when DNA replication and the formation of the new chromosomes take place. The G2 stage is the stage when the cell continues to prepare for division.

In prophase of mitosis the chromosomes condense, the nuclear membrane and the nucleoli disappear, and spindle fibers form. In metaphase, the chromosomes align on a plane in the middle of the spindle fibers. In anaphase, the centromeres divide and separate and the two daughter chromosomes of each pair travel to opposite poles of the spindle. In telophase, new nuclear membranes form around each group of daughter chromosomes, the nucleoli appear, the chromosomes decondense and cytokenesis follows where the constricting ring closes down on the remains of the spindle fiber to form two daughter cells.

180. (C)   181. (D)   182. (E)   183. (A)   184. (B)

Serial homology (C) is the specialization of series of traits (e.g., vertebrae) within an individual. It is a special case of homology. Allometric growth results in the condition allometry (D), and is the differential rate of growth of body parts. The difference in the shape of the skull in humans and chimps could be due to differential growth. Homeosis (E) is the evolutionary alteration of the placement of parts and could explain the difference in numbers of petals on different species of flowers. Homoplastic traits look similar and have similar functions: the tail fin of a salmon (a fish) and the flipper of the seal (a mammal) both aid in propulsion during swimming. Analogy (B) is a similarity in function or behavior among organisms or their anatomical structures resulting from convergent evolution rather than common ancestry.

185. (B)

Neutrophils are phagocytic cells that are concerned primarily with local infections. They indicate localized infections such as appendicitis or abscesses in other parts of the body.

186. (C)

Eosinophils are blood cells that generate protein substances that inactivate foreign protein. High eosinophil counts may indicate allergic conditions or invasion of parasitic roundworms such as trichinella spiralis.

187. (E)

Basophils are leukocytes, which probably produce heparin, histamine, and serotonin.

188. (A)

Erythrocytes are cells that transport O_2 and CO_2. This ability is related to their ability to carry hemoglobin.

189. (D)

Lymphocytes are phagocytic cells that are concerned primarily with generalized infections. High lymphocyte counts are present in whooping cough and some viral infections.

190. (B)

According to the experiments ribosomes were not synthesized after infection. This was evidenced by the absence of "light" ribosomes.

191. (D)

The new protein labeled with heavy isotopes this means that new proteins were synthesized in the pre-existing ribosomes, in the heavy isotope medium.

Questions 192–195 concern the data below, regarding F_1 phenotypes.

$$\left.\begin{array}{l} + + v - 37 \\ ct\ y + - 35 \end{array}\right\} \text{parental chromosomes}$$

$$\left.\begin{array}{l} ct + + - 10 \\ +\ y\ v - \ \ 7 \end{array}\right\} \text{single recombinant class I}$$

$$\left.\begin{array}{l} ct\ y\ v = \ 4 \\ +\ +\ + = \ 4 \end{array}\right\} \text{single recombinant class II}$$

$$\left.\begin{array}{l} +\ y\ + = \ 2 \\ ct\ +\ v = \ 1 \end{array}\right\} \text{double recombinant or double crossover}$$

192. (A)

The two parental gametes have the highest frequency ($+ + V = 37/100$ and $ct\ y + = 35/100$) because the loci are linked. Only one of the parental gametes ($ct\ y +$), is contained in the given list. It is a parental gamete.

193. (C)

The difference between $\dfrac{+ + v}{+ y +}$ and $\dfrac{+ + +}{+ y v}$ is the change in position of y in single recombinant class I above. Therefore, allele y is on one end of the chromosome. The difference between $\dfrac{+ + v}{c + y +}$ and $\dfrac{c + y v}{+ + +}$ is the change in position of v. Therefore, gene v is on the other end of the chromosome. Hence, the arrangement of these three genes should be y ct v. The frequency of recombination for the y to ct interval is $(10 + 7 + 2 + 2) \div 100 = 0.2 = 20$ map units. The frequency of recombination for the ct to v interval is $(4 + 4 + 2 + 1) \div 100 = 0.11 = 11$ map units. The linkage map should be

194. (A)

The coefficient of coincidence

$$= \frac{\% \text{ actual double crossover}}{\% \text{ expected double crossover}}$$

$$= \frac{0.03}{(0.17 + 0.03)(0.08 + 0.03)} = 1.36$$

195. (A)

Interference $= 1 -$ coefficient of coincidence
$= 1 - 1.36 = -0.36$

196. (E)

From the diagrams of the curves it is obvious that *P. caudatum*'s growth activities were inhibited since instead of reproducing, its growth actually decreased.

197. (D)

When the two cultures were mixed, one continued growing and one stopped. This effect was most likely due to competition between species, with the most adapted one winning.

198. (D)

From the chart, it was observed that vica faba and the mouse suffered from the same damages in chromosomal aberrations and mutations.

199. (C)

Chromosomal aberrations and mutations can only be understood if screening tests and experiments are carried out. The outcome of such tests explains the causes and reasons for the aberration or mutation. The absence of a mutation or chromosomal aberration means that it is not present, not that the screening program can not detect it.

200. (E)

The most frequent type of chromosomal aberrations, not mutations, were deletions, duplications, and nondisjunction.

GRE BIOLOGY

Practice Test 2

This test is also on CD-ROM in our special interactive GRE Biology TEST*ware*®. It is highly recommended that you first take this exam on computer. You will then have the additional study features and benefits of enforced timed conditions and instant, accurate scoring. See page xi for guidance on how to get the most out of our GRE Biology software.

Answer Sheet

1. Ⓐ Ⓑ Ⓒ Ⓓ Ⓔ
2. Ⓐ Ⓑ Ⓒ Ⓓ Ⓔ
3. Ⓐ Ⓑ Ⓒ Ⓓ Ⓔ
4. Ⓐ Ⓑ Ⓒ Ⓓ Ⓔ
5. Ⓐ Ⓑ Ⓒ Ⓓ Ⓔ
6. Ⓐ Ⓑ Ⓒ Ⓓ Ⓔ
7. Ⓐ Ⓑ Ⓒ Ⓓ Ⓔ
8. Ⓐ Ⓑ Ⓒ Ⓓ Ⓔ
9. Ⓐ Ⓑ Ⓒ Ⓓ Ⓔ
10. Ⓐ Ⓑ Ⓒ Ⓓ Ⓔ
11. Ⓐ Ⓑ Ⓒ Ⓓ Ⓔ
12. Ⓐ Ⓑ Ⓒ Ⓓ Ⓔ
13. Ⓐ Ⓑ Ⓒ Ⓓ Ⓔ
14. Ⓐ Ⓑ Ⓒ Ⓓ Ⓔ
15. Ⓐ Ⓑ Ⓒ Ⓓ Ⓔ
16. Ⓐ Ⓑ Ⓒ Ⓓ Ⓔ
17. Ⓐ Ⓑ Ⓒ Ⓓ Ⓔ
18. Ⓐ Ⓑ Ⓒ Ⓓ Ⓔ
19. Ⓐ Ⓑ Ⓒ Ⓓ Ⓔ
20. Ⓐ Ⓑ Ⓒ Ⓓ Ⓔ
21. Ⓐ Ⓑ Ⓒ Ⓓ Ⓔ
22. Ⓐ Ⓑ Ⓒ Ⓓ Ⓔ
23. Ⓐ Ⓑ Ⓒ Ⓓ Ⓔ
24. Ⓐ Ⓑ Ⓒ Ⓓ Ⓔ
25. Ⓐ Ⓑ Ⓒ Ⓓ Ⓔ
26. Ⓐ Ⓑ Ⓒ Ⓓ Ⓔ
27. Ⓐ Ⓑ Ⓒ Ⓓ Ⓔ
28. Ⓐ Ⓑ Ⓒ Ⓓ Ⓔ

29. Ⓐ Ⓑ Ⓒ Ⓓ Ⓔ
30. Ⓐ Ⓑ Ⓒ Ⓓ Ⓔ
31. Ⓐ Ⓑ Ⓒ Ⓓ Ⓔ
32. Ⓐ Ⓑ Ⓒ Ⓓ Ⓔ
33. Ⓐ Ⓑ Ⓒ Ⓓ Ⓔ
34. Ⓐ Ⓑ Ⓒ Ⓓ Ⓔ
35. Ⓐ Ⓑ Ⓒ Ⓓ Ⓔ
36. Ⓐ Ⓑ Ⓒ Ⓓ Ⓔ
37. Ⓐ Ⓑ Ⓒ Ⓓ Ⓔ
38. Ⓐ Ⓑ Ⓒ Ⓓ Ⓔ
39. Ⓐ Ⓑ Ⓒ Ⓓ Ⓔ
40. Ⓐ Ⓑ Ⓒ Ⓓ Ⓔ
41. Ⓐ Ⓑ Ⓒ Ⓓ Ⓔ
42. Ⓐ Ⓑ Ⓒ Ⓓ Ⓔ
43. Ⓐ Ⓑ Ⓒ Ⓓ Ⓔ
44. Ⓐ Ⓑ Ⓒ Ⓓ Ⓔ
45. Ⓐ Ⓑ Ⓒ Ⓓ Ⓔ
46. Ⓐ Ⓑ Ⓒ Ⓓ Ⓔ
47. Ⓐ Ⓑ Ⓒ Ⓓ Ⓔ
48. Ⓐ Ⓑ Ⓒ Ⓓ Ⓔ
49. Ⓐ Ⓑ Ⓒ Ⓓ Ⓔ
50. Ⓐ Ⓑ Ⓒ Ⓓ Ⓔ
51. Ⓐ Ⓑ Ⓒ Ⓓ Ⓔ
52. Ⓐ Ⓑ Ⓒ Ⓓ Ⓔ
53. Ⓐ Ⓑ Ⓒ Ⓓ Ⓔ
54. Ⓐ Ⓑ Ⓒ Ⓓ Ⓔ
55. Ⓐ Ⓑ Ⓒ Ⓓ Ⓔ
56. Ⓐ Ⓑ Ⓒ Ⓓ Ⓔ

57. Ⓐ Ⓑ Ⓒ Ⓓ Ⓔ
58. Ⓐ Ⓑ Ⓒ Ⓓ Ⓔ
59. Ⓐ Ⓑ Ⓒ Ⓓ Ⓔ
60. Ⓐ Ⓑ Ⓒ Ⓓ Ⓔ
61. Ⓐ Ⓑ Ⓒ Ⓓ Ⓔ
62. Ⓐ Ⓑ Ⓒ Ⓓ Ⓔ
63. Ⓐ Ⓑ Ⓒ Ⓓ Ⓔ
64. Ⓐ Ⓑ Ⓒ Ⓓ Ⓔ
65. Ⓐ Ⓑ Ⓒ Ⓓ Ⓔ
66. Ⓐ Ⓑ Ⓒ Ⓓ Ⓔ
67. Ⓐ Ⓑ Ⓒ Ⓓ Ⓔ
68. Ⓐ Ⓑ Ⓒ Ⓓ Ⓔ
69. Ⓐ Ⓑ Ⓒ Ⓓ Ⓔ
70. Ⓐ Ⓑ Ⓒ Ⓓ Ⓔ
71. Ⓐ Ⓑ Ⓒ Ⓓ Ⓔ
72. Ⓐ Ⓑ Ⓒ Ⓓ Ⓔ
73. Ⓐ Ⓑ Ⓒ Ⓓ Ⓔ
74. Ⓐ Ⓑ Ⓒ Ⓓ Ⓔ
75. Ⓐ Ⓑ Ⓒ Ⓓ Ⓔ
76. Ⓐ Ⓑ Ⓒ Ⓓ Ⓔ
77. Ⓐ Ⓑ Ⓒ Ⓓ Ⓔ
78. Ⓐ Ⓑ Ⓒ Ⓓ Ⓔ
79. Ⓐ Ⓑ Ⓒ Ⓓ Ⓔ
80. Ⓐ Ⓑ Ⓒ Ⓓ Ⓔ
81. Ⓐ Ⓑ Ⓒ Ⓓ Ⓔ
82. Ⓐ Ⓑ Ⓒ Ⓓ Ⓔ
83. Ⓐ Ⓑ Ⓒ Ⓓ Ⓔ
84. Ⓐ Ⓑ Ⓒ Ⓓ Ⓔ

85. Ⓐ Ⓑ Ⓒ Ⓓ Ⓔ
86. Ⓐ Ⓑ Ⓒ Ⓓ Ⓔ
87. Ⓐ Ⓑ Ⓒ Ⓓ Ⓔ
88. Ⓐ Ⓑ Ⓒ Ⓓ Ⓔ
89. Ⓐ Ⓑ Ⓒ Ⓓ Ⓔ
90. Ⓐ Ⓑ Ⓒ Ⓓ Ⓔ
91. Ⓐ Ⓑ Ⓒ Ⓓ Ⓔ
92. Ⓐ Ⓑ Ⓒ Ⓓ Ⓔ
93. Ⓐ Ⓑ Ⓒ Ⓓ Ⓔ
94. Ⓐ Ⓑ Ⓒ Ⓓ Ⓔ
95. Ⓐ Ⓑ Ⓒ Ⓓ Ⓔ
96. Ⓐ Ⓑ Ⓒ Ⓓ Ⓔ
97. Ⓐ Ⓑ Ⓒ Ⓓ Ⓔ
98. Ⓐ Ⓑ Ⓒ Ⓓ Ⓔ
99. Ⓐ Ⓑ Ⓒ Ⓓ Ⓔ
100. Ⓐ Ⓑ Ⓒ Ⓓ Ⓔ
101. Ⓐ Ⓑ Ⓒ Ⓓ Ⓔ
102. Ⓐ Ⓑ Ⓒ Ⓓ Ⓔ
103. Ⓐ Ⓑ Ⓒ Ⓓ Ⓔ
104. Ⓐ Ⓑ Ⓒ Ⓓ Ⓔ
105. Ⓐ Ⓑ Ⓒ Ⓓ Ⓔ
106. Ⓐ Ⓑ Ⓒ Ⓓ Ⓔ
107. Ⓐ Ⓑ Ⓒ Ⓓ Ⓔ
108. Ⓐ Ⓑ Ⓒ Ⓓ Ⓔ
109. Ⓐ Ⓑ Ⓒ Ⓓ Ⓔ
110. Ⓐ Ⓑ Ⓒ Ⓓ Ⓔ
111. Ⓐ Ⓑ Ⓒ Ⓓ Ⓔ
112. Ⓐ Ⓑ Ⓒ Ⓓ Ⓔ

Continued

Answer Sheet

Continued

113. Ⓐ Ⓑ Ⓒ Ⓓ Ⓔ
114. Ⓐ Ⓑ Ⓒ Ⓓ Ⓔ
115. Ⓐ Ⓑ Ⓒ Ⓓ Ⓔ
116. Ⓐ Ⓑ Ⓒ Ⓓ Ⓔ
117. Ⓐ Ⓑ Ⓒ Ⓓ Ⓔ
118. Ⓐ Ⓑ Ⓒ Ⓓ Ⓔ
119. Ⓐ Ⓑ Ⓒ Ⓓ Ⓔ
120. Ⓐ Ⓑ Ⓒ Ⓓ Ⓔ
121. Ⓐ Ⓑ Ⓒ Ⓓ Ⓔ
122. Ⓐ Ⓑ Ⓒ Ⓓ Ⓔ
123. Ⓐ Ⓑ Ⓒ Ⓓ Ⓔ
124. Ⓐ Ⓑ Ⓒ Ⓓ Ⓔ
125. Ⓐ Ⓑ Ⓒ Ⓓ Ⓔ
126. Ⓐ Ⓑ Ⓒ Ⓓ Ⓔ
127. Ⓐ Ⓑ Ⓒ Ⓓ Ⓔ
128. Ⓐ Ⓑ Ⓒ Ⓓ Ⓔ
129. Ⓐ Ⓑ Ⓒ Ⓓ Ⓔ
130. Ⓐ Ⓑ Ⓒ Ⓓ Ⓔ
131. Ⓐ Ⓑ Ⓒ Ⓓ Ⓔ
132. Ⓐ Ⓑ Ⓒ Ⓓ Ⓔ
133. Ⓐ Ⓑ Ⓒ Ⓓ Ⓔ
134. Ⓐ Ⓑ Ⓒ Ⓓ Ⓔ

135. Ⓐ Ⓑ Ⓒ Ⓓ Ⓔ
136. Ⓐ Ⓑ Ⓒ Ⓓ Ⓔ
137. Ⓐ Ⓑ Ⓒ Ⓓ Ⓔ
138. Ⓐ Ⓑ Ⓒ Ⓓ Ⓔ
139. Ⓐ Ⓑ Ⓒ Ⓓ Ⓔ
140. Ⓐ Ⓑ Ⓒ Ⓓ Ⓔ
141. Ⓐ Ⓑ Ⓒ Ⓓ Ⓔ
142. Ⓐ Ⓑ Ⓒ Ⓓ Ⓔ
143. Ⓐ Ⓑ Ⓒ Ⓓ Ⓔ
144. Ⓐ Ⓑ Ⓒ Ⓓ Ⓔ
145. Ⓐ Ⓑ Ⓒ Ⓓ Ⓔ
146. Ⓐ Ⓑ Ⓒ Ⓓ Ⓔ
147. Ⓐ Ⓑ Ⓒ Ⓓ Ⓔ
148. Ⓐ Ⓑ Ⓒ Ⓓ Ⓔ
149. Ⓐ Ⓑ Ⓒ Ⓓ Ⓔ
150. Ⓐ Ⓑ Ⓒ Ⓓ Ⓔ
151. Ⓐ Ⓑ Ⓒ Ⓓ Ⓔ
152. Ⓐ Ⓑ Ⓒ Ⓓ Ⓔ
153. Ⓐ Ⓑ Ⓒ Ⓓ Ⓔ
154. Ⓐ Ⓑ Ⓒ Ⓓ Ⓔ
155. Ⓐ Ⓑ Ⓒ Ⓓ Ⓔ
156. Ⓐ Ⓑ Ⓒ Ⓓ Ⓔ

157. Ⓐ Ⓑ Ⓒ Ⓓ Ⓔ
158. Ⓐ Ⓑ Ⓒ Ⓓ Ⓔ
159. Ⓐ Ⓑ Ⓒ Ⓓ Ⓔ
160. Ⓐ Ⓑ Ⓒ Ⓓ Ⓔ
161. Ⓐ Ⓑ Ⓒ Ⓓ Ⓔ
162. Ⓐ Ⓑ Ⓒ Ⓓ Ⓔ
163. Ⓐ Ⓑ Ⓒ Ⓓ Ⓔ
164. Ⓐ Ⓑ Ⓒ Ⓓ Ⓔ
165. Ⓐ Ⓑ Ⓒ Ⓓ Ⓔ
166. Ⓐ Ⓑ Ⓒ Ⓓ Ⓔ
167. Ⓐ Ⓑ Ⓒ Ⓓ Ⓔ
168. Ⓐ Ⓑ Ⓒ Ⓓ Ⓔ
169. Ⓐ Ⓑ Ⓒ Ⓓ Ⓔ
170. Ⓐ Ⓑ Ⓒ Ⓓ Ⓔ
171. Ⓐ Ⓑ Ⓒ Ⓓ Ⓔ
172. Ⓐ Ⓑ Ⓒ Ⓓ Ⓔ
173. Ⓐ Ⓑ Ⓒ Ⓓ Ⓔ
174. Ⓐ Ⓑ Ⓒ Ⓓ Ⓔ
175. Ⓐ Ⓑ Ⓒ Ⓓ Ⓔ
176. Ⓐ Ⓑ Ⓒ Ⓓ Ⓔ
177. Ⓐ Ⓑ Ⓒ Ⓓ Ⓔ
178. Ⓐ Ⓑ Ⓒ Ⓓ Ⓔ

179. Ⓐ Ⓑ Ⓒ Ⓓ Ⓔ
180. Ⓐ Ⓑ Ⓒ Ⓓ Ⓔ
181. Ⓐ Ⓑ Ⓒ Ⓓ Ⓔ
182. Ⓐ Ⓑ Ⓒ Ⓓ Ⓔ
183. Ⓐ Ⓑ Ⓒ Ⓓ Ⓔ
184. Ⓐ Ⓑ Ⓒ Ⓓ Ⓔ
185. Ⓐ Ⓑ Ⓒ Ⓓ Ⓔ
186. Ⓐ Ⓑ Ⓒ Ⓓ Ⓔ
187. Ⓐ Ⓑ Ⓒ Ⓓ Ⓔ
188. Ⓐ Ⓑ Ⓒ Ⓓ Ⓔ
189. Ⓐ Ⓑ Ⓒ Ⓓ Ⓔ
190. Ⓐ Ⓑ Ⓒ Ⓓ Ⓔ
191. Ⓐ Ⓑ Ⓒ Ⓓ Ⓔ
192. Ⓐ Ⓑ Ⓒ Ⓓ Ⓔ
193. Ⓐ Ⓑ Ⓒ Ⓓ Ⓔ
194. Ⓐ Ⓑ Ⓒ Ⓓ Ⓔ
195. Ⓐ Ⓑ Ⓒ Ⓓ Ⓔ
196. Ⓐ Ⓑ Ⓒ Ⓓ Ⓔ
197. Ⓐ Ⓑ Ⓒ Ⓓ Ⓔ
198. Ⓐ Ⓑ Ⓒ Ⓓ Ⓔ
199. Ⓐ Ⓑ Ⓒ Ⓓ Ⓔ
200. Ⓐ Ⓑ Ⓒ Ⓓ Ⓔ

Practice Test 2

DIRECTIONS: Choose the best answer for each question and mark the letter of your selection on the corresponding answer sheet

1. Two phenotypic markers are found to be inherited according to Mendel's Law of Independent Assortment. What can you say about the corresponding gene loci?

   (A) Both loci correspond to the phenotypic wild type.
   (B) Both genes are dominant.
   (C) Both genes are recessive.
   (D) The two loci are on different chromosomes.
   (E) The two loci are on the same chromosome.

2. All of the following are autotrophs EXCEPT

   (A) Herbivores
   (B) Flowering plants
   (C) Diatom
   (D) Photosynthetic bacteria
   (E) Primary producers

3. In 3-point (trihybrid) gene mapping, the occurrence of the first crossover reduces the chance of a nearby second crossover nearby. This phenomenon is called

   (A) encroachment
   (B) enhancement
   (C) linkage reduction
   (D) interference
   (E) advancement

4. Basic drives, such as hunger, thirst, sex, and rage, as well as internal environmental parameters of blood pressure, heart rate, and body temperature, have all been linked to the functioning of

   (A) the basal ganglia
   (B) the adrenal gland
   (C) the pineal gland
   (D) the hypothalamus
   (E) the corpus callosum

5. Cellulose is a natural polymer composed of the monomer

   (A) glucagon
   (B) amino acids
   (C) glucose
   (D) amides
   (E) lipids and amino acids

6. The R-group of amino acids is always attached to a

   (A) nitrogen atom
   (B) carbon atom
   (C) oxygen atom
   (D) hydrogen atom
   (E) phosphorus atom

7. The protein transthyrein has 4 identical polypeptide subunits. This represents which level of protein structure?

   (A) primary
   (B) secondary
   (C) tertiary
   (D) quaternary
   (E) enzymatic

8. The ecological role played by trees on land is the same as that played by which organisms in the ocean?

   (A) Bacteria
   (B) Algae
   (C) Fish
   (D) Shrimp
   (E) Sharks

9. Amino acids complex with transfer RNA through

   (A) an aminoacyl link with the cytosine residue at the 5′ end of the chain
   (B) a phosphodiester bond
   (C) ionic and van der Waals attractions involving the phosphate and hydroxyl moieties of the nucleotide chain
   (D) hydrophobic interactions
   (E) an aminoacyl linkage with a 3′ terminal adenosine residue

10. Respiration consists of four distinct stages. The conversion of one glucose molecule to two molecules of pyruvic acid is called

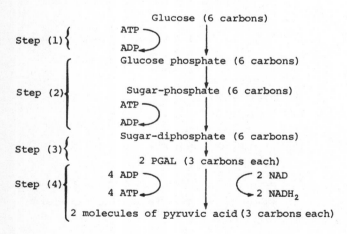

   (A) the Krebs cycle
   (B) the tricarboxylic acid cycle
   (C) glycolysis
   (D) gluconeogenesis
   (E) the citric acid cycle

11. During fertilization which of the following enzymes is released by the sperm cells?

   (A) calcium carbonase
   (B) hyaluronidase
   (C) acid phosphatase
   (D) chrondroitin sulfatase
   (E) chorionase

12. The process whereby cells and tissues become specialized during development is called

   (A) neutralization
   (B) differentiation
   (C) specialization
   (D) actualization
   (E) predestination

13. Wolves in the Mojave Desert were observed by two scientists to display peculiar behavior. When hunting a hare, which can outrun wolves, one wolf would chase the hare in a large circular pattern for approximately 10 minutes. It would then stop to rest and another wolf would continue the chase. These two wolves would alternate chasing the hare in this fashion until the hare was exhausted. The wolves would then catch the hare and share the meat. The type of behavior displayed by the wolves is typical of

   (A) communication
   (B) mimicry
   (C) altruism
   (D) cooperation
   (E) mutualism

14. All of the following are found in most plant cells EXCEPT

   (A) cell wall
   (B) endoplasmic reticulum
   (C) centrioles
   (D) centromere
   (E) nuclear membrane

15. Lactose consists of

    (A) two glucose molecules
    (B) one glucose molecule and one galactose molecule
    (C) one fructose molecule and one glucose molecule
    (D) one glucose molecule and one maltose molecule
    (E) one galactose molecule and one fructose molecule

16. The following list describes the functions of a particular hormone:

    I.   stimulates glycogen formation and storage
    II. inhibits formation of new glucose
    III. stimulates synthesis of protein and fat

    Which of the following hormones is primarily responsible for the functions described in the list above?

    (A) Epinephrine and/or norepinephrine
    (B) Glucocorticoids
    (C) Insulin
    (D) Glucagon
    (E) ACTH

17. Muscle fatigue is due, in part, to the accumulation of

    (A) lactic acid
    (B) citric acid
    (C) pyruvic acid
    (D) ACTH
    (E) ATP

18. The apical meristem is responsible for growth and differentiation in the

    (A) stem, root tips, and branches
    (B) vascular cambium
    (C) stem, sepal, and root tips
    (D) blastula
    (E) flower

19. Double-stranded DNA, when exposed to high temperature, changes in pH, or denaturing agents such as urea or formamide, may be induced to significantly increase its absorption of ultraviolet radiation at 260 nm. This effect is due to

    (A) an unwinding of the helix into unpaired, random coils
    (B) tautomerization of guanine and cytosine bases
    (C) disruption of the tertiary structure of the chain
    (D) the formation of thymine dimers
    (E) a change in the chirality, or "handedness," of the coils in the superhelix

Questions 20 and 21 refer to the graph below, which shows the carbon dioxide content over a large temperate forest.

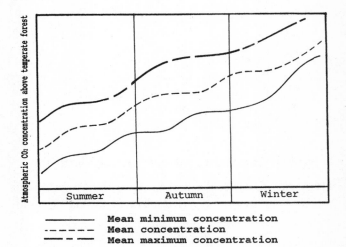

Mean minimum concentration
Mean concentration
Mean maximum concentration

20. The CO_2 concentration observed in the summer can be attributed to

    (A) low rates of DNA replication in forest animals
    (B) adaptation of plants to cold temperatures
    (C) high uptake of CO_2 by the trees
    (D) high numbers of autotrophs
    (E) loss of carbon from the forest animals

21. The consistent pattern of atmospheric CO_2 over the large temperate forest shown in the figure can be attributed to

    (A) seasonal differences in tree growth
    (B) seasonal changes in the distance of the earth from the sun
    (C) migration of zooplankton to the deep waters of the oceans
    (D) migration of birds to areas south of the forest
    (E) increase in automobile exhaust in winter

22. Protozoans can reproduce in a number of ways; they are, however, incapable of

    (A) sporulation
    (B) binary fission
    (C) sexual reproduction
    (D) viviparity
    (E) budding

23. What is the derived characteristic for vertebrates?

    (A) Notochord
    (B) Pharyngeal gill slits
    (C) Dorsal hollow nerve cord
    (D) Neural crest cells
    (E) Post-anal tail

Question 24 refers to the following diagram.

The Y-axis is either the death rate of the prey (DR) or the birthrate of the prey (BR). The X-axis is the numbers of prey, or prey density.

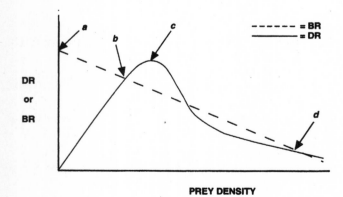

PREY DENSITY

24. In the diagram, at what point(s) will the size of the prey population be stable?

    (A) a and c
    (B) b and d
    (C) b, c, and d
    (D) a, b, and d
    (E) c

25. Which is NOT a currently accepted example of an evolutionary event?

    (A) An individual in a population synthesizes a new variation of an enzyme.
    (B) Giraffes develop longer necks as they strain to reach leaves at the tops of trees.
    (C) Individuals leave their dwelling and establish a new feeding and breeding ground.
    (D) An outside group joins an established population.
    (E) Antelopes that can run extremely fast survive to reproductive age.

Questions 26–29 concern the diagram.

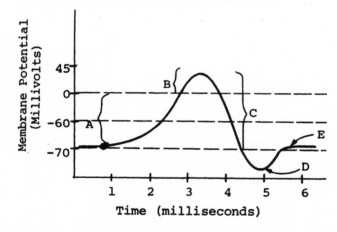

Time (milliseconds)

26. Part A of the curve represents a condition called

    (A) depolarization
    (B) reversed potential
    (C) hyperpolarization
    (D) destabilization
    (E) restabilization

27. If the threshold potential is −60 millivolts, what would be the result when a stimulus changed the membrane potential to −66 millivolts?

    (A) Depolarization will be effected
    (B) Na$^+$ ions will flow out of the system
    (C) An action potential will not be created
    (D) Hyperpolarization would be effected
    (E) Repolarization would be effected

28. Part C of the curve is due to the

    (A) outflow of Na$^+$
    (B) outflow of K$^+$
    (C) influx of Na$^+$
    (D) influx of K$^+$
    (E) outflow of Na+ and influx of K$^+$

29. Parts A and B of the curve are due to the

    (A) outflow of Na$^+$
    (B) influx of Na$^+$.
    (C) outflow of K$^+$
    (D) influx of K$^+$
    (E) influx of K$^+$ and outflow of Na$^+$

30. Of the following, which represents the most logical evolutionary advantage of multicellularity?

    (A) Multicellular organisms are able to become quite large.
    (B) Multicellular organisms are more efficient than single-celled.
    (C) Multicellular organisms can cope differently with the environment and its resources.
    (D) Multicellular organisms are better equipped to survive than single-celled.
    (E) Multicellular organisms can live on land.

31. The orienting of a grayling butterfly towards the sun, thus causing the pursuing predator to be partly blinded, is an example of

    (A) kinesis
    (B) tropism
    (C) taxis
    (D) instinctive behavior
    (E) conditioning

32. Bacteria may be classified into physiological groups according to the range of temperatures which will permit their growth. The type most suited for cold conditions are the

    (A) mesophiles
    (B) psychrophiles
    (C) thermophiles
    (D) thermophobes
    (E) poikilotherms

33. Osmoregulation in freshwater bony fishes is accomplished by

    (A) constant drinking, relatively impermeable skin, and active excretion of salts by specialized cells in the gills
    (B) rarely drinking, highly permeable skin, and production of copious amounts of dilute urine
    (C) rarely drinking, impermeable skin with scales, active absorption of salts through specialized cells in the gills, and production of copious, dilute urine
    (D) constant drinking, impermeable skin with scales, active excretion of salts through specialized cells in the gills, and production of relatively little isotonic urine
    (E) constant drinking, excretion of excess salt through specialized cells in the rectum, impermeable skin with scales, and production of copious dilute urine

34. The sources of genetic variation in a population include all of the following EXCEPT

    (A) mutation
    (B) natural selection
    (C) fertilization of egg by sperm
    (D) crossing-over
    (E) meiosis

35. Xylem is one type of vascular tissue found in plants. Which of the following best describes its functions in flowering plants?

    (A) Its exclusive function is transport of water and dissolved substances upward in the plant.
    (B) It functions exclusively to carry water and solutes in both directions throughout the plant.
    (C) It carries water and solutes in both directions through the plant and gives mechanical support to aerial parts of the plant.
    (D) It transports organic materials, particularly sugars and amino acids, in both directions.
    (E) It transports carbohydrates, amino acids, and other organic materials from the leaves to other parts of the plant and lends mechanical support.

36. All of the following traits are true about dicot roots EXCEPT

    (A) the xylem radiates like spokes with the phloem between the spokes
    (B) the xylem and phloem surround the pith
    (C) there is no vascular cambian
    (D) the xylem and phloem are contained in the stele
    (E) there is a pericycle layer present

37. The model of antigenic stimulation known as clonal selection states that

    (A) antigens react with a lymphocyte of the proper specificity, stimulating proliferation of a clone of antigen-specific lymphocytes
    (B) interaction with an antigen confers the appropriate specificity on a lymphocyte
    (C) antigens are generally able to react with a large variety of lymphocytes
    (D) antigens bind only antibodies produced by a lymphocyte of the correct specificity
    (E) binding of antigen prevents the formation of clones of incorrect specificity

38. In the Jacob-Monod operon theory, an inducer is

    (A) a low molecular weight compound which binds to the promoter enhancing polymerase binding
    (B) a molecule, often the translated product of the operon, which binds to the repressor and inactivates it
    (C) a protein which binds to the promoter, displacing the previously bound repressor
    (D) a small molecule which binds to the regulator gene, preventing transcription of the repressor
    (E) a peptide which binds to the operator

39. Which of the following molecules is thought to block the myosin (cross bridge) binding site on actin when a muscle fiber is not contracting?

    (A) Calcium
    (B) Troponin
    (C) Tropomyosin
    (D) ATP
    (E) ATP + Pi

40. Gram-negative bacteria have all of the following characteristics EXCEPT

    (A) they accept the secondary red stain
    (B) penicillin has no effect on them
    (C) the cell walls are only about 10 nanometers thick
    (D) they have an outer layer of lipopolysaccarides
    (E) they do not contain peptidoglycan

41. The frequencies of the alleles for blood type in the European population are $I_A = 0.46$, $I_B = 0.14$ and $i = 0.40$. What is the frequency of the A phenotype in this population?

    (A) 0.21
    (B) 0.18
    (C) 0.46
    (D) 0.06
    (E) 0.39

42. Assuming a frequency of a recessive allele of 0.5 and random segregation, what proportion of the individuals of a polyploidy population will be homozygous recessive in a stable population?

    (A) 0.25
    (B) 0.5
    (C) 0.0625
    (D) 0.125
    (E) 1.0

43. A eutrophic aquatic system is one where

    (A) plankton is at a very low density
    (B) there is no input of phosphate and sodium detergents
    (C) there is high productivity
    (D) plants die due to lack of nutrients
    (E) the biomass is steadily decreasing

44. A famous experiment by Hershey and Chase may be summarized as follows:

    Bacteriophage was grown on a medium containing radioactive sulfur (^{35}S) and radioactive phosphorus (^{32}P). This radioactive bacteriophage was then introduced to normal, nonradioactive bacteria. Some time later, before the phage was able to reproduce, both the phage and the bacteria were put into a blender. The empty "ghosts" of the bacteriophage were dislodged from the bacterial surface and were separated by centrifugation. It was discovered that all of the radioactive sulfur was located in the empty bacteriophage, while all the radioactive phosphorus was found inside the infected bacteria. What did this experiment indicate?

    (A) That DNA is the transforming principle and carrier of heredity in microbial genetics
    (B) That the phages were not inactivated by the radioactivity they absorbed
    (C) That bacteria may be transformed by viral infection
    (D) That bacteriophage was capable of injecting both proteins and nucleic acid into bacteria
    (E) That the transformed bacteria produced viral specific proteins

45. Which of the following vertebrate tissues come from mesoderm?

    (A) Epidermis
    (B) Notochord
    (C) Jaws
    (D) Liver and pancreas
    (E) Brain and spinal column

46. Passing from the nucleus, through the cytoplasm and into the internal matrix of a mitochondrion involves passing through how many membranes (without using pores in the nuclear envelope)?

    (A) 1
    (B) 2
    (C) 3
    (D) 4
    (E) 5

47. Consider the following Mendelian model. Pure breeding red flowers are crossed with pure breeding white flowers. The progeny of this cross are all red. Subsequently they are used in two further crosses:

    I.   The first filial generation above (F_1) is crossed with itself to form a second filial generation (F_2).
    II. The F_1 plants are crossed with true-breeding white plants (i.e., a testcross).

    The results of the two crosses are:

    (A) the F_2 generation is 3:1 (red:white) flowers while the testcross results in a 1:1 ratio of red and white plants
    (B) the F_2 generation is 3:1 (red:white) and the testcross gives only red flowers
    (C) the F_2 generation is 3:1 (red:white) as are the testcross progeny
    (D) the F_2 generation is 3:1 (white:red) as are testcross progeny
    (E) the F_2 generation is uniformly red and the testcross gives a 1:1 ratio of the types crossed

48. The experiment in the previous question illustrates

    (A) Mendel's Law of Independent Assortment
    (B) Mendel's Law of Segregation
    (C) the concept of linkage
    (D) the occurrence of crossovers
    (E) pleiotropy

49. Now consider a slightly more involved Mendelian pattern. A parental cross is made whereby true-breeding red, tall plants are crossed with true-breeding white, short plants. The resulting F_1 generation of plants are all tall and red. As in the monohybrid cross, the F_1 plants are crossed with each other to obtain the F_2 generation. Out of every 16 plants in the second filial generation there are

    (A) 12 tall red plants, 2 short red, 1 short white, 1 tall white
    (B) 12 tall red plants, 2 short white, 1 tall white, 1 short red
    (C) 9 short red, 3 tall white, 3 tall red, 1 short white
    (D) 9 tall red, 3 short red, 3 tall white, 1 short white
    (E) 9 tall white, 3 short red, 3 short white, 1 tall red

50. An undifferentiated cell can be induced to follow the developmental patterns of a different species by

    (A) replacing the original nucleus with one from another (chosen) species
    (B) introducing cytoplasm from another species
    (C) removing the native nucleus
    (D) no known means
    (E) adding mitochondria from another species

51. As each electron completes its movement back and forth across the mitochondrial membrane, starting at the $FADH_2$ level, the chemiosmotic differential is enriched by

    (A) one H^+
    (B) two H^+
    (C) three H^+

    (D) four H^+
    (E) five H^+

52. In flowering plants, the female gametophyte is the

    (A) stigma
    (B) style
    (C) ovary with ovules
    (D) embryo sac in ovule
    (E) fruit

53. Consider the sense strand in a length of genomic DNA. Which of the following represents the correct order of structures travelling in the direction taken by a functioning polymerase?

    (A) 3′-terminus-antileader-TAC-structural gene-antitrailer-terminator-5′ terminus
    (B) 5′-terminus-antileader-TAC-structural gene-antitrailer-terminator-3′ terminator
    (C) 5′-terminus-leader-AUG-gene-trailer-3′ terminus
    (D) 5′-terminus-leader-TAC-structural gene-trailer-terminator-3′ terminus
    (E) 5′-terminus-promoter-leader-AUG-gene-antitrailer-terminator-5′ terminus

54. The so-called Pribnow box (TATAATG consensus sequence) is described best as

    (A) a prokaryotic terminator, a signal for polymerase and transcriptase to dissociate
    (B) a eukaryotic promoter sequence
    (C) a promoter found in *E. coli.*, an upstream binding site for RNA polymerase
    (D) the leader terminus in eukaryotes responsible for signalling the addition of a 7-methylguanosine cap structure to the transcription product
    (E) a structure found at intron-exon boundaries

55. Which of the following are NOT aspects of promoter function?

    (A) Binding site for RNA polymerase
    (B) Unwinding of DNA into single stranded regions accessible to polymerase
    (C) Establishing the correct sense and antisense strands

(D) Base pairing with rRNA and stabilizing mRNA-ribosomal complex in the proper geometry

(E) Specifying the correct direction for transcription to proceed

56. You are startled by a large bear, which suddenly rushes into the campsite where you are quietly reading. Your heart rate increases, your pupils dilate, and blood is shunted away from the digestive tract and peripheral vessels. Changes such as these are directed by

(A) the parasympathetic system of the autonomic nervous system

(B) the parasympathetic system of the somatic nervous system

(C) the sympathetic system of the autonomic nervous system

(D) the sympathetic system of the somatic nervous system

(E) the parasympathetic and sympathetic systems of the somatic nervous system

57. An individual is characterized as being heterozygous with respect to ABO locus; specifically he is I^A/I^B. This means that he

(A) can accept transfusions from any donor but cannot produce any of the antibodies associated with blood typing

(B) can accept transfusions from type O donors only, but cannot produce any of the antibodies involved

(C) can accept transfusions from type O donors only and can produce anti-A and anti-B

(D) can accept transfusions from type A or type O donors and can produce anti-B

(E) can accept transfusions from type B or type O donors and can produce anti-A

58. Two lengths of DNA along the same chromosome encode very similar products. They produce enzymes identical in catalytic properties and substrate selectivity which are very similar in electrophoretic mobility and other physical properties. These genes are

(A) isozymes

(B) isologous

(C) allelic variants

(D) epistatic

(E) pleiotropic

59. An individual is diagnosed as possessing a sickle cell trait. This individual's hemoglobin is correctly described as

(A) having 2 α chains and 2 S chains in each tetramer

(B) possessing 2 α chains, one S chain, and one β chain

(C) having half of the hemoglobin being composed of 2 α chains and 2 β chains and the other half being composed of 2 α and 2 S chains

(D) having each tetramer being composed of 22 β chains, 1 α, and 1 S chain

(E) half being composed of 2 β chains and 2 S chains, while the other half is found to be composed of 2 β chains and 2 α chains

60. Numerous antibiotics work by interfering with protein synthesis. An example is cycloheximide, which

(A) inhibits the peptidyl transferase activity of 60 S ribosomal subunits in eukaryotes

(B) binds to the 30 S subunit and inhibits binding of aminoacyl-tRNAs (prokaryotes)

(C) causes premature chain termination by acting as an analogue of aminoacyl-tRNA (prokaryotes and eukaryotes)

(D) inhibits translocation by binding to 50 S ribosomal subunit (prokaryotes)

(E) inhibits initiation and causes misreading of mRNA in prokaryotes

61. Offspring showing Down's syndrome are trisomic for chromosome 21. In the following crosses, 21 represents a normal 21 chromosome, 14 a normal 14 chromosome and 14^{21} a translocation heterozygote carrying part of chromosome 21 attached to chromosome 14. Which of the crosses would produce an offspring with Down's syndrome?

	Gamete of Parent 1	Gamete Parent 2
I	14, 21	14, 21
II	14, 21	14^{21}
III	14, 21	14^{21}, 21
IV	14, 21	14

(A) I and II
(B) II and III
(C) III only
(D) IV only
(E) II, III and IV

62. All of the following are characteristics of eukaryotic chromatin EXCEPT

(A) DNA wound around a protein core
(B) five histones each with large amounts of arginine and lysine
(C) a small nucleiod in the centre
(D) nucleosomes connected by stretches of DNA
(E) looped or folded domains

63. All of the following are characteristics of eukaryotic chromatin EXCEPT

(A) nucleosome core particles are composed of approximately 2×10^8 base pairs
(B) histone (H1) is attached to the outside of the nucleosome where the DNA enters and leaves
(C) in the chromatosomes, approximately two full turns of the left-handed superhelix wraps around the nucleosome core
(D) the term mononucleosome refers to a chromatosome and associated linker DNA
(E) at physiological salt concentrations, the nucleosomes form a folded structure consisting of approximately six nucleosomes per turn

64. In *E. coli*, the enzyme primase functions in DNA replication by

(A) catalysing the energy-dependent unwinding of the duplex using 2 ATPs per base pair separated
(B) stabilizing the single stranded DNA thus formed
(C) catalysing polymerization of ribonucleoside 5'-triphosphates to form 3'-5' phosphodiester bonds with a concomitant release of diphosphate
(D) catalysing addition of deoxyribonucleoside to 3' ends of the chains formed in (C)
(E) multicellular organisms can live on land

65. Of the following, which represents the most logical evolutionary advantage of multicellularity?

(A) Multicellular organisms are able to become quite large.
(B) Multicellular organisms are more efficient than single-celled organisms.
(C) Multicellular organisms can cope differently with the environment and its resources.
(D) Multicellular organisms are better equipped to survive than single-celled organisms.
(E) Of the above, none are logical evolutionary advantages.

66. A genetic engineering process that is usually done in the laboratory and which involves the transfer of genetic material from one bacterium to another using a bacteriophage as the carrier is

(A) transduction
(B) transcription
(C) plasmid transfer
(D) transformation
(E) translocation

67. The most significant adaptation of succulent plants to arid environments is the

(A) low surface-to-volume ratio of above-ground organs
(B) accumulation of H2O in the cells
(C) formation of highly dissected leaves
(D) absence of stomata
(E) production of nonpigmented cuticles

68. The following Lineweaver-Burke plot represents the kinetics of an enzyme-mediated reaction and illustrates the course of the reaction without any inhibitor present and in the presence of increasing concentrations of inhibitor.

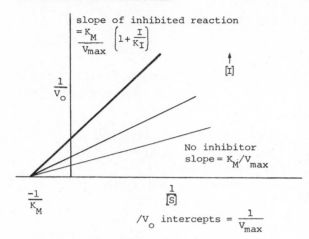

These idealized kinetics reflect which of the following mechanisms of inhibition?

(A) Competitive inhibition
(B) Steric inhibition
(C) Noncompetitive inhibition
(D) Irreversible inhibition
(E) Reversible inhibition

69. A certain bacterial infection is characterized by extreme virulence due to toxins from the bacteria. The toxin cannot be inactivated by heating or by decreasing the pH. It may be concluded that

(A) the bacteria is probably gram positive and is elaborating an exotoxin
(B) the bacteria is probably gram negative and is producing an endotoxin
(C) the bacteria is probably gram positive, the virulence is due to an endotoxin
(D) the symptoms are caused by a gram negative bacteria and its exotoxin
(E) the bacteria is undergoing rapid mitosis

70. An important function of gibberellins is

(A) elongation of the stem
(B) delay of senescene
(C) branching of roots

(D) opening and closing of stomata
(E) water transport within the plant

71. During the human menstrual cycle, peak levels of estrogen and lutenizing hormone are associated with

(A) the flow phase
(B) the early part of the follicular phase
(C) the latter part of the follicular phase
(D) the early part of the luteal phase
(E) the latter part of the luteal phase

72. The age distribution of a population is

(A) always a summary of the proportional representation of the various age classes
(B) always a good indicator of population growth rate
(C) is a good measure of age-specific survivorship
(D) always a good indicator of population growth rate
(E) a good measure of fecundity (birth rate)

73. Communication systems in animals vary and are sometimes elaborate. One important type of communication is identification of individuals of the same species. Which of the following is an example of communication among individuals to bring the individuals together for mating.

(A) The dance of bees
(B) The call of frogs in spring
(C) The roaring of male mule deer
(D) Electric signals from fish
(E) Songs of humpback whales

74. Cholecystokinin (CCK) stimulates the release of digestive enzymes from the pancreas. It is best described as

(A) a hormone produced in the duodenum
(B) a neurotransmitter produced in the duodenum
(C) an neurosecretion secretion from the goblet cells of the stomach
(D) an endocrine secretion from the chief cells of the stomach
(E) a hormone produced in the pancreas

75. Plant cells are able to withstand a much wider fluctuation in the osmotic pressure of the surrounding medium than animal cells. This is due to the plant cells'

    (A) stomata
    (B) lipid membrane
    (C) cell wall
    (D) chloroplasts
    (E) mitochondria

76. All of the following are statements about the reflex arc EXCEPT

    (A) there is only one sensory neuron in a pathway
    (B) the cell body of a sensory neuron is always outside the spinal cord in the dorsal root ganglia
    (C) the axons of motor neurons always enter the spinal cord dorsally
    (D) the cell body of the interneuron is in the white matter of the spinal cord
    (E) a minimum of three cells are involved in a given pathway

77. All of the following statements are true about the light reactions of photosynthesis EXCEPT

    (A) ATP and NADPH are both produced during non cyclic electron flow
    (B) ATP is produced in noncyclic and cyclic photophosphorylation
    (C) photosystem I and photosystem II are required for noncyclic electron flow
    (D) the electron transport chain transfers electrons from photosystem II to photosystem I
    (E) the light reactions occur in the stroma of the chloroplast

78. Lambda phages are being used as a vector to deliver recombinant DNA into bacteria for the purpose of DNA cloning. Which of the following is or are the advantage(s) of choosing λ phage as a vector?

    I. There is no limitation on the length of foreign DNA being inserted into λ genome.
    II. λ phage infects bacteria with a high frequency.

    III. Large segments of the λ genome are not essential for its lytic and lysogenic life cycles.

    (A) I
    (B) II
    (C) III
    (D) I and II
    (E) II and III

79. Which of the following occurs during the dark reaction of the photosynthetic process of an angiosperm?

    (A) Production of carbohydrates
    (B) Production of ATP
    (C) Production of O_2
    (D) Production of NADPH
    (E) Production of CO_2

80. Most of the ATP produced during cellular respiration

    (A) is in the cystol during glycolysis
    (B) is in the mitochondral matrix during the Krebs cycle
    (C) is in the inner membrane during electron transport
    (D) is in the inner membrane during oxidative phosphorylation
    (E) is in the intermembrane space during production of acetyl CoA and Krebs cycle

81. Walking along a fence behind which stand several houses, you find yourself being barked at loudly by a German shepherd. When the animal realizes that the fence will not allow him to get at you, he bites at a stick with exaggerated ferocity. This last behavior is an example of

    (A) habituation
    (B) intentional movements
    (C) redirected activity
    (D) imprinting
    (E) operant conditioning

82. Two individuals, A and B, live in marshes. If only A is present, it is found throughout the marsh, similarly for B. However, when A and B are both present, A lives in the water and B lives in the emergent vegetation. The words that describe the new position of B are

    (A) fundamental niche
    (B) producer component
    (C) consumer component
    (D) realized niche
    (E) selected habitat

83. The organism chlamydomonas is characterized by

    (A) possession of a macronucleus as well as a micronucleus
    (B) a life cycle generally based on asexual reproduction but occasionally on sexual reproduction under conditions of low ambient nitrogen
    (C) a life cycle in which the prevalent multicellular, motile form is under conditions such as low ambient nitrogen, replaced by a unicellular, amoeboid form
    (D) sexual reproduction which alternates between isogamy and anisogamy
    (E) diploid spores are produced during sexual reproduction

84. Cyclic AMP (5'-3' cyclic adenine monophosphate) has been found to play a role as a "second messenger" and regulator in all of the following cases EXCEPT

    (A) phosphorylation of a protein kinase as part of a reaction cascade triggered by the binding of epinephrine to membrane receptors which leads to the hydrolysis of glycogen.
    (B) binding to CAP (catabolite gene activator protein), thereby stimulating the transcription of a number of inducible catabolic operons (such as the Lac operon in *E. coli*).

    (C) the aggregation of amoeboid cells into a pseudoplasmodium.
    (D) the stimulation of lipolysis by the phosphorylation of a protein kinase which in turn activates a lipase.
    (E) the formation of a steroid receptor complex which difuses into the nucleus and initiates the production of regulatory proteins.

85. Which of the following DOES NOT accurately describe selection and adaptation?

    (A) Adaptations occur in individuals, not populations.
    (B) Adaptations are new traits that arise as a result of selection.
    (C) Selection is on individuals but populations evolve.
    (D) Selection usually results in the acquisition of positive traits.
    (E) Not all changes in alleles or genotypes are adaptations.

86. Consider a hypothetical cell in a young woody plant. Proceeding inwardly from an intercellular space to the interior of the cell, which list correctly gives the order in which the given structures would be encountered?

    (A) middle lamella—primary cell wall—plasma membrane—secondary cell wall
    (B) middle lamella—secondary cell wall—primary cell wall—plasma membrane
    (C) plasma membrane—secondary cell wall—primary cell wall—middle lamella
    (D) middle lamella—primary cell wall—secondary cell wall—plasma membrane
    (E) middle lamella—secondary cell wall—plasma membrane—primary cell wall

Question 87 refers to the following figure.

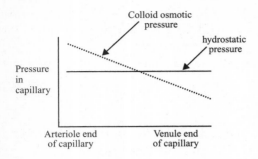

87. The figure shows the hydrostatic and colloidal osmotic pressure in a capillary from the arteriole to venule end. Which of the following statements satisfies the conditions shown in the figure?

(A) Reabsorbtion occurs at the arteriole end of the capillary.
(B) Fluid moves out of the capillaries when the colloid osmotic pressure is greater than the hydrostatic pressure.
(C) There is a net uptake of fluids from the intracellular space to the capillaries.
(D) Hydrostatic pressure decreases from the arteriole to venule end due to filtration.
(E) The colloid osmotic pressure is due primarily to the presence of small ions in the blood.

88. Which of the following best explains possible relationships between genotypes and phenotypes?

(A) Different genotypes never produce the same phenotype.
(B) The same genotypes can occur in different phenotypes.
(C) The genotype is equivalent to the phenotype.
(D) The genotype is unrelated to the phenotype.
(E) The genotype and phenotype are the same in all siblings.

89. If a sperm cell in which nondisjunction of the sex chromosomes has occurred fertilizes a normal human female egg, how many chromosomes will the resulting zygote contain?

(A) 47
(B) 24
(C) 23
(D) 46
(E) 22

Questions 90–91 refer to the following table, which shows the number of individuals of five different species (species 1 to 5) in five different communities (a, b, c, d, and e).

Community	Species				
	No. 1	No. 2	No. 3	No. 4	No. 5
a	18	21	22	20	19
b	150	0	0	75	5
c	47	48	0	0	46
d	100	3	5	4	5
e	6	0	6	0	0

90. Which one of the communities would have the highest species diversity index?

(A) a
(B) b
(C) c
(D) d
(E) e

91. Which one of the communities has the lowest evenness?

(A) a
(B) b
(C) c
(D) d
(E) e

92. An investigator wishes to determine the intracellular location of a protein P via immunoflorescence. Purified P is injected into a rabbit and anti-P is obtained from rabbit blood serum. The investigator uses the indirect method of immunoflorescence. Therefore, he completes all of the steps EXCEPT

(A) injection of rabbit immunoglobulins into a goat
(B) incubation of the cells to be studied with nonflorescent anti-P
(C) brief fixation of cells to be studied in formaldehyde, and extraction with acetone

(D) linking anti-P to a fluorescent marker
(E) collection of goat anti-rabbit serum (GAR)

93. The following proteins or protein subunits have all been isolated from various filamentous or tubular systems and play some contractile and/or cytoskeletal role. Of these, which has been shown to possess endogenous ATPase activity?

(A) Heavy meromyosin
(B) α-Tubulin
(C) β-Tubulin
(D) Microtubule associated proteins
(E) Actin

94. All of the following are true about the life cycle of plants EXCEPT

(A) in some plants, haploid and diploid stages occur as independent, free-living forms
(B) the zygote is the diploid stage
(C) gametes may be produced by mitosis
(D) spores may be produced by meiosis
(E) the haploid stage develops into fruit

95. Density-dependent factors are those that

(A) cause the population size to decrease to zero
(B) allow birthrates to increase as population size decreases
(C) cause birthrates to increase as population size increases
(D) allow death rates to increase as population size decreases
(E) are always due to predation

96. The growth of a population depends on the initial population size and which of the following parameters?

(A) Births, deaths, immigrations, and emigrations
(B) Maturity, parity, and fecundity
(C) Fertility, deaths, immigrations, and dispersal
(D) Fitness, deaths, immigrations, and spacing
(E) Fecundity, deaths, immigrations, and dispersal

97. In *Drosophila*, color and wing size are linked. Gray color and normal wings are wild characteristics but can mutate to black color and vestigial wings. The resulting alleles are C+ for gray, C for black, W+ for normal wings and W for vestigial wings. The following cross between gray normal-winged flies and black vestigial-winged flies produced the following percents of the different offspring.

C+CW+W crosses with CCWW

Offspring genotype	C+CW+W	C+CWW	CCW+W	CCWW
Percent of offspring	41%	7%	9%	43%

Which of the following statements best explains the result?

(A) Color is sex-linked to the Y chromosome.
(B) Some crossing over occurred during meiosis producing recombinants.
(C) Color and wing size sort independently.
(D) Wing size is sex-linked to the X chromosome.
(E) There was non-random mating among the *Drosophila*.

98. A gotwit is a shorebird. As the chick develops to an adult, the length of the beak increases faster than the length of the skull. The result is an adult gotwit with a long beak relative to the head—a characteristic that is advantageous in obtaining food from shallow water. This is an example of

(A) allometric growth
(B) natural selection
(C) gene flow
(D) heterochrony
(E) paedomorphosis

99. Which of the following is a correct explanation for the feeling of dizziness after being rotated in a swivel chair?

(A) The movement of hair cells with respect to the movement of perilymph in the cochlea of the inner ear causes the hair cells to send impulses to the cerebellum.
(B) The movement of hair cells with respect to the movement of endolymph in the ampulla of the inner ear causes the hair cells to send impulses to the cerebellum.

(C) The movement of perilymph with respect to the movement of hair cells in the ampulla of the inner ear causes the hair cells to send impulses to the cerebrum.

(D) The movement of endolymph with respect to the movement of perilymph in the ampulla of the inner ear causes the former to send impulses to the cerebellum.

(E) The movement of hair cells with respect to the movement of perilymph in the cochlea of the inner ear causes the hair cells to send impulses to the cerebrum.

100. Which of the following is a realistic chart illustrating the relationship between enzymatic activity and pH?

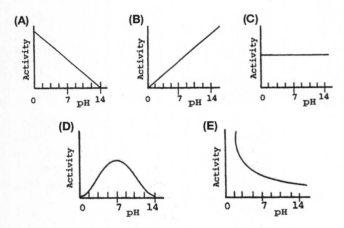

Questions 101 through 105 refer to the following diagrams illustrating early embryonic development in amphioxus.

Amphioxus is a small marine chordate; its egg cells are characterized by a very small amount of yolk. Identify the structures indicated; if there is no specific structure specified then give the appropriate developmental stage.

101.

(A) Blastula
(B) Morula
(C) Animal pole
(D) Vegetal pole
(E) Gastrula

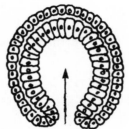

102.

(A) Blastopore
(B) Animal pole
(C) Vegetal pole
(D) Archenteron
(E) Blastocoel

103.

(A) Early neurula
(B) Blastula
(C) Late neurula
(D) Gastrula
(E) Fetula

104.

(A) Animal pole
(B) Yolk plug
(C) Presumptive notochord
(D) Vegetal pole
(E) Blastopore

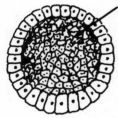

105.

(A) Blastocoel
(B) Coelom
(C) Gut
(D) Blastopore
(E) Archenteron

106. Locomotion in protozoans is accomplished using

I. Axopodia
II. Pseudopodia
III. Rhizopoda
IV. Flagella
V. Cilia

(A) I, II, and III
(B) IV and V
(C) I, II, IV, and V
(D) II, IV, and V
(E) I, II, III, IV, and V

107. Which of the following gives the correct composition of the thin myofilaments that are attached to the Z lines of a sarcomere?

(A) Actin, myosin, troponin
(B) Myocin, tropomyosin, troponin
(C) Actin, troponin, tropomyosin
(D) Myocin, actin, tropomyosin
(E) Tropomyosin, myocin, troponin

108. Ovum and follicle development are stimulated by

(A) increased FSH production
(B) low estrogen levels
(C) high estrogen levels
(D) high progesterone levels
(E) low progesterone levels

109. The following numbered list corresponds to characteristics of various plant types. Decide which of the lettered choices best describes the categories of plants to which they apply.

I. Flower parts arranged in threes
II. Principal veins of leaves parallel to each other.
III. Root system fibrous, no taproot
IV. Stem characterized by vascular bundles distributed irregularly through pith tissue

(A) Monocots
(B) II and III characterize monocots, whereas I and IV pertain to dicots
(C) II, III, IV describe monocots but I pertains to dicots
(D) I, III, IV describe dicots, II applies to monocots
(E) I, II, and IV describe monocots while III applies to some but not all monocots and some but not all dicots

110. Consider a biochemical reaction $X \rightleftharpoons Y$. Which of the graphs below correctly describes the changes in free energy during both the catalyzed and the uncatalyzed conversion of X to Y?

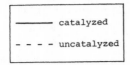

(A)

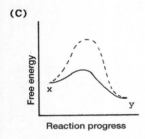

(B)

(C)

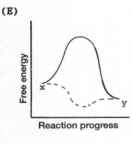

(D)

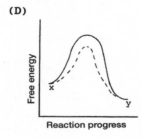

(E)

111. Certain derivatives of minerals represent small inorganic molecules that are bound to enzymes during chemical reactions to make the enzymes functional. These are called

(A) hydrophobic determinants
(B) cofactors
(C) coenzymes
(D) quarternary groups
(E) catalysts

112. Which one of the following is an angiosperm?

(A) Fern
(B) Mushroom
(C) Oak tree
(D) Pine tree
(E) Spruce tree

113. Neanderthals lived in Europe, the Middle East, and parts of Asia beginning about 130,000 years ago. They are believed to be direct descendants of

(A) *Australopithecus afarensis*
(B) *Australopithecus africenus*
(C) *Homo habilus*
(D) *Homo erectus*
(E) *Homo sapiens*

114. The acquisition of bipedalism is first documented for

(A) *Australopithecus*
(B) *Homo erectus*
(C) *Homo sapiens*
(D) *Purgatorius*
(E) *Aegyptopithecus*

115. Trilobites were arthropods in the Paleozoic. They are useful index fossils because

(A) they were abundant in shallow seas which dried up at the end of the Paleozoic
(B) they had a segmented exoskeleton
(C) their appendages showed little structural variation
(D) they were less specialized than existing arthropods
(E) they can be aged using carbon 14

116. In the human excretory system, blood components such as glucose and amino acids are returned to the blood from the filtrate by

(A) facililated diffusion
(B) pinocytosis
(C) simple diffusion
(D) osmosis
(E) passive transport

117. Ligation of the islets of Langerhans of the pancreas will deprive the circulatory system of

    (A) insulin
    (B) trypsin
    (C) serotonin
    (D) bile
    (E) pepsin

118. Sugar passes through the plasma membrane by combining with protein carrier molecule (a permease). This combination is then able to pass through the membrane from a region of high to one of low concentration of sugar, without the use of cellular energy. This process is called

    (A) filtration
    (B) active transport
    (C) facilitated diffusion
    (D) dialysis
    (E) simple diffusion

119. During muscle contraction, the Ca^{++} that is released combines with

    (A) troponin
    (B) actomyosin
    (C) tropomyosin
    (D) fibrinogen
    (E) myosin

120. The two main variables used to classify terrestrial biomes are

    (A) longitude and altitude
    (B) climate and topography
    (C) soil type and precipitation
    (D) precipitation and temperature
    (E) latitude and soil type

121. Mycorrhizae are best described as

    (A) interwoven mats of hyphae
    (B) mutualistic associations between fungi and plant roots
    (C) mutualistic associations between fungi and algae
    (D) strong odors produced by truffles
    (E) a group called imperfect fungi

122. Fungi secrete enzymes that break down the food extracellularly and then absorb the food molecules. Fungi that obtain food from fallen logs are called

    (A) autotrophs
    (B) parasites
    (C) predators
    (D) saprobes
    (E) symbionts

123. Which statement most accurately describes the circulatory patterns in humans and fish?

    (A) In humans, blood is returned to the heart after being aerated prior to being sent through the tissues; whereas in fish aerated blood is pumped directly to the tissues.
    (B) In humans, blood is pumped directly to the tissues after being aerated, whereas in fish the blood is routed through the heart.
    (C) In both fish and humans, blood is sent through the heart immediately after aeration and before systemic circulation.
    (D) In both fish and humans, blood is routed directly to the tissues after aeration.
    (E) In humans, aerated blood is first routed through the heart; in fish either circulatory pattern may occur.

124. Which one of the following factors contributes to the tendency of compartment formation for lipid bilayers?

    (A) Kinetic
    (B) Energetic
    (C) Equilibrium
    (D) Static
    (E) Ionic

125. The ecosystem characterized by coniferous trees is the

    (A) temperate forest
    (B) tundra
    (C) taiga
    (D) chaparral
    (E) tropical rain forest

Question 126 refers to the following diagram.

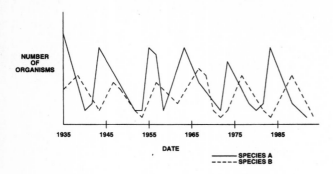

126. Species A and species B are members of the same community. From this graph we can conclude that

   (A) species A eats species B
   (B) species A increases in number every five years
   (C) species B eats species A
   (D) species B can increase in number when A is low
   (E) both species are competing for the same food source

127. The digestive action of bile salts is to

   (A) emulsify fats
   (B) act as an enzyme in the breakdown of lactose
   (C) hydrolyse fatty acids
   (D) esterify cholesterol
   (E) hydrolyse fats to fatty acids and glycerol

128. The major mechanism driving the long-distance transport of water up the xylem is

   (A) root pressure
   (B) cohesion of the water molecule to each other
   (C) adhesion of the water molecules to the hydrophilic walls of the xylem
   (D) pressure gradient between the atmosphere and the air spaces in the mesophyll of the leaves
   (E) turgor pressure changing in the guard cells

129. Flowering plants generally meet their needs for fixed nitrogen by absorbing

   (A) amino acids and nucleotides
   (B) ammonium ions
   (C) nitrates
   (D) nitrites
   (E) diazonium salts

130. Increased blood pressure is detected by the baroreceptors in the carotid sinus, which results in a reflex reduction in cardiac output. This reduction is the result of

   (A) an increase in heart rate
   (B) an increase in systolic pressure in the heart.
   (C) an increase in peripheral vascular resistance
   (D) an increase in cardiac sympathetic nerve stimulation
   (E) a decrease in heart rate

131. Animals with radial symmetry are diploblastic and lack

   (A) archenteron
   (B) true tissues
   (C) endoderm
   (D) mesoderm
   (E) ectoderm

132. For many generations, giraffes have stretched their necks in order to reach the leaves on the higher branches of trees. This stretching effort has caused their necks to grow longer; as an individual giraffe acquires a longer neck by stretching it, he is able to pass on this quality of a longer neck to his progeny. After continued small growth in neck length from one generation to the next, giraffes have evolved into exceptionally long necked animals.

   The above reasoning illustrates the theory of

   (A) Charles Darwin
   (B) Jean Baptiste de Lamarck
   (C) Gregor Mendel
   (D) Louis Pasteur
   (E) A. R. Wallace

133. One of the simplest kinds of behavior is the knee jerk, in which a tap below the knee causes the leg to jerk up. This behavior requires as a minimum which of the following combination of structures?

    (A) A motor neuron and a muscle
    (B) A receptor and at least two segments of the spinal cord
    (C) A receptor neuron, a motor neuron, and a muscle
    (D) An intact spinal cord and a brain
    (E) A receptor neuron connected to a muscle

134. All of the following is true about deserts EXCEPT that

    (A) they receive very little annual rainfall (10 inches or less)
    (B) they experience dramatic fluctuations in temperature from night to day
    (C) some deserts are produced and maintained by high mountain ranges that block coastal precipitation
    (D) they exist on all continents except Europe and Antarctica and tend to be created along the 30° north and south latitude lines
    (E) they are always sand-covered and largely shaped by wind

135. Assuming that all of the offspring survive, which of the following individuals is considered to have the greatest fitness?

    (A) A female bird that mates with one male and produces two eggs every year for three years.
    (B) A male bird that mates with two females and each female produces three eggs every year for three years.
    (C) A male bird that mates with five females and each female produces one egg every year for three years.
    (D) A female bird that lives for ten years.
    (E) A female bird that mates with four males and produces four eggs every year for three years.

136. Fermentation reactions may produce any of the following products EXCEPT

    (A) CO_2
    (B) O_2
    (C) ethyl alcohol
    (D) lactic acid
    (E) CO

137. The end products of the light reactions of photosynthesis are

    (A) H_2O, ADP, and NADP
    (B) PGAL, ADP, and ribulose
    (C) O_2, ATP, and NADPH
    (D) CO_2, PGAL, and $2H^+$
    (E) O_2, ATP, and ribulose

138. Which of the following statements defines meiosis?

    (A) The number of chromosomes in the diploid nucleus is reduced by half.
    (B) The fusion of two haploid nuclei to form a diploid nucleus.
    (C) The process by which four haploid cells or nuclei are transformed into a diploid cell.
    (D) The separation of chromosomes.
    (E) The fusion of chromosomes.

139. The movement of young salamanders from the pond where they were born to another pond is an example of

    (A) dispersal
    (B) energy flow
    (C) migration
    (D) nutrient cycling
    (E) spacing

Question 140 and 141 refer to the following situation.

While traveling to various islands in the Pacific, you observed that tortoises are common. To your surprise, one of the islands had a different habitat and the tortoises found there had longer necks.

140. Finding longer-necked tortoises on one island and not on the other islands is an example of

 I.   directional selection
 II.   stabilizing selection
 III. disruptive selection
 IV.   heterochrony
 V.   allometry

 (A) I and V
 (B) II and IV
 (C) III
 (D) I and IV
 (E) III and V

141. The habitat of this island was

 (A) open grassland with few trees or shrubs
 (B) tropical forest with little growth on the forest floor
 (C) thick shrubs shading out growth on the floor
 (D) rocky with little vegetation
 (E) an active volcanic island with little land around the volcano

142. As energy is passed through a food web

 (A) decomposers cannot break down organisms' wastes
 (B) herbivores dominate the last links of the web by numbers of organisms
 (C) less useful energy remains available at each successive feeding level
 (D) producers feed on remaining nutrients unused by other trophic levels
 (E) species numbers increase from link to link in the web

143. In an alga-minnow-bass-bear food chain, organisms are listed in successive trophic levels. The alga is a producer followed by successive consumer levels. The bass behave as a

 (A) decomposer
 (B) fourth level consumer
 (C) producer
 (D) secondary consumer
 (E) tertiary consumer

Use the following graph to answer question 144.

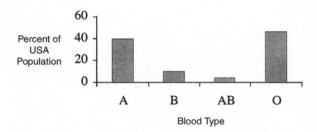

144. The graph shows the frequency of blood types in the United States. This frequency distribution suggests that blood type is

 (A) homozygous
 (B) polymorphic
 (C) normally distributed
 (D) polygenic
 (E) incompatible with the Hardy-Weinberg equilibrium

145. Which one of the following situations is most likely to result in an evolutionary change?

 (A) Athletic ability by extensive training
 (B) Maintenance of health by correct diet
 (C) Prompt medical attention when exposed to disease
 (D) Increased resistance to a disease because of a mutation
 (E) Vitamin supplements

146. During muscle contraction, the protein that also serves as an enzyme that breaks down ATP → ADP + Pi is

    (A) tropomyosin
    (B) fibrinogen
    (C) actomyosin
    (D) troponin
    (E) myosin

Questions 147 and 148 are concerned with the following dissociation curves.

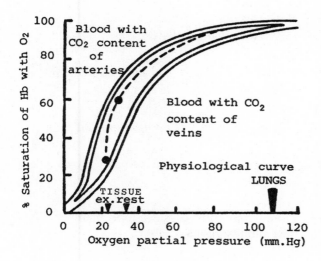

147. Upon examination of the graph, one notices that the hemoglobin is about 98 percent saturated with the oxygen at the partial pressure of oxygen typical of the lungs (108 mm), while it is only about 58 percent saturated at the partial pressure of oxygen typical of the tissues at rest (32 mm). What, therefore, does the 40 percentage-point difference represent?

    (A) The percentage of carbon dioxide released by the tissues
    (B) The log phase between oxygen uptake and carbon dixoide release
    (C) The approximate percent of carbon carried by hemoglobin in the form of carboxyhemoglobin
    (D) The approximate percentage of oxygen carried by hemoglobin that is actually released to the tissues

    (E) The relative amounts of oxygen and carbon dioxide in venous blood

148. The two different dissociation curves for human hemoglobin show the shift to the right due to an increase in the concentration of CO_2. This means that

    (A) a higher partial pressure of oxygen is needed to unload oxygen from the hemoglobin
    (B) the blood concentration becomes more alkaline, thus increasing the affinity of hemoglobin for oxygen
    (C) the oxyhemoglobin releases oxygen more readily
    (D) hemoglobin carries oxygen and carbon dioxide
    (E) the presence of CO_2 results in a reduction in the amount of carbonic acid in the blood

Question 149 refers to the following.

Stephen J. Gould suggests that there are three tiers of time to consider when discussing evolutionary change. These are

    I.   Tens of thousands of years resulting in changes in population genetic processes
    II. Millions of years resulting in speciation and extinction that is measured among groups of organisms
    III. Tens to hundreds of millions of years resulting in changes such as mass extinction

149. Which of the above apply/applies to macroevolution?

    (A) I and II
    (B) III
    (C) I and III
    (D) II and III
    (E) I, II, and III

150. Natural selection depends upon all of the following EXCEPT

    (A) more individuals are born in each generation than will survive and reproduce
    (B) there is genetic variation among individuals

(C) some individuals have a better chance of survival than others

(D) some individuals are reproductively more successful

(E) certain acquired traits can be passed on to the next generation of organisms with the phenotypes

DIRECTIONS: For each group of questions below, match the numbered word, phrase or sentence to the most closely related lettered heading and mark the letter of your selection on the corresponding answer sheet. A lettered heading may be chosen as the answer once, more than once, or not at all for the question in each group.

Questions 151–155

(A) T antigen
(B) Specialized transduction
(C) Generalized transduction
(D) Arc gene product
(E) Burkitt's lymphoma

151. SV-40

152. Occurs through defects in mechanism for cutting and packaging concatamers

153. Occurs through errors in excision mechanism

154. Herpes virus

155. Protein kinase required for oncogenic transformation

Questions 156–160

(A) Stamen
(B) Pistil
(C) Embryo sac
(D) Pollen grain
(E) Calyx

156. Style, stigma, ovary

157. Seven-celled, 8 nucleate, haploid

158. Haploid, dinucleate, thick walled

159. Anther, filament

160. Sepals

Questions 161–165

(A) Apical meristem
(B) Cortex
(C) Phloem
(D) Vascular cambium
(E) Parenchyma cells

161. Surrounds the large leaf veins and contains most of the chloroplasts of leaves

162. This is a thin layer of undifferentiated cells

163. This structure is made up of parenchyma and collenchyma

164. Tissue consisting of sieve-tube members, sieve-tube and companion cells

165. These cells divide actively and are of the growing tip of a root or shoot

Questions 166–170

(A) Colchicine
(B) Vinblastine
(C) Cytocholasin B
(D) Sodium azide
(E) Vanadate ions (Vo_4^{3-})

166. Transition state analog for phosphoryl group hydrolysis

167. Inhibits formation of actin filaments

168. Uncoupling agent

169. Specifically inhibits mitosis at metaphase, prevents distribution of chromosomes to daughter cells without interfering with chromosome replication, and can therefore be used in making polyploid cells

170. Inhibits formation of microtubles and spindle fibers by catalyzing intracellular crystallization of tubulin

Questions 171–175

    (A) Taiga
    (B) Tundra
    (C) Deciduous forests
    (D) Grasslands
    (E) Tropical rain forests

171. Most pronounced degree of vertical stratification

172. Relatively few trees, but the ground is largely blanketed with mosses, lichens, and grasses

173. Dominated by coniferous forests and characterized by small lakes, ponds, and bogs

174. Temperate regions with relatively long and warm summers and abundant rainfall

175. Undergoes annual warm-cold or wet-dry cycles in temperate and tropical zones, respectively

Questions 176–179

    (A) Cryptic coloration
    (B) Aposematic appearance or coloration
    (C) Batesian mimicry
    (D) Müllerian mimicry
    (E) Geographic isolation

176. Industrial melanism

177. Characterized by gaudy coloration and highly conspicuous appearance

178. Two or more inedible or unpleasant species evolve to resemble each other, each species serving as both model and mimic.

179. A prey species without any characteristics noxious to potential predators takes on the appearance of a species which has such characteristics

Questions 180–183

    (A) Cline
    (B) Population
    (C) Species
    (D) Subspecies (or race)
    (E) Parallel evolution

180. All of the yellow perch in a small lake

181. Largest unit of population within which gene flow can occur

182. Isolated populations with recognizably different traits but believed to be potentially capable of interbreeding

183. Gradual geographic variation of a given trait within a species

Questions 184–185

    (A) Preadaptation
    (B) Speciation
    (C) Adaptation
    (D) Natural Selection
    (E) Adaptive Radiation

184. The presence of lungs in a primitive fish

185. The formation of a new species

Questions 186–189

    (A) Habituation
    (B) Classical conditioning
    (C) Operant conditioning
    (D) Autonomic learning
    (E) Instinctive response

186. A rat is placed in a cage equipped with a small lever. As the rat becomes hungry, it begins a random exploration of its cage and accidentally depresses the lever. As the lever is depressed, a food pellet is thrust into the cage. At first the rat shows no signs of associating those two events, but in time his searches become less random and he proceeds more directly to the lever. Eventually, the rat spends most of its time just sitting and pressing the bar.

187. Rats were paralyzed with a curare derivative. Its effect was to allow the animals to remain fully conscious but unable to move any skeletal muscles voluntarily. Positive reinforcement was given to the rats via an electrode implanted in the pleasure center of their brains. Negative reinforcement was given by administering a mild shock.

The minor physiological variations that occur naturally in the body were the starting points for the experiment. If the scientists wanted to teach the animal to slow its heart rate, they stimulated the pleasure center of the brain whenever the heart rate naturally began to slow down. Whenever the heart rate increased, the animal was given a mild shock. Under these circumstances, the researchers found that in some cases the heart rate soon remained slow. Conversely, when a rapid heart rate was rewarded, the heart rate increased.

188. Russian biologist Ivan Pavlov conducted the following experiment. A dog was harnessed and had a tube attached to its mouth to measure saliva. On successive occasions, a light was flashed five seconds before food was dropped onto a feeding tray under the dog's nose. After a number of such trials, Pavlov discovered that the dog would begin to salivate upon seeing the light.

189. A reef fish, establishing its territory, repeatedly chases after its immediate neighbors, attempting to drive them away. Eventually, it accepts the nearby presence of familiar fish, and only chases after strange fish wandering through the area.

DIRECTIONS: The following groups of questions are based on laboratory or experimental situations. Choose the best answer for each question and mark the letter of your selection on the corresponding answer sheet.

Questions 190–192

A couple desiring genetic counseling undergoes genetic typing; the couple's human leukocyte antigen (HLA) karyotypes are:

$$
\begin{array}{llll}
A1 & B7 & C4 & D8 \\
A3 & B7 & C3 & D4 \\
\\
\text{and} \quad A3 & B5 & C2 & D9 \\
A3 & B1 & C7 & D6 \\
\end{array}
$$

The A3 allele has been associated with multiple sclerosis (MS). Assume independent assortment of the HLA alleles.

190. What percent of this couple's children can be expected to develop MS?

(A) 0
(B) 25
(C) 50
(D) 75
(E) 100

191. In later years, it is found that none of the offspring develop MS despite their homozygosity of the recessive allele. This is an illustration of

(A) incomplete penetrance
(B) mutation
(C) selection
(D) chance
(E) incomplete dominance

192. The following F_1 recombinations occur at the given percentages:

$$
\begin{array}{lllll}
A: A1 & B1 & C7 & D8 & 27\% \\
B: A3 & B1 & C3 & D4 & 17\% \\
C: A3 & B7 & C2 & D9 & 14\% \\
D: A3 & B5 & C4 & D6 & 42\% \\
\end{array}
$$

Which alleles have a recombination frequency of 41%?

(A) A and B
(B) B and C
(C) C and D
(D) A and C
(E) B and D

193. Consider the following experiment performed using a frog embryo. In normal development of the zygote, the first cleavage runs through the gray crescent. If the two daughter cells are experimentally separated after the first cleavage each develops into a normal tadpole. If, on the other hand, the plane of cleavage is altered (by tying off a portion of the zygote) so that all the gray crescent material is partitioned into one daughter cell, the cell with the gray crescent material develops normally while the other cell gives rise to an undifferentiated ball of cells. This shows that

(A) the first cleavage is determinate
(B) the first cleavage is indeterminate
(C) gray crescent material is required for normal development within the first few divisions
(D) B and C
(E) A and C

Questions 194–196

194. An in vitro protein synthesizing system was prepared using radioactively labelled amino acids, synthetic components from reticulocytes and mRNA from human placenta. The predominant transcript was specific for human placental lactogen (HPL), a hormone secreted by the placenta. Rat liver microsomes derived from either smooth membranes (SER) or rough endoplasmic recticulum (RER) were added to the sample; the samples were then analyzed using SDS-PAGE (sodium dodecyl sulfate polyacrylamide gel electropheresis), as illustrated in lanes a and b of the figure below. Lanes c and d illustrate the results obtained when the proleolytic enzymes trypsin and chymotrypsin were added to the mixture. In both cases the origin is at the top of the figure. The results show that

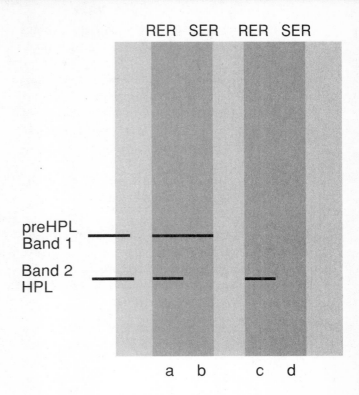

I. the translated product (pre HPL) is larger than the final, secreted form of the molecule (HPL)

II. the processing of pre HPL to form HPL is dependent on insertion of pre HPL into lumen of endoplasmic reticulum

III. the processing depends on components exclusively associated with RER and not with SER

IV. translocation of proteins is mediated by excision of a hydrophobic N-terminal sequence in accordance with the "signal hypothesis"

(A) I, II, III, and IV
(B) I, II, and III, but not IV
(C) I only
(D) I, III and IV, but not II
(E) I and IV

195. Which of the following is the most accurate interpretation of the experimental results?

(A) The experiment supports the signal hypothesis more than it does the membrane trigger hypothesis.

(B) The experiment supports the trigger hypothesis more than it does to the signal hypothesis.

(C) The experiment supports both hypotheses equally well (or equally poorly).

(D) The experiment shows that the binding of polysomes to endoplasmic reticulum is dependent on a protein receptor in the membrane of RER.

(E) The experiment shows that the mRNA of the human placenta is short-lived.

196. Suppose another type of mRNA had been used and that this message specified a protein similar in molecular weight and acid-base properties to pre HPL and was native to the cytoplasm of rat liver hepatocytes, you would expect that

(A) a single band would be found in all lines, at approximately the level where HPL was found

(B) a single band would be found in all lines, at approximately the level where pre HPL was found

(C) a single band would be found in lanes a and b at about the level of pre HPL, and no bands would be found in lanes c and d

(D) two bands would be found in lane a as with HPL and pre HPL, one band would be found in lane b at the pre HPL level, one band would be found in lane c at the same level and no bands would appear in lane d

(E) two bands would be found in lane a as above, one in lane b at the HPL level and no bands would be found in lanes c or d

197. Bacterial cells were incubated in a medium containing glucose, lactose and galactose. At regular intervals, cells were removed and plated. Colonies were counted, and the figure below illustrates the fashion in which the population increased over time:

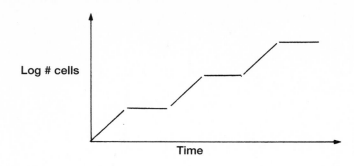

What can you say about the organisms involved?

(A) All enzymes required for fermentation of any of the sugars are adaptive.

(B) All enzymes required for fermentation of each sugar are constituitive.

(C) Lactose hydrolysis is repressible.

(D) Synthesis of β-galactosidase is adaptive.

(E) Lactose hydrolysis is not adaptive.

Questions 198–200

Carla, curious as to the cause of disease in her mother's tobacco plants, designed the following experiment. She crushed the infected leaves with sterile water. Then, the solution was filtered once through a funnel filter and once through a 0.4-micron filter. Both the filtrate and residue of each filter were kept and tested by rubbing each, individually, onto an uninfected plant. The table below shows the results.

	Infection Ensues	No Infection
initial filtrate	x	
initial residue	x	
second filtrate	x	
second residue		x

198. Which of the following is the most likely pathogen?

(A) virus
(B) bacteriophage
(C) bacterium
(D) fungus
(E) actinomycetes

199. The second filtrate was used to infect a healthy plant. This plant then underwent the same procedure as the plant that yielded the first set of filtrates and residues. The process was repeated four times; the degree of infection remained constant each time. It is assumed that the cause of disease is a pathogen and not a toxin because

    (A) the initial filtrate remains infectious
    (B) the initial residue remains infectious
    (C) the second filtrate remains infectious
    (D) the second residue remains uninfectious
    (E) the short duration of the experiment

200. Which of the following tests corroborates the diagnosis given in question 198?

    (A) Isolation and growth of the pathogen on essential media similar in chemical constitution to that of the plant
    (B) Isolation and growth of the pathogen on agar (nutrient deficient) media
    (C) Lack of growth of the pathogen on essential media similar in chemical constitution to that of the plant
    (D) Lack of growth of the pathogen on agar (nutrient deficient) media
    (E) There is no test available for corroboration

Answer Key

1. (D)	35. (C)	69. (B)	102. (D)	135. (B)	168. (D)
2. (A)	36. (B)	70. (A)	103. (B)	136. (B)	169. (A)
3. (D)	37. (A)	71. (C)	104. (A)	137. (C)	170. (B)
4. (D)	38. (B)	72. (A)	105. (A)	138. (A)	171. (E)
5. (C)	39. (C)	73. (B)	106. (C)	139. (A)	172. (B)
6. (B)	40. (E)	74. (A)	107. (C)	140. (A)	173. (A)
7. (D)	41. (E)	75. (C)	108. (A)	141. (C)	174. (C)
8. (B)	42. (C)	76. (D)	109. (A)	142. (C)	175. (D)
9. (E)	43. (C)	77. (E)	110. (C)	143. (D)	176. (A)
10. (C)	44. (A)	78. (E)	111. (B)	144. (B)	177. (B)
11. (B)	45. (B)	79. (A)	112. (C)	145. (D)	178. (D)
12. (B)	46. (D)	80. (D)	113. (D)	146. (C)	179. (C)
13. (D)	47. (A)	81. (C)	114. (A)	147. (D)	180. (B)
14. (C)	48. (B)	82. (D)	115. (A)	148. (C)	181. (C)
15. (B)	49. (D)	83. (B)	116. (A)	149. (D)	182. (D)
16. (C)	50. (A)	84. (E)	117. (A)	150. (E)	183. (A)
17. (A)	51. (B)	85. (D)	118. (C)	151. (A)	184. (A)
18. (A)	52. (D)	86. (D)	119. (A)	152. (C)	185. (B)
19. (A)	53. (A)	87. (D)	120. (D)	153. (E)	186. (C)
20. (C)	54. (C)	88. (B)	121. (B)	154. (E)	187. (D)
21. (A)	55. (D)	89. (A)	122. (D)	155. (D)	188. (B)
22. (D)	56. (C)	90. (A)	123. (A)	156. (B)	189. (A)
23. (D)	57. (A)	91. (B)	124. (B)	157. (C)	190. (C)
24. (B)	58. (B)	92. (D)	125. (C)	158. (D)	191. (A)
25. (B)	59. (C)	93. (A)	126. (C)	159. (A)	192. (D)
26. (A)	60. (A)	94. (E)	127. (A)	160. (E)	193. (A)
27. (C)	61. (C)	95. (D)	128. (D)	161. (E)	194. (B)
28. (B)	62. (C)	96. (A)	129. (C)	162. (D)	195. (A)
29. (B)	63. (A)	97. (B)	130. (E)	163. (B)	196. (C)
30. (C)	64. (C)	98. (A)	131. (D)	164. (C)	197. (D)
31. (C)	65. (C)	99. (B)	132. (B)	165. (A)	198. (A)
32. (B)	66. (A)	100. (D)	133. (C)	166. (E)	199. (D)
33. (C)	67. (A)	101. (B)	134. (E)	167. (C)	200. (C)
34. (B)	68. (C)				

Detailed Explanations of Answers

1. (D)

Phenotypic markers tend to follow the Law of Independent Assortment if and only if they correspond to loci on separate chromosomes. It is actually the chromosomes themselves that assort independently during meiosis.

2. (A)

An autotroph is by definition capable of manufacturing food from simple, low-energy sources; plants are a good example. Herbivores ingest plants. All of the remaining choices use photosynthesis to obtain sugars (organic nutrients).

3. (D)

In 3-point gene mapping the occurrence of the first crossover reduces the chances of a second crossover nearby via interference.

4. (D)

The hypothalamus exerts significant control over various aspects of the body by monitoring feedback from the Autonomic Nervous System and by releasing hormones to regulate the hormonal secretions of the pituitary.

5. (C)

Glucose is the monomer that makes up both cellulose and starch.

6. (B)

The R group is always attached to a carbon atom. Also attached to that carbon are an amino group which has a positive charge, a carboxyl group with a negative charge, and a hydrogen atom.

7. (D)

Quaternary structure is the joining of 2 or more polypeptide chains into a single, functional protein. Primary structure is the sequence of amino acids. Secondary structure is the helix or pleated sheet due to hydrogen bonds between amino acids on the same polypeptide chain. Tertiary structure is the 3-dimensional shape of a single polypeptide chain. Enzymes are proteins but do not decribe protein structure.

8. (B)

For this question you must know what role trees play in the ecosystem and determine what organism plays the same role in the ocean. Trees are primary producers. They have the ability to make organic molecules from inorganic molecules using sunlight as the energy source. In other words, they are photosynthetic organisms. Algae also carry out photosynthesis and, therefore, play the same role as trees in their ecological community. Fish, shrimp, sharks, and bacteria are consumers and/or decomposers.

9. (E)

Structure of aminoacyl tRNA complex

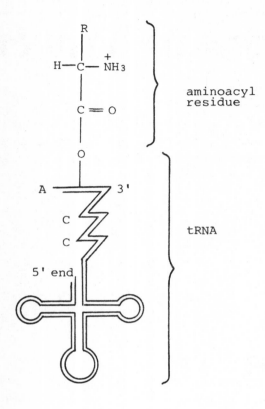

10. (C)

The four distinct stages of respiration are glycolysis, the conversion of pyruvate to acetyl-CoA, Krebs cycle, and oxidative phosphorylation. The tricarboxylic acid and citric acid cycles are alternative names for the Krebs cycle. Gluconeogenesis is not a cycle but a process by which glucose is produced from non-carbohydrate molecules.

11. (B)

During fertilization, when the sperm reaches the vicinity of the egg, the sperm releases the enzyme hyaluronidase to penetrate the corona radiata to enable contact with the egg.

12. (B)

During development, differentiation is the stage at which the cells and tissues become specialized.

13. (D)

Cooperative behavior occurs when two or more animals act toward their mutual benefit. When the animals belong to the same species, it is called intraspecific cooperation; when they belong to different species, it is known as interspecific cooperation. In the case of the two wolves (intraspecific), each wolf benefits by helping the other catch the rabbit and each sacrifices one-half of the rabbit to the other in return for help.

Communication occurs in many forms and in all kinds of behavior—friendly, hostile, aggressive. Mimicry is the adaptation of the appearance, behavior, or smell by a species of another species. Altruism differs from cooperation in that it involves an activity which benefits another organism but at the individual's own expense. A typical example of altruism is that of parents protecting their offspring.

14. (C)

Most plant cells do not have centrioles (also called basal bodies). Centrioles, which are important in organizing spindle fibers, are found in most animal cells. Plants have cell walls, endoplasmic reticulum, centromeres and a nuclear membrane.

15. (B)

Lactose consists of one glucose molecule covalently linked to one galactose molecule.

16. (C)

Epinephrine, norepinephrine, and especially glucagon all oppose insulin in their effect upon blood glucose levels. The net result of all the effects listed is the blood glucose levels decrease, while reserves of glycogen, fat, and protein increase. Glucocorticoids such as cortisone tend to stimulate the formation of polysaccharides from protein and fat, initially raising glycogen levels.

17. (A)

A muscle that has contracted strenuously and repeatedly and has exhausted its stored supply of organic phosphates and glycogen, will accumulate lactic acid. This lactic acid is a product of glycolysis and fermentation. The muscle has incurred what is known as an oxygen debt. When the violent activity is over, the muscle cells consume large quantities of oxygen as they convert lactic acid into pyruvic acid. Pyruvic acid is oxidized via the Krebs cycle and electron–transport process. The cells utilize the energy obtained to resynthesize glycogen from the lactic acid that remains.

18. (A)

The apical meristem near the tips of root, stem, and branches is important in growth differentiation in these areas of a plant. The meristems in the buds of a stem are responsible for outgrowths of the stem, and the roots give rise to cells that eventually differentiate into all of the cell types present in the cell.

19. (A)

The denaturing conditions disrupt the forces that maintain the double helix. These forces include hydrophobic interactions which keep the largely nonpolar bases in the inside of the molecule, hydrophilic interactions involving the phosphate moieties, hydrogen bonding between paired bases, and the combination of permanent depolar effects and van der Waals forces comprising the so-called "stacking energy" which stabilizes a parallel alignment of the bases.

20. (C)

In a temperate forest, the greatest growth occurs during the spring and summer. This growth requires energy and the energy is obtained from sugars produced during photosynthesis. Photosynthesis uses CO_2 and thus there is a lower concentration of CO_2 in the atmosphere during summer than autumn or winter. Increased growth would result in increased—not decreased—rates of DNA replication. The temperate forest plants may be adapted to cold temperatures, but this does not relate to summer levels of atmospheric CO_2. Autotrophs produce CO_2 and thus high numbers would act to increase atmospheric CO_2. Carbon lost from the animals could be tied up in organic matter or released into the atmosphere, thus increasing the amount of atmospheric CO_2.

21. (A)

There is a definite seasonal growth pattern in trees from temperate forests that is reflected in growth rings. The growth is directly related to photosynthetic activity and thus CO_2 fixation. The result is a seasonal pattern of atmospheric CO_2 with atmospheric CO_2 inversely related to CO_2 fixation and thus photosynthetic activity and growth. The greatest CO_2 fixation is in spring. The distance between the earth and sun varies over a year (B) but this has no effect on seasons or plant growth. Plankton (C) includes both plants and animals and thus there is a potential effect on both CO_2 fixation and production. Birds produce CO_2 in respiration and thus the migration of the birds away from the forest would not cause the changes observed (D). There is little or no vehicle traffic in a large forest and thus (E) is incorrect.

22. (D)

Protozoans are single-celled animals whose cells are often highly specialized containing many organelles. They can reproduce both sexually and asexually. However, they cannot give birth to live progeny (viviparity), in the way that mammals can.

23. (D)

The notochord, pharyngeal gill slits, dorsal hollow nerve cord, and post anal tail are the characteristics that define chordates. Neural crest cells produced from ectoderm during neurulation migrate to different locations and form various structures unique to vertebrates.

24. (B)

The prey population will be stable when the birth rate equals the death rate. This occurs at b and d. At a, the birth rate is greater than the death rate, and at c, the death rate is greater than the birth rate.

25. (B)

The process of change in the allele frequencies in the gene pool of a population over time is called evolution. Evolution occurs through the processes of mutation, genetic drift, migration, and natural selection. An individual who synthesizes a new enzyme has had a mutation in one of his genes so that a new, mutant gene product is produced. The Lamarckian view that evolution occurs through the use and disuse of body parts is illustrated in the example with giraffes. This viewpoint has been overwhelmed by contrary evidence.

26. (A)

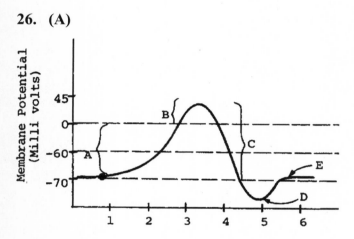

Part A of the curve represents a condition called depolarization.

27. (C)

The threshold potential is required in order for an action potential to occur. Therefore, no action potential will occur.

28. (B)

Part C of the curve is due to the outflow of K+.

29. (B)

Parts A and B of the curve are due to the influx of Na+.

30. (C)

The most logical evolutionary advantage of multicellularity is that multicellular organisms can cope differently with the environment and its resources.

In multicellular organisms there is a very high degree of specificity. This allows for a high level of efficiency which allows the multicellular organism to cope differently with the environment.

31. (C)

Taxis is a term which refers to the movement of a free-living organism toward the source of an external stimulation. The orientation of a grayling butterfly towards the sun thus causing the pursuing predator to be partly blinded is a typical example of taxis. Tropism is used usually to refer to plant movement toward a stimulus.

32. (B)

Psychrophiles may grow at 0°C or lower, although optimal temperatures range from 15–30°C.

33. (C)

All known living organisms are hypertonic with respect to freshwater. Freshwater fish thus tend to take in too much water and have difficulty retaining salt. They are able to compensate for this by seldom drinking, active absorption of salts, and the excretion of large volumes of dilute urine.

34. (B)

This question tests your understanding of evolution and, in particular, knowing what will increase genetic variation. Mutations may involve changes in particular genes that result in new forms of the gene being produced. Meiosis is a process that involves producing gametes with one of each chromosome of a homologous pair. The chromosomes for each egg or sperm are selected at random and, thus, result in each gamete having its own combination of chromosomes. During meiosis, crossing over may occur between homologous chromosomes. This allows new combinations of alleles to be located near each other and inherited as a unit on the new chromosome. The process of fertilization also increases genetic variability, since each egg and sperm contains its own combination of chromosomes. Natural selection is not a source of genetic variation, but rather acts on the genetic variation that already exists. Natural selection could, in fact, lead to decreased genetic variability.

35. (C)

Most xylem cells are essentially hollowed-out cell walls which serve two important functions: first, they provide a passageway through which water and dissolved substances can be transported up and down the tree; second, they provide structural support for the tree itself (xylem cells comprise what is commonly called wood). Organic materials are transported through the other kind of plant vascular tissue: the phloem.

36. (B)

Although there is a pith at the core of a dicot stem, there is no pith in a dicot root. All of the other traits are present.

37. (A)

The clonal selection theory proposes that lymphocytes become specialized early in embryonic development. When these lymphocytes begin to move through the bloodstream, they learn to "recognize" their parent organism, so that they can later distinguish a foreign substance from the body's own molecules.

Later, when the viable individual does encounter a foreign substance–an antigen–the lymphocyte with the antibody specific for that antigen binds to the antigen and begins to proliferate, thus producing a great number of antigen-specific antibodies to neutralize the invading foreign substance.

38. (B)

Binding of the inducer to the repressor inactivates it, thus opening up the site on the promoter for binding of polymerase.

39. (C)

Tropomyosin, a linear protein, attaches to actin and covers the myosin binding sites when the muscle is not contracting.

40. (E)

Both gram-positive and gram-negative bacteria have peptidoglycan. However, in gram-negative bacteria, the peptidoglycan is encased in the lipopolysaccharides. All of the other characteristics are typical of gram-negative bacteria.

41. (E)

The phenotype A results from AA or AO. The frequency of AA is $(0.46 \times 0.46 = 0.21)$ and the frequency of AO is $(0.46 \times 0.40 = 0.18)$. The combined frequency is $0.21 + 0.18$ or 0.39. Answer (A) is the frequency of AA. Answer (B) is the frequency of AO. Answer (C) is the frequency of allele IA. Answer (D) is the frequency of AB.

42. (C)

In the polyploidy condition, the number of chromosomes is doubled and each homologous pair is represented two times. If the frequency of each allele is 0.5, according to the Hardy-Weinberg equilibrium, the frequency of one homozygous recessive pair is $(0.5)^2$ or 0.25. The frequency of two homozygous recessive pairs is $(0.25)^2$ or 0.0625.

43. (C)

Eutrophic lakes are very productive with high nutrient levels and high biomass. Therefore, plankton densities are high (not low), nutrients are available (not limiting), and biomass is not decreasing. Phosphate is frequently a limiting nutrient and input of phosphate in detergents can enhance eutrophication; their absence would not be related to eutrophication.

44. (A)

The experiment relied on the fact that proteins contain sulfur, but no phosphorus; nucleic acids contain phosphorus, but no sulfur. Thus, the proteins and nucleic acids in the bacteriophage labeled themselves by incorporating radioactive sulfur and phosphorus, respectively. Since the infected bacteria was shown to contain only phosphorus, the bacteriophage must have injected only nucleic acids into the bacteria. Given the mechanism of phage reproduction, in which an infected host cell produces hundreds of new phages and then ruptures, the nucleic acid must be the genetic material of the cell.

45. (B)

Of the tissues given, only the notochord is produced from mesoderm. Epidermis is produced from somativc ectoderm, jaws from neural crest ectoderm, and brain and spinal column from neural tube ectoderm. The liver and pancreas are produced from endoderm.

46. (D)

The nucleus and the mitochondria each have both an inner and an outer membrane, so that transport from the nucleus to the internal matrix of a mitochondrion requires passage across four membranes.

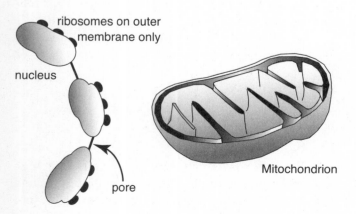

47. (A)

Since the F_1 generation consists entirely of red plants, it is obvious that the red allele is dominant to the white. To solve this problem, it may be helpful to draw a diagram of two Punnet squares. As always, indicate dominant traits with uppercase letters and recessive traits with lowercase letters:

Parental Cross RR × rr → F_1 3:1 red:white

F_1 cross

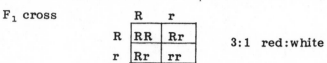

	R	r
R	RR	Rr
r	Rr	rr

3:1 red:white

Testcross Rr × rr 1:1 red:white

	R	r
r	Rr	rr
r	Rr	rr

48. (B)

Mendel's Law of Segregation describes simple dominant-recessive genetics, in which there are different versions of a gene called alleles. Each diploid individual has two copies of a gene, one from each parent. The Law of Independent Assortment and the occurrence of linkage and crossovers concern genetics with more than one gene and therefore are not an issue. Pleitropy is the ability of a gene to affect an organism in many ways.

49. (D)

Proceed as before using a Punnet square. Note that each of the two traits follows the same 3:1 ratio as obtained in the monohybrid cross, and the genotypes are 1:2:1 as in the monohybrid cross.

	RT	RT	rT	rt
RT	RRTT	RRTT	RrTT	RrTt
Rt	RRTt	RRTt	RrTt	Rrtt
rT	RrTT	RrTT	rrTT	rrTt
rt	RrTt	RrTt	rrTt	rrtt

50. (A)

The nucleus is the control center of the cell. Experiments have sucessfully demonstrated that the replacement of one nucleus with that of another species is sufficient for that cell to develop characteristically like the other species. It is therefore concluded that all of the necessary hereditary information is contained in the nucleus.

51. (B)

Most of the electrons from the citric acid cycle are carried as NADH, each electron will pump three hydrogen ions out against the gradient. One of these reactions, however, passed a pair of hydrogens directly to a lower energy FAD coenzyme. This FAD coenzyme now $FADH_2$ passes its hydrogen to the cycle further down the line. These lower energy hydrogens which do not participate in the first active transport reaction will contribute only two hydrogen ions each to the chemiosmotic differential.

52. (D)

The embryo sac in the ovules is the gametophyte. The stigma, style, and ovary with ovule are the female reproductive parts of the sporophyte. The fruit contains the fertilized zygote.

53. (A)

The convention is to describe genes by proceeding in the $5' \rightarrow 3'$ direction along the antisense strand, allowing one to preserve the $5' \rightarrow 3'$ orientation of the message while being able to read from left to right. Here we are describing structures along the transcribed strand in the order that a polymerase would encounter them, which is from a $3'$ to $5'$ direction.

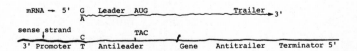

54. (C)

Centered at about the -10 region of a DNA template is one of the two known promoter sites in *E. coli*. The other site is further upstream at about -30 nucleotides. These promoters have an importance extending well beyond *E. coli* insofar as over 50 prokaryotic promoters have been sequenced and all contain similiar sequences in these two locations.

55. (D)

Promoters function with respect to transcription. The translation-related activities described in (D) are characteristic (in prokaryotes) of the Shine-Dalgarno sequence. The latter is about 5 to 10 base pairs upstream from the AUG codon and is homologous to the $3'$ terminus of 16 S rRNA. Note that specifying transcriptional direction (E) is essentially the same as choice of sense strand (C).

56. (C)

On being suddenly chased by the large bear the system of the body immediately comes into play with the fight-or-flight mechanism being activated. This causes an increase in heart rate, the pupils dilate, blood is shunted away from the digestive tract and peripheral vessels. These changes are all wrought by the sympathetic system of the autonomic nervous system.

57. (A)

Since I^A and I^B alleles are codominant, the individual produces glycosyl tranferases corresponding to both A and B antigens. The production of both antigens causes the individual to produce neither the two kinds of antibody. Since the primary risk in transfusions arises from agglutination of the transfused erythrocytes by circulating antibodies in the recipient's bloodstream and not by agglutination of the host erythrocytes by injecting antibodies, the AB individual may receive transfusions from any of the bloodtypes (he should not, however, donate to any but other AB recipients). AB types are known as "universal acceptors," O types are known as "universal donors."

297

58. (B)

The genes are isologous. The translated products are isozymes. Allelic forms refer to alternate sequences which occur at the same position along a given chromosome. A gene is epistatic to another gene if gene A can mask or enhance the effects of gene B (i.e., B produces a pigment while A regulates its deposition). Pleiotropy is the translated product of a single gene manifesting itself as more than one phenotypic trait.

59. (C)

Hemoglobin S results from the substitution of a single amino acid in the β chain. Normal adult hemoglobin (i.e., hemoglobin A) is a tetramer which may be represented as $\alpha 2S2$. Hemoglobin S, the sickling form, is thus $\alpha 2S2$. The individual with sickle cell trait is heterozygous for the β producing allele; the hemoglobin is composed of 50% hemoglobin A and 50% hemoglobin S.

60. (A)

The correct choice is A. The other effects listed correspond respectively to Tetracycline, Puromycin, Erythromycin, and Streptomycin.

61. (C)

The offspring from III has the gametes from both parents (14, 14^{21}, 21, 21) and thus has three copies of chromosome 21 and is trisomic. This offspring would show the symptoms of Down's syndrome. The offspring of cross I (14, 14, 21, 21) is a normal. The offspring of cross II (14, 14^{21}, 21) is phenotypically normal because the 21 is on the 14 and thus there are two of chromosome 21. There is no offspring from cross IV because it is a haploid and thus not viable.

62. (C)

A nucleiod is the small circular strand of DNA in prokaryotes. All of the other responses represent characteristics of eukaryotic chromatin.

63. (A)

The nucleosome core is composed of about 145 DNA base pairs. The other answers reflect the increasing levels of packing DNA in chromosomes.

64. (C)

Primase gets its name from its role in forming the RNA primers needed in DNA replication. Unlike RNA polymerization which can proceed de novo, there is a requirement in DNA synthesis for an exposed 3' end to add onto and hence the need for RNA primers. The other functions listed are also necessary and are performed, respectively, by helicase (A), single strand binding protein (B), DNA polymerase III holoenzyme (C), and DNA polymerase I.

65. (C)

The most logical evolutionary advantage of multicellularity is that multicellular organisms can cope differently with the environment and its resources. In multicellular organisms there is a very high degree of specificity. This means a high level of efficiency which allows the multicellular organism to cope differently with the environment.

66. (A)

Transduction is usually done in the laboratory. In this process genetic material from one bacterium is transferred to another bacterium using a bacteriophage as the carrier.

67. (A)

The first priority of succulent plants adapting to arid environments is conservation of water. Since most water loss will occur in the hot, dry air above ground, the plant must adapt by reducing the area exposed to this air. The less surface area exposed to the air, the less water the plant will lose.

68. (C)

Recall the definitions of noncompetitive and uncompetitive inhibition. In the latter, the inhibitor does not combine with free enzyme or substrate but combines with the enzyme-substrate complex. A Lineweaver-Burke plot of such a reaction is characterized by retaining a constant slope as inhibitor concentration increases for fixed substrate concentrations. Noncompetitive inhibition occurs when the inhibitor can complex with either free enzyme or enzyme-substrate complex, the inhibitor tends to bind at some site other than the active site and increasing substrate concentration is ineffectual in reversing inhibition. Noncompetitive inhibition may be recognized through Lineweaver-Burke plots of the type shown (i.e., different slopes but a common 1/[S] intercept for reactions run in various concentrations of inhibitor). Only enzyme inhibition reactions that are slowed by competitive inhibitors are reversible.

69. (B)

Exotoxins are primarily excreted by gram positive organisms, whereas gram negative bacteria are typically the source of endotoxins. The former are proteins, while the latter are lipopolysaccharides. A toxin which cannot be inactivated by either heat treatment or acidification is clearly not a protein. Mitosis does not occur in bacteria.

70. (A)

One important function of gibberellins is the elongation of the stem. The other choices are functions of other plant hormones.

71. (C)

As follicles grow they release increasing amounts of estrogen. The high estrogen levels (or possibly the slowdown in estrogen buildup as the peak is reached) elicit a sharp, spiked surge of luteinizing hormone from the pituitary. This, in turn, triggers the opening of the follicle and ovulation.

72. (A)

The age distribution of a population is a summary of the proportional representation of the various age classes. It may be an indicator of the potential for growth of future populations, but it is not a good indicator of population growth rate (B) because the data are only given for one time period. Although the age distribution may be dominated by one age category (C), this is not always the case, particularly in a distribution shown by a stable population. The distribution gives the number of individuals in each age group at a particular time but it does not indicate the number of births for each category. Therefore, it is not a good indicator of age-specific survivorship (D). The initial population size is not known and thus fecundity is unknown; therefore, the age distribution is not a measure of fecundity (E).

73. (B)

The male frogs call to attract females to the breeding pond. Bees have an elaborate dance (A) that shows other bees the direction to a new food source. Male mule deer roar (C) to establish territories and warn other males to stay away. Electric signals from fish (D) are for stunning prey or communicating information about approaching predators or navigation directions. Humpback whales have a common song (E) for the pod of whales that move as a group.

74. (A)

Cholecystokinin or CCK is a hormone that is produced in the duodenum in response to amino or fatty acids in the duodenum. Its action is in the pancreas (and gallbladder). It is not produced in the stomach (C and D) or pancreas (E).

75. (C)

Stomata are pores in plant stems and leaves and are not structures of a cell. The lipid membrane is common to both animals and plants. Chloroplasts, though structures in plant cells, are involved in the process of photosynthesis, not in the regulation of osmotic pressure.

The osmotic pressure of a plant cell is regulated by the cell wall. Consider the separate fates of an animal cell and a plant cell placed into very hypotonic solutions. The animal cell will take in water, causing it to swell, and if the original difference in osmotic pressure is great enough, the cell may take in more water and build up more pressure than its membrane can withstand, in which case the cell would burst (this is called lysis).

This lysis would not occur in a plant cell. A plant cell placed in a hypotonic medium would have water enter it, causing it to swell. However, an upper limit as to how much water can enter is imposed by the cell wall. As the cell swells, its plasma membrane exerts what is called turgor pressure on the cell wall. The wall exerts an equal and opposing pressure on the swollen membrane. Mature cell walls can be stretched only to a minute amount. When the pressure exerted by the cell wall is so great that further increase in cell size is not possible, water will cease to enter the cell. Thus, plant cells will only absorb a certain amount of water, even in an extremely dilute medium.

76. (D)

If present, interneurons are wholly contained in the gray matter of the spinal column. A reflex arc consists of a minimum of three cells: a sensory cell, a motor neuron and an effector cell (E). Only one sensory neuron is present (A) and the cell body of the sensory neuron is outside the spinal cord in the dorsal root ganglia (B). The axon of the motor neurons enter the spinal cord dorsally (C) and leave ventrally.

77. (E)

The light reactions occur in thylakoid membrane not the stroma or inside compartment.

78. (E)

Recombinant DNA has to be on the range of length from 75% to 105% of a unit λ genome in order to be encapsidated. Therefore, the length of the foreign DNA insert is not unlimited. λ phage infects bacteria at a higher frequency than plasmid transformation into bacteria. Large segments of the λ genome can be replaced with foreign DNA leaving its lytic and lysogenic life cycles unaffected.

79. (A)

Water is used in the light, not the dark, reaction. In the dark reaction, ATP and NADPH are used to fix the CO_2 into carbohydrates.

80. (D)

Oxidative phosphorylation (D), which occurs in the inner membrane of the mitochondria, generates about 90% of the total ATP produced during cellular respiration. Gycolysis (A), Krebs cycle (B), conversion of pyruvate to acetyl CoA (E), and electron transport (C) are other stages of cellular respiration and are required for oxidative phosphortylation to occur but they do not produce large quantities of ATP.

81. (C)

Habituation is the process of gradually learning not to respond to a stimulus. Intentional movements are component parts of an implied imminent action, such as a man clenching his fist but not striking. Redirected activity is transference of an action from its intended yet unavailable object to an available object. In this case, the dog's anger and desire to bite is redirected from your flesh to the stick. Imprinting (D) is a learned behavior frequently acquired immediately after hatching (chicks follow their mother). Operant conditioning (E) is a type of trial and error learning.

82. (D)

The ecological position, or niche, of species B is changed because of competition. The new niche is the realized niche. The niche in absence of any interactions is the fundamental niche (A). Species B lives in the emergent vegetation that is the producer component (B) but living in vegetation does not make the individual a producer; the type of individual is not given. Similarly, because the type of individual is not known it cannot be assumed to be a consumer (C). Because species B lived in the whole pond, the new habitat is not selected (E).

83. (B)

Chlamydomonas is characterized by asexual reproduction, although Chlamydomonas can sexually reproduce when there is low ambient nitrogen. During asexual reproduction, haploid not diploid (E) zoospores are produced. Possession of a macronucleus as well as a micronucleus (A) is characteristic of Paramecia and various cellular slime molds such as Dictyostelium. These organisms undergo amoeboid as well as motile multicellular stages, the change from free living amoebae occurring under conditions of diminished local food supply (C). Clamydomonas gametes are similar and the resulting fusion is called isogamy. Anisogamy is the fusion of gametes of different sizes.

84. (E)

Steroid hormones diffuse across the plasma membrane. It combines with a receptor that is found only in target cells. The steroid hormone/receptor complex diffuses into the nucleus and initiates the production of regulatory proteins. There is no involvement of cyclic AMP. All of the other cases involve cyclic AMP.

85. (D)

All of the statements except (D) are true. Selection can be against a particular trait. For instance, in stabilizing selection the extremes of the phenotype are selected against; there is no new trait.

86. (D)

The middle lamella is the pectin-based layer formed when the walls of adjoining plant cells come into contact. The primary wall is the first to be laid down, is stretchable, and remains the only wall present as long as the cell continues growing. After ceasing to grow, cells in the more rigid, wooden sections of the plant may lay down a secondary wall, lying internal to the primary wall (and external to the plasma membrane).

87. (D)

Hydrostatic pressure pushes fluids out of the capillary, and osmotic pressure pulls water in. The difference results in water movement. At the arteriole end, the hydrostatic pressure is greater than colloid osmotic pressure and water moves out. This is filtration, not reabsorption (A). At the venule end, reabsorption takes place: the colloid osmotic pressure is greater than the hydrostatic pressure and fluids move into the capillaries, not out (B). Therefore, as filtration occurs from arteriole to venule end, the hydrostatic pressure decreases (D). Colloid osmotic pressure remains constant because it is due almost exclusively to the blood proteins, not small ions (E). The part of the figure above the colloid osmotic pressure is greater than that below and thus filtration exceeds reabsorption. Therefore, there is a net loss of fluids from the capillary, not a net uptake (C).

88. (B)

The same genotypes can occur in different phenotypes (B) due to environmentally induced changes during development. Because these acquired conditions are not inherited, this possible difference in phenotypes is important in morphological comparisons among individuals in evolutionary studies. The genotype is the allele composition of a gene. The phenotype is the physical or chemical expression of the gene. Therefore, the genotype is not equivalent to the phenotype (C) but it is related to the phenotype (D). Different genotypes can produce the same phenotype (A). For instance, if W is the dominant allele and w the recessive allele, Ww and WW will produce the same phenotype. Siblings are not identical and do not have the same genotypes or phenotypes.

89. (A)

Nondisjunction refers to the failure of chromosomes to separate after synapsis. Synapsis is a phase of meiosis in which homologous chromosomes wrap themselves around each other. When nondisjunction occurs, both chromosomes migrate into one of the sperm cells. If this sperm cell fertilizes an egg, the resulting zygote will have 47 chromosomes. The normal number of chromosomes for humans is 46.

90. (A)

Species diversity is an index that measures both the number of species and the relative numbers of each species (evenness). The more evenly the species are distributed, the greater the evenness. Therefore, the community with the greatest number of species and the most even distribution of the species has the highest species diversity (community a). Community b has the most individuals but only three species that are unevenly distributed. Community c has even distribution but only three species. Community d has five species, but species number 1 dominates. Community e has only two species. Community a has five species evenly distributed.

91. (B)

Evenness is the relative number of each of the species, regardless of the total number of species. The lowest evenness is found in the community with the most different numbers of individuals of each species. Community a has similar numbers of each of the five species. Although community c has only three species, they are present in similar numbers. Community d has a dominance of one species but the remaining four species are present in similar numbers. Community e has few individuals but the two species have the same number of individuals. Note community b with 150–0–0–75–5.

92. (D)

Linking anti-P to a fluorescent marker is used only in the direct method. All of the other steps are required in the indirect method.

93. (A)

The ATP-ase activity in contraction of skeletal muscle is associated with the myosin thick filaments; more specifically it is localized in the S-1 heavy meromyosin fragment. The ATP-ase activity in microtubule activity is associated with neither of the tubulin subunits but rather with the dynein "arms" that connect adjacent doublets. The MAPs are known to bind to microtubules during polymerization/depolymerization cycles and are presumably functionally involved in initiation, elongation, and stabilization of microtubule assembly but possess no known ATP-ase activity.

94. (E)

Most common plants undergo "alternation of generations" with both diploid and haploid multicellular stages. The relative prominence of these stages varies from one plant type to another. The most primitive plants are characterized by the absence of any diploid phase other than the zygote, which meiotically divides to produce spores. The fruit develops after fertilization and contains the diploid seeds.

A: Rudimentary plant

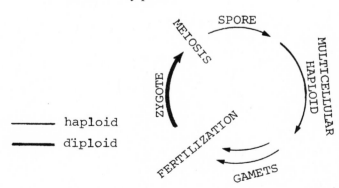

B: More advanced multicellular plant

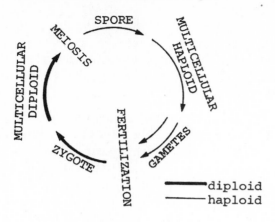

95. (D)

Density-dependent factors are those whose effect increases with population density. The effect may be to increase the death rate or decrease the birth rate as the population size increases. Of these two possible effects, only an increase in death rate was given as an option. There is an inverse relationship between birth rate and population density, and thus the birth rate would not increase as the population size increased (C) nor would the birth rate decrease as the population size decreased (A). The effect of density-dependent factors increases with increasing density but does not cause the populations to decrease to zero (B). Predation may increase as population density increases but density-dependent factors are not due to predation (E).

96. (A)

Many different factors affect the size, distribution, and growth of populations, but the growth depends on the initial population size and births, deaths, immigrations, and emigrations. Maturity, parity, and fecundity (B) are the factors that determine fitness or the contribution of an individual to future generations but they do not determine the growth of the population. Dispersal and spacing contribute to population distributions, and fertility is potential births. Both fertility and dispersal are included in (C) and both fitness and spacing are included in (D). Fecundity (E) could be considered as births, but dispersal is incorrect.

97. (B)

The genes are linked and thus are on the same chromosome and should appear together in the offspring. The offspring should all be like either of the parents (B). There is no indication of the sex of the different parents and thus sex-linked genes are not a solution (A and D). The four phenotypes should be equally represented if there was independent assortment (C). There is no reason to expect non-random mating as the flies were able to breed randomly in jars (E).

98. (A)

Allometric growth is the differential growth of different parts of the body, in this case the beak and the skull. Paedomorphosis (E) is a special case of heterochrony (D) and both refer to changes in the timing and rate of developmental processes which is different than growth rates of individual body parts. Gene flow (C) and natural selection (B) affect the frequency of genes in a population. The allometric growth may be a consequence of natural election but it is not the process.

99. (B)

The labyrinth of the inner ear has three semicircular canals, each consisting of a semicircular tube connected at both ends to the utriculus. Each canal lies in a plane perpendicular to the other two. At the base of each canal, where it leads into the utriculus, is a bulb-like enlargement (the ampulla) containing tufts of hair cells similar to those in the utriculus and sacculus, but lacking otoliths. These cells are stimulated by movements of the fluid (endolymph) in the canals. When a person's head is rotated, there is a lag in the movement of the endolymph in the canals.

Thus, the hair cells on the ampulla attached to the head rotate, in effect, in relation to the fluid. This movement of the hair cells with respect to the endolymph stimulates the former to send impulses to the cerebellum of the brain. There, these impulses are interpreted and a sensation of dizziness is felt.

100. (D)

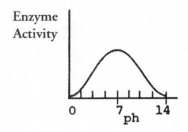

As the pH rises, enzyme activity rises rapidly until an optimum pH of about 7.00 is reached. Then, it begins to drop just as rapidly as it rose.

101. (B)   102. (D)   103. (B)   104. (A)   105. (A)

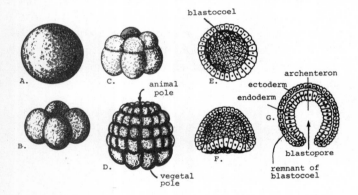

A.   Zygote
B.   First cleavage
C.   Morula
D.   Blastula
E.   Longitudinal section through a blastula
F.   Longitudinal section through early gastrula
G.   Longitudinal section through late gastrula

Note the following characteristics: The early cleavage stages occur without cytoplasmic growth, resulting in cell clusters no larger than the original zygote.

Early cleavages are oriented in such fashion that the embryo is composed of an animal and a vegetal hemisphere, with the cells in the former being noticeably smaller. Animals with greater amounts of yolk in the eggs show greater variation in size than amphioxus.

In addition the more yolk present, the more the pattern of embryological development diverges from amphioxus. The first cleavages produce a grapelike cell cluster known as the morula. Subsequent cleavages result in a blastula stage. The blastula is a semispherical aggregate of cells (blastomeres) which secrete fluids into a central cavity, the blastocoel.

Gastrulation typically begins when the blastula is composed of approximately 500 cells. The process begins with invaginatiori at the vegetal pole and continues as more cells move to the point of invagination and fold inward, enlarging the invagination. Subsequently the invaginated cell layer comes to lie almost flush against the surface layer, filling in the old blastocoel and forming a 2-tiered cup-like structure. The interior of the cup forms the archenteron and opens to the outside through the blastopore.

106. (C)

Rhiopoda (III) is a phylum which includes the amoebas, it is not a form of locomotion. The amoebas move using pseudopodia (II). Axopodia (I) are the "feet" of Axtinopoda. Flagella (IV) are used by zooflagellates. Cilia (V) are used by ciliates. The remaining choices are structures found in protozoans that are used in locomotion

107. (C)

The thin myofilaments attached to the Z lines of a sarcomere contain three proteins: a thin protein called actin; a globular protein called troponin which attaches to actin and contains binding sites for Ca++; and tropomyosin, a linear protein which is also attached to actin and which covers the myosin binding sites when the muscle is at rest but uncovers them during contraction.

108. (A)

Increased production of follicle stimulating hormone (FSH) stimulates the development of ovum and follicle.

109. (A)

	Dicots	**Monocots**
No. of seed leaves (cotyledons)	2	1
Venation of leaves	Pinnate, palmate (i.e., branched network)	Parallel
Type of root system	Woody, taproot (i.e., large primary root with branch roots growing from it)	fibrous, no taproot, all roots approximately equal in size
Arrangement of vasculature in stem	Vascular bundles in a single ring	Vascular bundles irregularly distributed through pith tissue
Cambium growth rings	New growth ring each year or growing season	Stem and root devoid of cambium, no growth rings
Arrangement of flower parts	Arranged in twos, fours or most often fives	Arranged in threes

Question 109 listed exclusive traits of monocots.

110. (C)

A catalyst lowers the energy of activation of a reaction, thus facilitating transition from reactants to products. The catalyst does not, however, erase the energy barrier altogether, and the reactants and products start and arrive at the same energy levels as in the uncatalyzed reaction.

In graph C, the energy of activation is significantly lowered in the catalyzed reaction, while the reactants and products begin and end at the same levels in both the catalyzed and uncatalyzed reactions.

Graph B has the catalyzed and uncatalyzed reactions reversed.

In graph A, the catalyzed reaction shows an energy of activation which is lower than the final energy state of the products—this is impossible. Graph E shows the same thing—an energy of activation lower than the energy of the products—and also has the catalyzed reaction needing more energy than the uncatalyzed reaction. In graph D, the catalyzed and uncatalyzed reactions are reversed, and the energy levels of reactants and products before and after the reaction are not the same for both reactions.

111. (B)

Cofactors are the nonprotein part of enzymes. They are derivatives of minerals represented by small inorganic molecules that are bound with enzymes during chemical reactions to make the enzymes functional.

112. (C)

Ferns are lower vascular plants. They bear fronds that produce spores. Mushrooms belong to the Kingdom Fungi. The pine tree and spruce are both gymnosperms. They bear cones as reproductive structures. The oak tree is an angiosperm. It bears flowers and "hidden" seeds.

113. (D)

Neanderthals were descendants of *Homo erectus*. They are named for the location of the first fossils in the Neander Valley, Germany.

114. (A)

Australopithecus, *Homo erectus*, and *Homo sapiens* are all bipedal but *Australopithecus* (A) fossils date to 4.5 million years ago. *Homo erectus* (B) fossils are only about 1.8 million years old. *Homo sapiens* (C) are younger than both of the others. Neither *Purgatorius* (oldest known primates) (D) nor *Aegyptopithecus* (oldest known apes) (E) are bipedal.

115. (A)

Index fossils are fossils that are abundant and widespread for a defined period of time. The trilobites were abundant in the Paleozoic but became extinct when the seas dried up at the end of the Paleozoic. They meet the criteria of index fossils. They cannot be aged by carbon 14 because they are too old. Their physical characteristics are not important as an index fossil.

116. (A)

In the human excretory system large blood components such as glucose and amino acids are returned to the blood from the filtrate by facilitated diffusion.

117. (A)

If the islets of Langerhans of the pancreas are ligated or damaged in any way the flow of insulin from the pancreas to the circulatory system is affected.

If the pancreatic duct is ligated, the flow of trypsinogen to the small intestine will be hindered.

118. (C)

Sugar passes through the plasma membrane by combining with a protein carrier molecule (a permease). This combination is then able to pass through the membrane from a region of higher to one of lower concentration of sugar without the use of cellular energy. This process is called facilitated diffusion.

119. (A)

During the initiation of muscular contraction the action potentials traveling down the T-tubules and stimulate the sarcoplasmic reticulum to release its stored calcium, which binds to the troponin binding sites.

120. (D)

Biomes refer primarily to terrestrial ecosystems and are defined on the basis of the main vegetation types, which are dictated primarily by temperature and precipitation. Longitude and latitude affect the temperature and precipitation, which are products of the climate and topography. Soil type is also important in dictating the type of terrestrial ecosystem present but neither soil type and precipitation, nor soil type and latitude, combine as prime variables.

121. (B)

Mycorrhizae are associations between fungi and plant roots. Interwoven mats of hyphae are called mycelium. Mutualistic associations between fungi and algae are called lichens. Strong odors are produced by truffles but there is no particular name for these odors. Imperfect fungi are not part of mycorrhizal associations.

122. (D)

Saprobes obtain food from dead plants (including fallen logs) or animals.

123. (A)

Fish retain a primitive type of heart, with one atrium and one ventricle. Blood is routed directly from the gills to the tissues in fish.

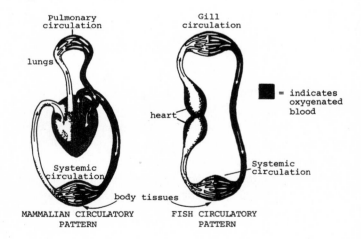

124. (B)

Lipid bilayers tend to close on themselves so that no hydrocarbon chains are exposed to the aqueous medium. As the lipid bilayer closes, water molecules are released from the hydrocarbon chains and a favorable increase in entropy occurs in the system. Van der Waals forces between hydrophobic tails and electrostatic interactions between the polar head groups lower the energy content of the system which is also favorable.

125. (C)

This question asks you to recognize the ecosystem that is dominated by evergreen trees. This is the taiga, an area of extremely cold winters. The tundra, a region north of the taiga, has a permanently frozen subsoil (permafrost) and cannot support any large plants. The temperate forests are farther south and contain deciduous plants (plants that lose their leaves) that need more rainfall and a relatively long, warm summer. Tropical rain forests have abundant rainfall and are characterized by tremendous species diversity, with up to ten dominant trees. A chaparral is a region dominated by drought-resistant and fire-resistant small trees and shrubs because of the cool, rainy winters and hot, dry summers.

126. (C)

This question tests your ability to interpret experimental data given in graphical form. This graph shows the changes in abundance of two species over a period of time. Species A shows a sharp increase in number every 9 to 10 years, followed by a sharp decrease. Species B increases in number following the same cycle—except that the maximum abundance is about one year later than that for species A. Since species B peaks after species A, it is most likely that species B is eating species A. B is the predator, and A is the prey. When there are more species A organisms, there is more food for species B, and B increases. Larger numbers of B feed on A and cause the decline of A. As A decreases, there is less food for B, and B dies off.

127. (A)

Bile salts have no enzymatic action as they do not change any substrate into another component. Their function is to emulsify fats so that lipases can hydrolyse them.

128. (D)

All of the answers contribute to the movement of water up the xylem, but pressure gradient between the atmosphere and the air spaces in the mesophyll of the leaves is the driving mechanism.

129. (C)

Some higher plants, especially grasses and trees, pick up quanties of ammonium ion and subsequently incorporate the nitrogen into more complex molecules. However, most flowering plants absorb required nitrogen in the form of nitrates.

130. (E)

When the baroreceptors detect an increase in pressure, the signal is passed to the cardiac medullary cardiovascular center and then to the autonomic parasympathetic motor neurons, not the sympathetic (D) motor neurons. This results in a decrease in heart rate (E), not an increase (A), and a decrease in the force of cardiac systole, not an increase (B), which act to reduce cardiac output and reduce peripheral vascular resistance, not increase it (C).

131. (D)

Animals such as Cnidaria (jellyfish, corals, anemones) have radial symmetry and only two germ layers: the endoderm and the ectoderm. They have true tissue and during development the archenteron is present.

132. (B)

Lamarck believed that evolution occurred through physical changes during the lives of individuals who struggled to adapt to their environments. This theory is now known to be wrong. We now know that DNA directs the formation of the phenotype, and that DNA does not accept instruction back from the phenotype. Lamarck's theory has been replaced by Darwin's theory of natural selection, which proposes that changes in the phenotype of a certain type of organism occur as individuals with characteristics that favor survival pass those characteristics on to successive generations, while individuals with characteristics less suited to survival are not as successful at passing on their genes; hence, characteristics that are less suited to survival fade out and eventually disappear.

133. (C)

Highly developed animals can react extremely quickly to certain types of stimuli. This can be important when the animal comes into contact with harmful stimuli, such as fire or sharp objects piercing the skin. The faster the animal responds, the less damage is done.

These kinds of responses do not involve the brain, for conscious input is unnecessary and would slow the response. Instead, the impulse is carried from the receptor neuron directly to the spinal cord and then back out through a motor neuron to the appropriate muscle. Thus, the receptor neuron, the motor neuron, and the muscle are the only structures necessary for the reflex action to occur.

134. (E)

It is a common misconception that all deserts are covered by sand and shaped by wind.

135. (B)

Fitness is the contribution of alleles an individual makes to the next generation. These alleles are contained in surviving offspring. The greater the number of offspring that survive, the greater the contribution of alleles will be to the next generation. In (B) the male has his alleles in 18 offspring. This is obtained by multiplying the number of matings each year (2) times the number of eggs (3) produced by the females times the number of years that eggs are produced (3). For the females, the number of mating is only important in the total number of eggs produced. The length of time the bird lives does not affect the contribution of alleles to the next generation.

136. (B)

Fermentation is the breaking down of glucose in the absence of oxygen into alcohol and carbon dioxide.

137. (C)

The light reaction phase of photosynthesis that occurs in the thylakoid concludes with the yielding of the following products: O_2, ATP, and NADPH.

138. (A)

Meiosis is the process by which the number of chromosomes in the diploid cell or nucleus is reduced by half.

139. (A)

Dispersal is the movement of organisms away from the breeding site. The movement of the young salamanders from the pond where they were born to a new pond is dispersal. Migration involves movement but not from a specific place, the birth location. Energy flow is the movement of energy. Similarly, nutrient cycling is the movement of nutrients. Spacing is one result of dispersal.

140. (A)

The necks of the tortoises are longer and, thus, the mean length of the tortoise neck is increased. This is directional selection. The increased growth could easily be explained by differential growth of the cervical vertebrae relative to the other vertebrae. This is allometry. Stabilizing selection would not result in a change in the length of the neck. Disruptive selection would result in an abundance of very short and very long necks. Heterochrony is a differential rate of development.

141. (C)

The long-neck variation had an advantage on the island with thick shrubs shading out growth on the floor. The tortoises could browse the shrubs with their long necks. The tropical forest, rocky islands, and active volcanic islands did not support low vegetation or the terrain was not suitable for tortoises. Tortoises did not need long necks to survive on the open grassland island. This is an example of Darwin's theory of natural selection, the theory that explains why evolution occurs. Because each species produces more offspring than can survive (overpopulation), there is competition (struggle for existence). In every population, organisms have different traits (variations), and these traits can be inherited. The individuals with favorable variations in a particular environment will most likely live longer and produce more offspring (survival of the fittest). The long-neck tortoise is an example of natural selection and is a culmination of all the principles that explain survival of the fittest.

142. (C)

Producers and the herbivores that feed on them start a food web. Each step of food chain energy conversion leads to dissipation of some energy into useless heat. This limits numbers that can be supported at each successive link, as less energy to support them becomes available. Decomposers act on organisms and their wastes throughout.

143. (D)

The bass is a second-level feeder, or secondary consumer. It is not a producer starting the chain, as is the alga. The minnow is an herbivore and a primary consumer. It eats the alga. The bear is a third-order consumer.

144. (B)

Polymorphic characteristics are discrete phenotypes—as are the four blood types (B).

145. (D)

This question focuses on one of the basic factors required for evolutionary change to occur. If change is to occur, there must be genetic variation for natural selection to act upon. Athletic training, correct diet, medical care, and vitamin supplements may all improve the

physical well-being of the individual, but the improvements would not be passed on to the next generation. Mutation provides the raw material for evolution.

146. (C)

During muscular contraction, the myosin cross bridges bind to actin to form a functional protein called actinomyosin, which serves as an enzyme in the breakdown of ATP to ADP + Pi.

147. (D)

As stated in the set-up for this item, the partial pressure of oxygen at the lungs is 108 mm, while it is only 32 mm at tissues at rest. The hemoglobin is about 98% saturated with oxygen at the lungs and only 58% saturated at the partial pressure of oxygen typical of the tissues at rest. The 40-point difference between the percentage of the former and that of the latter represents the consumption of oxygen by the tissues themselves and, thus, the approximate percentage of oxygen carried and released to the tissues by hemoglobin.

148. (C)

In the figure two different dissociation curves are given for human hemoglobin, one for the carbon dioxide concentration in arterial blood and the other for the carbon dioxide concentration in venous blood. An increase in CO_2 concentration shifts the curve to the right thus indicating that a higher partial pressure of oxygen is needed to load the hemoglobin and that oxyhemoglobin releases oxygen more readily. The release of O_2 is facilitated by the waste of CO_2, which is picked up by the blood in the capillary beds of the tissues at the same time O_2 is released. Hemoglobin does carry small amounts of carbon dioxide but this does not affect the shift of the dissociation curve. Also, the presence of CO_2 results in the formation of carbonic acid and thus the amount of carbonic acid increases.

149. (D)

Macroevolution is evolution on a large scale and examines the origin of new structures, evolutionary trends, adaptive radiation, and phylogenetic relationships among species and mass extinction. These are large scale events that are shown in the fossil record (II and III). Microevolution is the change in allele frequencies that ultimately leads to new species. This is studied on a smaller time scale (I).

150. (E)

This question deals with the assumption upon which evolution is based. Certain individuals have a better chance of survival because of their genetic makeup. Since more individuals are born than can survive and reproduce, those individuals with the genetic advantage will have greater reproductive success. Acquired traits will not be passed on to the next generation because there is no genetic base for them.

151. (A)   152. (C)   153. (E)   154. (E)   155. (D)

In the early region of the SV-40 genome (so named for its being the first portion to be transcribed) there is a length of DNA which codes for two products: t antigen and T antigen. The latter is a protein kinase which activates the replication of viral DNA by host cell machinery. The former becomes associated with the cell membrane and may be involved in the loss of contact inhibition. Both are thought to be required for the establishment and maintenance of the transformed state.

Specialized transduction is that in which fragments of chromosomal DNA break off with viral DNA when the prophage is induced to come out of its lysogenic state. The included host DNA will then be transferred along with the viral genome; the transfer of genetic information is specialized insofar as only sequences close to phage attachment site can be transduced.

The converse to specialized transduction, generalized transduction, occurs by an entirely different mechanism. Recall that viral DNA is often replicated in the form of long concatamers; the latter is then spliced into fragments of the appropriate length and packaged into a capsid. Transducing particles are formed when pieces of host DNA, generally of the same length as the phage chromosome, are mistakenly packaged into a phage coat and subsequently released with the phage progeny when the cell lyses. Note that such a transducing parti-

cle contains only host-specific DNA instead of bacterial and viral DNA as in specialized transduction; note also the lack of constraints on gene loci transported—hence the apellation "generalized."

The src gene is a component of the RNA oncovirus ASV (Avian Sarcoma Virus). A protein kinase, it is significant as the sole proteinaceous product found in cells transformed by ASV but not in cells which are ASV infected and nontransformed. Hence, expression of this gene seems to be the critical step in the oncogenic transformation mediated by ASV.

Burkitt's lymphoma is a cancer found in African children. It is thought to be caused by the Epstein-Barr virus, which is a type of herpes virus. E.B. virus is also implicated in mononucleosis and nasopharyngeal carcinomas.

156. (B)   157. (C)   158. (D)   159. (A)   160. (E)

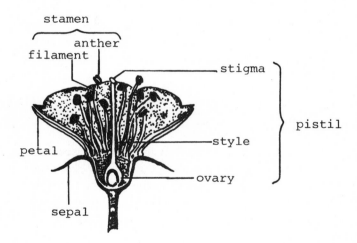

The stamen comprises the male reproductive organs (sporophylls), of which the two primary structures are the stalk-like filament and the terminal pollen producing structure known as the anther.

The female reproductive organs correspond to the pistil which, as illustrated, is composed of the ovary, the style and the stigma. The embryo sac is the female gametophyte; the pollen grain, of course, corresponds to the male gametophyte.

The calyx is the structure made up of all the sepals (corolla is the term for the corresponding petal structure).

161. (E)

Parenchyma cells surround the leaf veins and are important in photosynthesis, therefore contain chloroplasts.

162. (D)

The vascular cambium is a thin layer of undifferentiated cells in vascular tissue.

163. (B)

The cortex is made up of parenchyma and collenchyma and is found just outside the phloem.

164. (C)

The phloem tissue composes of sieve tube members and sieve tubes and also companion cells, phloem fibers, and phloem parenchyma.

165. (A)

The root tips contain the apical meristem tissue, which is located just above the root cap. The apical meristem is an area of small, rapidly dividing cells.

166. (E)   167. (C)   168. (D)   169. (A)   170. (B)

Colchicine and vinblastine are both specific inhibitors of mitosis and both exert their effects by interacting with tubulin. Colchicine binds tubulin dimers and forms colchicine-tubulin dimers that associate with the (+) end of growing microtubule structures (i.e., at the fast growing, fast disassembling end). This association prevents the continued formation of microtubules; thus, spindle fibers are not formed and mitosis is arrested at the metaphase stage. More stable microtubule structures, such as those in flagellae, do not seem to be affected.

Vinblastine arrests mitosis by the crystallization of tubulin and so prevents formation of spindle microtubules.

Cytocholasin B is a specific inhibitor of cytokinesis and has no effect upon mitosis. Eukaryotic cells incubated with the alkaloid show altered shapes and inhibited motility. The effects are produced by a specific interaction with one end of actin microfilaments which prevents further assembly of actin filaments.

Sodium azide (NaN_3) is a metabolic poison which inhibits respiratory ATP synthesis. It prevents energy produced by oxidation of Krebs cycle intermediates from being used to drive the phosphorylation of ADP.

The presence of vanadate ions disrupts numerous processes that require ATPase activity (such as sodium-potassium pumping). The pentavalent vanadium ion binds a bipyramidal array of ligands in a conformation similar to that around a phosphorus atom during the hydrolysis of a phosphate bond.

171. (E)   172. (B)   173. (A)   174. (C)   175. (D)

Taiga (also known as boreal forest) comprises a wide zone in North America and Eurasia, south of the tundra. It is characterized by large coniferous forests interspersed with small bodies of fresh water.

Although the subsoil is frozen for much of the year and the winters are quite cold, the subsoil thaws during the summers. Consequently, there is an abundance of vegegation.

In contrast to the above, the primary characteristic of tundra is the permanently frozen subsoil. The word "tundra" is derived from a Siberian expression for "North of the timberline," and trees are in fact relatively rare. The dominant vegetation consists of mosses and grasses.

Deciduous forests cover large areas of the Eastern U.S. and are characterized by summers that are warmer and longer than those of the taiga as well as abundant rainfall.

Grasslands occur both in temperate and tropical zones, and are typically regions where relatively low annual rainfall or seasonally uneven rainfall prevent the establishment of forests but allow for abundant grasses. Those grasslands in tropical zones (savannas) are characterized by wet-dry cycles as opposed to the annual temperature cycles of those in the temperate zones.

Tropical rain forests contain tall, closely-packed trees whose foliage forms continuous canopies which absorb large quantities of sunlight and leave the forest floor relatively dark throughout the day. These trees catch rain during storms and allow the water to continue to percolate down to the ground long after the rain has ceased. Furthermore, lower levels are shielded from wind, which results in a decreased rate of evaporation and leaves the forest floor more humid than the upper levels. All these factors lead to an abundant variety of life forms found at different elevations.

176. (A)   177. (B)   178. (D)   179. (C)

"Cryptic coloration" refers to the similarity in appearance between certain organisms and their natural environment. This similarity enhances the organisms' survival by making it difficult for predators to see them.

A classic example of cryptic coloration occurred in the 19th century in Manchester, England, where a species of moth composed almost entirely of white individuals evolved into a species composed almost entirely of black individuals. This change in color, which was called industrial melanism, was brought about by the industrial revolution, in which Manchester was centrally involved. Before the industrial buildup, the trees around Manchester were covered with spotty white lichens; hence, spotty white moths were difficult to see on the trees, and therefore had a better chance of escaping death from the moths' main predators—the birds. Black moths, on the other hand, stood out conspicuously against the lichens and therefore had a lesser chance of survival. A black moth was considered a rare and valuable catch to the avid British moth collector.

With the industrial revolution came profuse pollution, and the trees, like everything else, became covered with dark soot. This gave the black moth the survival advantage, and eventually it achieved a predominant frequency of 98% in Manchester and its immediate surrounding areas. In rural areas, the white moth remained predominant. Note that this evolutionary process of cryptic coloration follows precisely the principles of Darwin's natural selection.

Aposematic appearance is the highly noticeable, flashy appearance of certain organisms which possess qualities noxious to predators. While most organisms wish to hide from predators, these noxious organisms rely upon their unpleasant characteristics being remembered and subsequently avoided by predators. For this reason, it is advantageous for the species as a whole to possess physical characteristics which are easily distinguished from the appearance of animals without the same noxious qualities.

Batesian mimicry is the adaptation of an innocuous species to the appearance of a similar but noxious species. This is highly advantageous to the innoxious organism, for it escapes the predation of those animals which have learned to avoid the noxious species.

Müllerian mimicry is the mutually beneficial evolution of a similar appearance by two or more species possessing qualities which are noxious to their predators. Each of the species involved benefits because the predator has only one avoidance response to learn, instead of two or more. Each species is afforded greater protection from predators because the repellent qualities are more frequently advertised.

180. (B)   181. (C)   182. (D)   183. (A)

A population is all of the individuals of one species (yellow perch) in a given area (small lake). Individuals of the same species can interbreed and thus gene flow can interbreed. A cline is a graduation in an inherited trait along a geographic continuum.

184. (A)   185. (B)

Preadaptation (A) is the presence of a trait or character in an organism before it is used in it present function. An adaptation (C) is a trait that results from natural selection (D). Speciation (B) is the formation of a new species and subsequent species that arise from an ancestral specie is adaptive radiation (E).

186. (C)

In operant conditioning, reinforcers follow the behavioral response. It is based on the fact that if a pattern is rewarded, the probability of that pattern reappearing is increased. In this example, the rat continues to press the lever because of the reward in the form of food.

187. (D)

Autonomic learning is the ability of the body to learn to control bodily functions that are generally regulated by the autonomic nervous system. This is also called "visceral learning."

188. (B)

Pavlov's experiments are very well known. The kind of learning illustrated in this example is called "classical conditioning" because the dog's salivation became conditional on the light. The light became associated with the original stimulus—the meat—and eventually became the stimulus itself.

189. (A)

Habituation is, in a sense, learning not to respond to a stimulus. While it is generally beneficial for an animal to be wary and to respond to any stimulus which presents a possible threat, the animal would waste a great deal of time and energy if it continued to respond to stimuli which consistently did not attack, threaten or compete. Thus, the animal learns to ignore stimuli which has proved to be harmless by experience and instead reserves his energy to respond to extraordinary stimuli which may pose real danger.

190. (C)

This is a simple Mendelian cross concerning the A allele. Consider Al to be the dominant allele, A3 to be the recessive. A standard cross yields 50% heterozygous (no MS) and 50% homozygous recessive.

191. (A)

Penetrance is the percentage of individuals in a population who carry a gene in the correct combination for expression and who express the gene phenotypi-

cally. Incomplete penetrance occurs when genes do not express themselves as expected.

192. (D)

When given the frequency of recombination for alleles as a whole, relative allelic recombination may be determined by adding the total recombination frequencies. For the given genotypes, A and C recombine at frequencies of 27% and 14%, which equal 41%.

193. (A)

In the present context an indeterminate cleavage is one where the daughter cells produced retain their full spectrum of developmental potentialities. A determinate cleavage is one where asymmetric partitioning of cytoplasmic components limits the development of one or more daughter cells.

194. (B)   195. (A)   196. (C)

The results clearly indicate that preHPL is of greater molecular weight than HPL as the latter travels further in the gel than the former.

Clearly, some factor in the membrane of RER which is not present in SER is required for processing, as shown by the fact that only the unprocessed band appears in lane b. It is also obvious that the processing is dependent upon entry into the ER, as shown by lanes c and d.

However, the experiment shows nothing that indicates that the excised portion is a "hydrophobic, N-terminal sequence." The experiment supports the signal hypothesis more than the trigger hypothesis because it does show that some sequence is being excised and that translocation occurs during translation. The experiment does not attempt to show what factor(s) cause binding to be limited to RER. The experiment gives no information on the extend of protein formation or the turnover rate. Thus, the longevity of the mRNA is not known.

If the experiment was repeated with mRNA specifying a protein that had an electrophoretic mobility similar to preHPL but was found in the cytoplasm, the protein would not be processed or inserted into the microsomes. Hence, single bands would be found in lanes a and b at the preHPL level but no protein would escape digestion in lanes c and d.

197. (D)

The (triauxic) growth pattern indicates that one fermentation pathway is constituitive, while the other two are adaptive. Since lactose is the disaccharide composed of glucose and galactose, fermentation of glucose or galactose must be constituitive, but both cannot be constituitive. If both were constituitive, there would be a maximum of one growth plateau (which would occur if lactose hydrolysis was adaptive). The presence of two lag phases indicates that lactose hydrolysis and at least one of the fermentation pathways from the monosaccharides are adaptive. Note that this sort of pattern would occur in an organism with constituitive glucose catabolism and inducible lactose and galactose utilization.

198. (A)

We see that even the final filtrate causes disease in the plants. Therefore, the pathogen must be smaller than 0.4 microns. Of the choices given, only viruses and bac-

teriophages are this small. Bacteriophages are bacteria-specific viruses. This question asks about plant viruses so that of the two choices, virus is the more correct.

199. (D)

The process described is a form of dilution. If the toxin were present in the second filtrate, then it would be assumed that the toxin-producing agent had been filtered out and, therefore, no new toxin is being released. This limiting amount of toxin would show decreased activity with increased dilution. However, the degree of infection remains constant with each dilution. This can only mean that a pathogenic agent is present and that its activity remains constant through each dilution so that ensuing second residues remain uninfectious. An effect was observed in some of the treatments and thus the length of the experiment is not a factor.

200. (C)

We would normally expect growth on fully nutritive media from bacteria or fungi. However, virus particles require not nutrient media, but genetic vectors (i.e., bacteria) for their growth.

GRE BIOLOGY

Practice Test 3

Answer Sheet

1. Ⓐ Ⓑ Ⓒ Ⓓ Ⓔ
2. Ⓐ Ⓑ Ⓒ Ⓓ Ⓔ
3. Ⓐ Ⓑ Ⓒ Ⓓ Ⓔ
4. Ⓐ Ⓑ Ⓒ Ⓓ Ⓔ
5. Ⓐ Ⓑ Ⓒ Ⓓ Ⓔ
6. Ⓐ Ⓑ Ⓒ Ⓓ Ⓔ
7. Ⓐ Ⓑ Ⓒ Ⓓ Ⓔ
8. Ⓐ Ⓑ Ⓒ Ⓓ Ⓔ
9. Ⓐ Ⓑ Ⓒ Ⓓ Ⓔ
10. Ⓐ Ⓑ Ⓒ Ⓓ Ⓔ
11. Ⓐ Ⓑ Ⓒ Ⓓ Ⓔ
12. Ⓐ Ⓑ Ⓒ Ⓓ Ⓔ
13. Ⓐ Ⓑ Ⓒ Ⓓ Ⓔ
14. Ⓐ Ⓑ Ⓒ Ⓓ Ⓔ
15. Ⓐ Ⓑ Ⓒ Ⓓ Ⓔ
16. Ⓐ Ⓑ Ⓒ Ⓓ Ⓔ
17. Ⓐ Ⓑ Ⓒ Ⓓ Ⓔ
18. Ⓐ Ⓑ Ⓒ Ⓓ Ⓔ
19. Ⓐ Ⓑ Ⓒ Ⓓ Ⓔ
20. Ⓐ Ⓑ Ⓒ Ⓓ Ⓔ
21. Ⓐ Ⓑ Ⓒ Ⓓ Ⓔ
22. Ⓐ Ⓑ Ⓒ Ⓓ Ⓔ
23. Ⓐ Ⓑ Ⓒ Ⓓ Ⓔ
24. Ⓐ Ⓑ Ⓒ Ⓓ Ⓔ
25. Ⓐ Ⓑ Ⓒ Ⓓ Ⓔ
26. Ⓐ Ⓑ Ⓒ Ⓓ Ⓔ
27. Ⓐ Ⓑ Ⓒ Ⓓ Ⓔ
28. Ⓐ Ⓑ Ⓒ Ⓓ Ⓔ

29. Ⓐ Ⓑ Ⓒ Ⓓ Ⓔ
30. Ⓐ Ⓑ Ⓒ Ⓓ Ⓔ
31. Ⓐ Ⓑ Ⓒ Ⓓ Ⓔ
32. Ⓐ Ⓑ Ⓒ Ⓓ Ⓔ
33. Ⓐ Ⓑ Ⓒ Ⓓ Ⓔ
34. Ⓐ Ⓑ Ⓒ Ⓓ Ⓔ
35. Ⓐ Ⓑ Ⓒ Ⓓ Ⓔ
36. Ⓐ Ⓑ Ⓒ Ⓓ Ⓔ
37. Ⓐ Ⓑ Ⓒ Ⓓ Ⓔ
38. Ⓐ Ⓑ Ⓒ Ⓓ Ⓔ
39. Ⓐ Ⓑ Ⓒ Ⓓ Ⓔ
40. Ⓐ Ⓑ Ⓒ Ⓓ Ⓔ
41. Ⓐ Ⓑ Ⓒ Ⓓ Ⓔ
42. Ⓐ Ⓑ Ⓒ Ⓓ Ⓔ
43. Ⓐ Ⓑ Ⓒ Ⓓ Ⓔ
44. Ⓐ Ⓑ Ⓒ Ⓓ Ⓔ
45. Ⓐ Ⓑ Ⓒ Ⓓ Ⓔ
46. Ⓐ Ⓑ Ⓒ Ⓓ Ⓔ
47. Ⓐ Ⓑ Ⓒ Ⓓ Ⓔ
48. Ⓐ Ⓑ Ⓒ Ⓓ Ⓔ
49. Ⓐ Ⓑ Ⓒ Ⓓ Ⓔ
50. Ⓐ Ⓑ Ⓒ Ⓓ Ⓔ
51. Ⓐ Ⓑ Ⓒ Ⓓ Ⓔ
52. Ⓐ Ⓑ Ⓒ Ⓓ Ⓔ
53. Ⓐ Ⓑ Ⓒ Ⓓ Ⓔ
54. Ⓐ Ⓑ Ⓒ Ⓓ Ⓔ
55. Ⓐ Ⓑ Ⓒ Ⓓ Ⓔ
56. Ⓐ Ⓑ Ⓒ Ⓓ Ⓔ

57. Ⓐ Ⓑ Ⓒ Ⓓ Ⓔ
58. Ⓐ Ⓑ Ⓒ Ⓓ Ⓔ
59. Ⓐ Ⓑ Ⓒ Ⓓ Ⓔ
60. Ⓐ Ⓑ Ⓒ Ⓓ Ⓔ
61. Ⓐ Ⓑ Ⓒ Ⓓ Ⓔ
62. Ⓐ Ⓑ Ⓒ Ⓓ Ⓔ
63. Ⓐ Ⓑ Ⓒ Ⓓ Ⓔ
64. Ⓐ Ⓑ Ⓒ Ⓓ Ⓔ
65. Ⓐ Ⓑ Ⓒ Ⓓ Ⓔ
66. Ⓐ Ⓑ Ⓒ Ⓓ Ⓔ
67. Ⓐ Ⓑ Ⓒ Ⓓ Ⓔ
68. Ⓐ Ⓑ Ⓒ Ⓓ Ⓔ
69. Ⓐ Ⓑ Ⓒ Ⓓ Ⓔ
70. Ⓐ Ⓑ Ⓒ Ⓓ Ⓔ
71. Ⓐ Ⓑ Ⓒ Ⓓ Ⓔ
72. Ⓐ Ⓑ Ⓒ Ⓓ Ⓔ
73. Ⓐ Ⓑ Ⓒ Ⓓ Ⓔ
74. Ⓐ Ⓑ Ⓒ Ⓓ Ⓔ
75. Ⓐ Ⓑ Ⓒ Ⓓ Ⓔ
76. Ⓐ Ⓑ Ⓒ Ⓓ Ⓔ
77. Ⓐ Ⓑ Ⓒ Ⓓ Ⓔ
78. Ⓐ Ⓑ Ⓒ Ⓓ Ⓔ
79. Ⓐ Ⓑ Ⓒ Ⓓ Ⓔ
80. Ⓐ Ⓑ Ⓒ Ⓓ Ⓔ
81. Ⓐ Ⓑ Ⓒ Ⓓ Ⓔ
82. Ⓐ Ⓑ Ⓒ Ⓓ Ⓔ
83. Ⓐ Ⓑ Ⓒ Ⓓ Ⓔ
84. Ⓐ Ⓑ Ⓒ Ⓓ Ⓔ

85. Ⓐ Ⓑ Ⓒ Ⓓ Ⓔ
86. Ⓐ Ⓑ Ⓒ Ⓓ Ⓔ
87. Ⓐ Ⓑ Ⓒ Ⓓ Ⓔ
88. Ⓐ Ⓑ Ⓒ Ⓓ Ⓔ
89. Ⓐ Ⓑ Ⓒ Ⓓ Ⓔ
90. Ⓐ Ⓑ Ⓒ Ⓓ Ⓔ
91. Ⓐ Ⓑ Ⓒ Ⓓ Ⓔ
92. Ⓐ Ⓑ Ⓒ Ⓓ Ⓔ
93. Ⓐ Ⓑ Ⓒ Ⓓ Ⓔ
94. Ⓐ Ⓑ Ⓒ Ⓓ Ⓔ
95. Ⓐ Ⓑ Ⓒ Ⓓ Ⓔ
96. Ⓐ Ⓑ Ⓒ Ⓓ Ⓔ
97. Ⓐ Ⓑ Ⓒ Ⓓ Ⓔ
98. Ⓐ Ⓑ Ⓒ Ⓓ Ⓔ
99. Ⓐ Ⓑ Ⓒ Ⓓ Ⓔ
100. Ⓐ Ⓑ Ⓒ Ⓓ Ⓔ
101. Ⓐ Ⓑ Ⓒ Ⓓ Ⓔ
102. Ⓐ Ⓑ Ⓒ Ⓓ Ⓔ
103. Ⓐ Ⓑ Ⓒ Ⓓ Ⓔ
104. Ⓐ Ⓑ Ⓒ Ⓓ Ⓔ
105. Ⓐ Ⓑ Ⓒ Ⓓ Ⓔ
106. Ⓐ Ⓑ Ⓒ Ⓓ Ⓔ
107. Ⓐ Ⓑ Ⓒ Ⓓ Ⓔ
108. Ⓐ Ⓑ Ⓒ Ⓓ Ⓔ
109. Ⓐ Ⓑ Ⓒ Ⓓ Ⓔ
110. Ⓐ Ⓑ Ⓒ Ⓓ Ⓔ
111. Ⓐ Ⓑ Ⓒ Ⓓ Ⓔ
112. Ⓐ Ⓑ Ⓒ Ⓓ Ⓔ

Continued

Answer Sheet

Continued

113. Ⓐ Ⓑ Ⓒ Ⓓ Ⓔ   135. Ⓐ Ⓑ Ⓒ Ⓓ Ⓔ   157. Ⓐ Ⓑ Ⓒ Ⓓ Ⓔ   179. Ⓐ Ⓑ Ⓒ Ⓓ Ⓔ
114. Ⓐ Ⓑ Ⓒ Ⓓ Ⓔ   136. Ⓐ Ⓑ Ⓒ Ⓓ Ⓔ   158. Ⓐ Ⓑ Ⓒ Ⓓ Ⓔ   180. Ⓐ Ⓑ Ⓒ Ⓓ Ⓔ
115. Ⓐ Ⓑ Ⓒ Ⓓ Ⓔ   137. Ⓐ Ⓑ Ⓒ Ⓓ Ⓔ   159. Ⓐ Ⓑ Ⓒ Ⓓ Ⓔ   181. Ⓐ Ⓑ Ⓒ Ⓓ Ⓔ
116. Ⓐ Ⓑ Ⓒ Ⓓ Ⓔ   138. Ⓐ Ⓑ Ⓒ Ⓓ Ⓔ   160. Ⓐ Ⓑ Ⓒ Ⓓ Ⓔ   182. Ⓐ Ⓑ Ⓒ Ⓓ Ⓔ
117. Ⓐ Ⓑ Ⓒ Ⓓ Ⓔ   139. Ⓐ Ⓑ Ⓒ Ⓓ Ⓔ   161. Ⓐ Ⓑ Ⓒ Ⓓ Ⓔ   183. Ⓐ Ⓑ Ⓒ Ⓓ Ⓔ
118. Ⓐ Ⓑ Ⓒ Ⓓ Ⓔ   140. Ⓐ Ⓑ Ⓒ Ⓓ Ⓔ   162. Ⓐ Ⓑ Ⓒ Ⓓ Ⓔ   184. Ⓐ Ⓑ Ⓒ Ⓓ Ⓔ
119. Ⓐ Ⓑ Ⓒ Ⓓ Ⓔ   141. Ⓐ Ⓑ Ⓒ Ⓓ Ⓔ   163. Ⓐ Ⓑ Ⓒ Ⓓ Ⓔ   185. Ⓐ Ⓑ Ⓒ Ⓓ Ⓔ
120. Ⓐ Ⓑ Ⓒ Ⓓ Ⓔ   142. Ⓐ Ⓑ Ⓒ Ⓓ Ⓔ   164. Ⓐ Ⓑ Ⓒ Ⓓ Ⓔ   186. Ⓐ Ⓑ Ⓒ Ⓓ Ⓔ
121. Ⓐ Ⓑ Ⓒ Ⓓ Ⓔ   143. Ⓐ Ⓑ Ⓒ Ⓓ Ⓔ   165. Ⓐ Ⓑ Ⓒ Ⓓ Ⓔ   187. Ⓐ Ⓑ Ⓒ Ⓓ Ⓔ
122. Ⓐ Ⓑ Ⓒ Ⓓ Ⓔ   144. Ⓐ Ⓑ Ⓒ Ⓓ Ⓔ   166. Ⓐ Ⓑ Ⓒ Ⓓ Ⓔ   188. Ⓐ Ⓑ Ⓒ Ⓓ Ⓔ
123. Ⓐ Ⓑ Ⓒ Ⓓ Ⓔ   145. Ⓐ Ⓑ Ⓒ Ⓓ Ⓔ   167. Ⓐ Ⓑ Ⓒ Ⓓ Ⓔ   189. Ⓐ Ⓑ Ⓒ Ⓓ Ⓔ
124. Ⓐ Ⓑ Ⓒ Ⓓ Ⓔ   146. Ⓐ Ⓑ Ⓒ Ⓓ Ⓔ   168. Ⓐ Ⓑ Ⓒ Ⓓ Ⓔ   190. Ⓐ Ⓑ Ⓒ Ⓓ Ⓔ
125. Ⓐ Ⓑ Ⓒ Ⓓ Ⓔ   147. Ⓐ Ⓑ Ⓒ Ⓓ Ⓔ   169. Ⓐ Ⓑ Ⓒ Ⓓ Ⓔ   191. Ⓐ Ⓑ Ⓒ Ⓓ Ⓔ
126. Ⓐ Ⓑ Ⓒ Ⓓ Ⓔ   148. Ⓐ Ⓑ Ⓒ Ⓓ Ⓔ   170. Ⓐ Ⓑ Ⓒ Ⓓ Ⓔ   192. Ⓐ Ⓑ Ⓒ Ⓓ Ⓔ
127. Ⓐ Ⓑ Ⓒ Ⓓ Ⓔ   149. Ⓐ Ⓑ Ⓒ Ⓓ Ⓔ   171. Ⓐ Ⓑ Ⓒ Ⓓ Ⓔ   193. Ⓐ Ⓑ Ⓒ Ⓓ Ⓔ
128. Ⓐ Ⓑ Ⓒ Ⓓ Ⓔ   150. Ⓐ Ⓑ Ⓒ Ⓓ Ⓔ   172. Ⓐ Ⓑ Ⓒ Ⓓ Ⓔ   194. Ⓐ Ⓑ Ⓒ Ⓓ Ⓔ
129. Ⓐ Ⓑ Ⓒ Ⓓ Ⓔ   151. Ⓐ Ⓑ Ⓒ Ⓓ Ⓔ   173. Ⓐ Ⓑ Ⓒ Ⓓ Ⓔ   195. Ⓐ Ⓑ Ⓒ Ⓓ Ⓔ
130. Ⓐ Ⓑ Ⓒ Ⓓ Ⓔ   152. Ⓐ Ⓑ Ⓒ Ⓓ Ⓔ   174. Ⓐ Ⓑ Ⓒ Ⓓ Ⓔ   196. Ⓐ Ⓑ Ⓒ Ⓓ Ⓔ
131. Ⓐ Ⓑ Ⓒ Ⓓ Ⓔ   153. Ⓐ Ⓑ Ⓒ Ⓓ Ⓔ   175. Ⓐ Ⓑ Ⓒ Ⓓ Ⓔ   197. Ⓐ Ⓑ Ⓒ Ⓓ Ⓔ
132. Ⓐ Ⓑ Ⓒ Ⓓ Ⓔ   154. Ⓐ Ⓑ Ⓒ Ⓓ Ⓔ   176. Ⓐ Ⓑ Ⓒ Ⓓ Ⓔ   198. Ⓐ Ⓑ Ⓒ Ⓓ Ⓔ
133. Ⓐ Ⓑ Ⓒ Ⓓ Ⓔ   155. Ⓐ Ⓑ Ⓒ Ⓓ Ⓔ   177. Ⓐ Ⓑ Ⓒ Ⓓ Ⓔ   199. Ⓐ Ⓑ Ⓒ Ⓓ Ⓔ
134. Ⓐ Ⓑ Ⓒ Ⓓ Ⓔ   156. Ⓐ Ⓑ Ⓒ Ⓓ Ⓔ   178. Ⓐ Ⓑ Ⓒ Ⓓ Ⓔ   200. Ⓐ Ⓑ Ⓒ Ⓓ Ⓔ

Practice Test 3

1. Which of the following best describes the society of a social insect?

    (A) The members of the society are very industrious.
    (B) Each society cooperates with other societies.
    (C) There is a division of labor among the various members.
    (D) There exists a social hierarchy.
    (E) Dominance hierarchy is practiced.

2. When two or more traits produced by genes located on two or more different chromosome pairs are expressed independently, there has been

    (A) mutation
    (B) independent segregation
    (C) crossing-over
    (D) independent assortment
    (E) cross linkage

3. Which of the following is NOT characteristic of heterotrophic organisms?

    (A) They obtain their energy from the oxidation of organic molecules.
    (B) There is an interdependent relationship between them and photosynthetic organisms.
    (C) Most heterotrophic organisms use aerobic respiration.
    (D) They can manufacture their own food.
    (E) They have well-developed enzyme systems.

4. The R-group of amino acids is always attached to a

    (A) nitrogen atom
    (B) carbon atom
    (C) oxygen atom
    (D) hydrogen atom
    (E) phosphorus atom

5. Which of the following statements defines meiosis?

    (A) The number of chromosomes in the diploid nucleus is reduced by half.
    (B) The fusion of two haploid nuclei to form a diploid nucleus.
    (C) The process by which four haploid cells or nucleiare transformed into a diploid cell.
    (D) The separation of chromosomes.
    (E) The fusion of chromosomes.

6. In the model for the formation of a replication bubble in prokaryotic DNA, the coiled DNA is first relaxed by the action of

    (A) gyrase
    (B) RNA primer
    (C) DNA primase
    (D) DNA helicase
    (E) initiator proteins

7. Which of the following explains why evergreen trees do not have to shed their leaves in winter?

    (A) Low transpiration rates
    (B) Conical shape
    (C) High transpiration rates
    (D) Only one species of trees in each forest
    (E) Great height

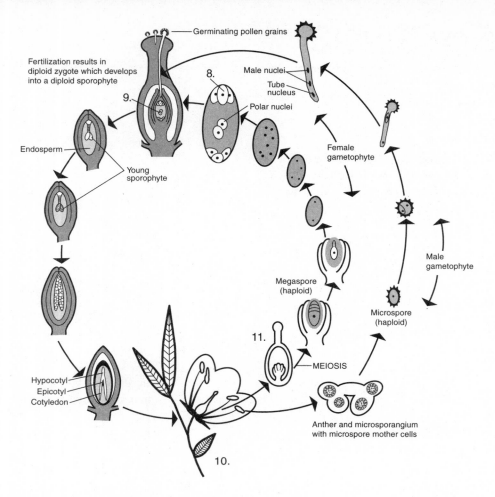

Germinating pollen grains

Fertilization results in
diploid zygote which develops
into a diploid sporophyte

Male nuclei

8.

Tube
nucleus

Polar nuclei

9.

Endosperm

Female
gametophyte

Young
sporophyte

Male
gametophyte

Megaspore
(haploid)

Microspore
(haploid)

11.

MEIOSIS

Hypocotyl
Epicotyl
Cotyledon

Anther and microsporangium
with microspore mother cells

10.

Questions 8–11 refer to the above schematic rendering of the life cycle of a flowering plant.

In each case, decide which of the words or phrases best corresponds to the illustrated object or process; unless a line is drawn to indicate a specific element, the entire structure is to be indicated.

8.

   (A) Egg (n)
   (B) Ovule
   (C) Embryo sac
   (D) Megaspore
   (E) Polar nuclei (n)

9.

   (A) Sperm nucleus (n)
   (B) Polar nucleus (n)
   (C) Polar nucleus (2n)

   (D) Endosperm nucleus (3n)
   (E) Fusion nucleus (2n)

10.

   (A) Sporophyte (2n)
   (B) Gametophyte (n)
   (C) Sporangium (2n)
   (D) Sporophyll (2n)
   (E) Sporophyte (n)

11.   (same structure in cross-section and outer view)

   (A) Ovule (2n)
   (B) Megaspore mother cell (2n)
   (C) Microsporangium (n)
   (D) Gametophyte (n)
   (E) Sporophyte (n)

12. The _____ microscope has the distinct advantage of producing the illusion of three-dimensional images with unusually great depth of field.

    (A) compound light
    (B) transmission electron
    (C) dissecting light
    (D) scanning electron
    (E) telescopic electron

13. A produce clerk sprayed her lettuce with a solution of salt in which the number of salt molecules equaled that in the cytoplasm of the lettuce cells. The spray would be _____.

    (A) hypotonic
    (B) isotonic
    (C) hypertonic
    (D) isokinetic
    (E) supertonic

14. Sea water evaporation ponds are used to concentrate sea water for salt production. Sometimes these ponds appear red due to the presence of certain organisms. Which group of organisms is most probably present?

    (A) Archaebacteria
    (B) Cyanobacteria
    (C) Clamydias
    (D) Rhodophyta
    (E) Proteobacteria

15. When a virus infects a bacterium, the material that enters the bacterium from the virus is

    (A) sulfur
    (B) nucleic acid
    (C) a mutagen
    (D) protein
    (E) glycogen

16. A gardener took cuttings from a short and stocky plant to grow and give her friends. One of the cuttings was left at the edge of the garden, in the shade, and grew very tall and spindly. The gardener liked the tall spindly form of the plant so she took cuttings from it and grew them in the usual sunny conditions. Unfortunately, all of the cuttings from the tall spindly plant were short and stocky. The best explanation for this result is

    (A) plants require direct sunlight for growth
    (B) plant characteristics acquired under different growing conditions are not hereditary
    (C) the tall spindly characteristic was recessive and only occurred in one of the plants
    (D) the short stocky plants were the dominant genotype
    (E) plant size is a polymorphic characteristic

17. Territories are areas that increase opportunities for food procurement, mating, and/or nesting sites. They are defended by the occupants against others, usually of the same species. Territories serve to

    (A) provide for equal distribution of food
    (B) increase the time that individuals spend fighting
    (C) regulate population size
    (D) increase physical contact among species members
    (E) decrease reproductive success

18. All microtubules are made up of a common protein. This protein is

    (A) tubulin
    (B) myosin
    (C) pectin
    (D) actin
    (E) actinomyosin

19. Which of the following are members of the monophyletic group chordates?

    (A) Hemichordates, urochordates, cephalochordates, and vertebrates
    (B) Cephalochordates and vertebrates
    (C) Urochordates, cephalochordates, and vertebrates
    (D) Hemichordates, cephalochordates, and vertebrates
    (E) Urochordates and vertebrates

20. The electron transport system of chloroplasts and mitochondria, even though similar in their use of electron carriers, are different in that
    I.   in electron transport system of chloroplasts CF1 particles project outward of the thylakoid whereas in the electron transport system of mitochondria stalked F1 particles project inwardly.
    II.   the Z scheme of photosynthesis began with the splitting of water into hydrogen ions, electrons, and molecular oxygen while in the mitochondrial transport system, the flow ended with the combining of hydrogen ions, electrons, and molecular oxygen into water.
    III. the photosynthetic electron transport chain started with NADP and the mitochondrial chain ended with NADP.
    IV.   chloroplasts turn energy into carbon compounds and oxygen while mitochondria turn carbon compounds and oxygen into energy.

    (A)   III only
    (B)   I, II, and III
    (C)   III and IV
    (D)   I, II, and IV
    (E)   I and II

21. A community of organisms interacting with abiotic environmental factors composes a/an

    (A)   biogeochemical cycle
    (B)   food chain
    (C)   ecosystem
    (D)   niche
    (E)   population

22. The kangaroo rat is able to live successfully in desert areas because it

    (A)   has a small surface-to-volume ratio
    (B)   has very few predators
    (C)   is only active at night
    (D)   excretes a very concentrated urine
    (E)   lives close to waterholes

23. Organisms which break down the compounds of dead organisms are called

    (A)   phagotrophs
    (B)   parasites
    (C)   saprophytes
    (D)   producers
    (E)   autotrophs

24. When the blood of an individual becomes hypertonic, the kidney will then

    (A)   secrete more antidiuretic hormone (ADH)
    (B)   excrete more water
    (C)   decrease its rate of filtration
    (D)   excrete a smaller volume of, but more concentrated urine
    (E)   increase its rate of filtration

Questions 25–30 refer to the following diagram of the human heart.

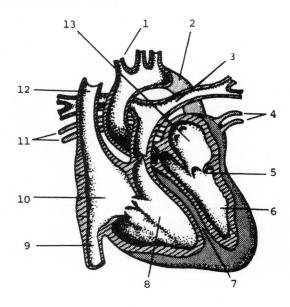

25. The pacemaker of the heart is found at
    (A)   7
    (B)   13
    (C)   5
    (D)   10
    (E)   8

26. Number 5 represents the

    (A) tricuspid valve
    (B) pulmonary valve
    (C) bicuspid valve (mitral valve)
    (D) aortic valve
    (E) semilunar valve

27. Deoxygenated blood returns to the heart by

    (A) 12 and 4
    (B) 11 and 4
    (C) 9 and 13
    (D) 3 and 9
    (E) 9 and 12

28. Number 1 is the

    (A) right subclavian artery
    (B) right common carotid artery
    (C) left subclavian artery
    (D) brachiocephalic artery
    (E) left common carotid artery

29. It is a general rule that all arteries carry oxygenated blood. Which artery is the exception to this rule?

    (A) 2
    (B) 12
    (C) 4
    (D) 3
    (E) 1

30. The blood pressure is maximum in which vessel?

    (A) 3
    (B) 2
    (C) 9
    (D) 1
    (E) 12

31. Progesterone, which is secreted by the corpus luteum, has the effect of

    (A) suppressing LH production
    (B) increasing LH production
    (C) stimulating FSH production
    (D) suppressing estrogen production
    (E) decreasing the thickness of the uterine lining

32. Interactions among species, independent of the abiotic environment, is a characteristic of a

    (A) niche
    (B) ecosystem
    (C) population
    (D) community
    (E) biome

33. In certain parts of the Middle East where agriculture is done on a very limited scale, many people suffer from poor bone growth and poor healing of wounds. This has been traced to a severe deficiency of fresh fruits. The underlying biochemical cause is a lack of

    (A) vitamin K
    (B) calcium
    (C) vitamin E
    (D) vitamin C
    (E) folic acid

34. A cell undergoing mitosis is observed to be devoid of a nuclear envelope and to have its double stranded chromosomes arranged so that each is attached by its centromere to spindle microtubules at the spindle equator. The cell is

    (A) in early anaphase
    (B) in telophase
    (C) in late prophase
    (D) in late anaphase
    (E) in metaphase

Question 35 refers to the following table.

a	A larval wasp that eats its insect host
b	A blackfly that sucks human blood
c	A roundworm in the gut of a carnivorous fish
d	A mink that eats fish
e	A bird that eats conifer seeds

35. Which one of the above is classified trophically as a secondary consumer and functionally as a parasitoid?

    (A) a
    (B) b
    (C) c
    (D) d
    (E) e

36. As one moves up an ecological pyramid, generally

    (A) the biomass increases
    (B) photosynthesis increases
    (C) available energy decreases
    (D) the number of organisms increases
    (E) productivity increases

37. Which of the following statements is correct?

    (A) Root tips show positive geotropism and positive phototropism.
    (B) Shoots show positive phototropism and positive geotropism.
    (C) Root tips show positive phototropism and negative geotropism.
    (D) Shoots show positive geotropism and negative phototropism.
    (E) Root tips show positive geotropism and negative phototropism.

38. The relationship between fungi and the algae in lichens is known as

    (A) mutualism
    (B) parasitism
    (C) commensalism
    (D) saprophytism
    (E) communalism

39. In males, which of the following hormones is responsible for secondary sexual characteristics?

    (A) Estrogen
    (B) Progesterone
    (C) Growth hormone
    (D) Prolactin
    (E) Testosterone

40. Which of the following pairs represents ecological equivalents?

    (A) Squirrel and rattlesnake
    (B) House cat and lion
    (C) Seagull and codfish
    (D) Wild horse and zebra
    (E) Phytoplankton and zooplankton

41. Membrane-associated proteins are studied by obtaining SDS-polyacrylamide gel patterns from "ghosts" and membrane fragments isolated under different conditions. A group of proteins shows the following characteristics:
    I. They may be extracted from the membranes by altering the ionic strength or pH of the medium.
    II. They are thoroughly digested by treating leaky ghosts with proteases, but remain unaffected if sealed ghosts or intact red cells are treated with the same enzymes.

    What can you conclude about these proteins?

    (A) They are peripheral proteins which are associated with the cytoplasmic face of the membrane.
    (B) They are integral proteins.
    (C) They are peripheral proteins associated with the outer face of the membrane.
    (D) They are relatively high molecular weight basic proteins.
    (E) They are relatively low molecular weight acidic proteins.

42. All of the following are true about alteration of generations in flowering plants EXCEPT

    (A) the haploid and diploid stages alternate
    (B) the diploid stage is called the sporophyte
    (C) the sporophyte produces spores
    (D) the haploid stage is called the gametophyte
    (E) the spores divide by meiosis to produce the gametophyte

43. The most significant adaptation of succulent plants to arid environment is the

    (A) low surface-to-volume ratio of above-ground organs
    (B) accumulation of H_2O in the cells
    (C) formation of highly dissected leaves
    (D) absence of stomata
    (E) production of nonpigmented cuticles

44. Sharks live in salt water but have body fluids that are slightly hypertonic to the water. This is because

    (A) sharks drink large quantities of salt water, which increases the ionic content of the body fluids
    (B) sharks secrete urea into the blood, which increases the ionic content of the blood
    (C) sharks absorb salts from the rectal gland that is associated with the gut
    (D) sharks absorb salts in specialized cells located in the gills called ionocytes
    (E) sharks do not produce antidiuretic hormone and thus they do not reabsorb water from the kidney tubule

45. Which of the following makes it possible for eukaryotic genes to be expressed by bacteria?

    (A) A universal genetic code
    (B) The cDNA
    (C) The same composition of 20 amino acids.
    (D) Plasmid vectors
    (E) tRNA

46. Protein synthesis in bacteria starts with

    (A) alanine
    (B) leucine
    (C) methionine
    (D) serine
    (E) cysteine

Use the following diagram and description to answer questions 47 to 49.

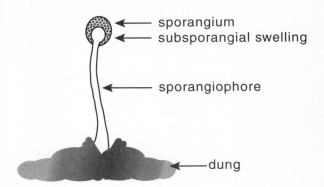

Pilobolus is a fungus that grows on dung. The fruiting body consists of a stalk called the sporangiophore. On top of the sporangiophore is a subsporangial swelling and a sporangium.

In the presence of light, two things happen. First, maximum growth occurs on the side of the sporangiophore that faces away from the light. Second, the turgor pressure in the subsporangial swelling increases, causing the swelling to burst and sending the sporangia flying onto blades of grass. The grass and sporangia are eaten by herbivores.

47. The response of the sporangiophore is

    (A) positive phototropism
    (B) positive thigmotropism
    (C) positive gravitropism
    (D) negative photoropism
    (E) thigmomorphogenesis

48. The subsporangial swelling bursts because

    (A) there is decreased movement of water into the vacuole
    (B) there is increased wall pressure of the vacuole
    (C) the increased growth of the sporangiophore pushes against the swelling
    (D) there is decresed osmotic pressure in the subsporangial swelling
    (E) there is increased movement of water out of the vacuole

49. Which of the following is true about the sporangia of *Pilobolus*?

    (A) The sporangia are produced in the initial dung during plasmogeny.
    (B) The sporangia are produced in new dung during karyogeny.
    (C) The sporangia release haploid spores.
    (D) The sporangia are dikaryotic.
    (E) The sporangia are only involved in asexual reproduction.

50. Which of the following blood cells produces anti-bodies?

    (A) Monocytes
    (B) B-lymphocytes
    (C) Eosinophil
    (D) Basophil
    (E) Thrombocytes

51. Stability in the helical structure of DNA is main-tained by

    (A) carbon bonding
    (B) phosphate bonding
    (C) hydrogen bonding
    (D) ATP
    (E) binding between sulphide groups in the side groups of some amino acids

52. Many single cells have evolved the following method for excreting cellular wastes:

    A vacuole containing material to be expelled trav-els to the cell membrane and fuses with it. After this fusion has been completed, the site of contact opens up and the contents of the vacuole are jet-tisoned out of the cell. This process is known as

    (A) endocytosis
    (B) phagocytosis
    (C) pinocytosis
    (D) exocytosis
    (E) facilitated diffusion

53. This neurotransmitter is found at neuromuscular junctions and central nervous system junctions.

    (A) Norepinephrine
    (B) Serotonin
    (C) Dopamine
    (D) Acetylcholine
    (E) Epinephrine

54. A given cell has a membrane which is permeable to water but impermeable to solute. What will happen to that cell if it contains 1% solute and is placed in a solution containing 3% solute?

    (A) The cell will swell.
    (B) The cell will shrink.
    (C) The cell will remain the same size but will change shape to minimize surface area.
    (D) The cell will remain the same size but will change shape to maximize surface area.
    (E) The material inside the cell will be diluted.

55. A universally accepted scientific principle states that in a closed system, energy can neither be cre-ated nor destroyed. This principle is contained in which of the following?
    I. the First Law of Thermodynamics
    II. the Second Law of Thermodynamics
    III. the Law of Conservation of Energy

    (A) I
    (B) II
    (C) III
    (D) I and II
    (E) I and III

Questions 56–57 refer to the following table.

The following table gives the mean monthly temperature (°C) and precipitation (mm). (Data from Environment Canada, 1990.)

Month	Mean Temperature (°C)	Mean Precipitation (mm)
January	2.5	153.8
February	4.6	114.7
March	5.8	101.0
April	8.8	59.6
May	12.2	51.6
June	15.1	45.2
July	17.3	32.0
August	17.1	41.1
September	14.2	67.1
October	10.0	114.0
November	5.9	150.1
December	3.9	182.4

56. In which month would precipitation in the form of snow be most likely?

  (A) January
  (B) November
  (C) April
  (D) October
  (E) December

57. Which biome would best describe these varying conditions of temperature and precipitation?

  (A) Arctic tundra
  (B) temperate forest
  (C) tropical alpine
  (D) tropical grassland
  (E) tropical forest

58. All of the following statements are true about phototropism EXCEPT

  (A) phototropism results from unequal rates of growth
  (B) phototropism results from a difference in cell division and cell elongation
  (C) phototropism occurs only when the tip of the stem is present
  (D) phototropism is due to auxin produced in the tip of the stem
  (E) phototropism occurs because the auxin hinders cell growth in the actively growing area

59. There are three main steps to blood coagulation. In the first step, a prothrombin activator is released. In the second, prothrombin is converted to its active form. In the third step, fibrinogen is converted to a smaller insoluble protein. Which of the following is the small insoluble protein?

  (A) Prothrombin
  (B) Thromboplastin
  (C) Fibrinogen
  (D) Fibrin
  (E) Thrombin

60. A resting, non-stimulated neuron is electrically _____ along the outside of its surface membrane and electrically _____ along the inside.

  (A) negative..positive
  (B) negative..neutral
  (C) neutral..positive
  (D) neutral..negative
  (E) positive..negative

61. Factors that play a role in determining an individual's rank in a dominance hierarchy include which of the following?
  I.   physical condition
  II. experience
  III. gender

  (A) I only
  (B) II only
  (C) I and II only
  (D) II and III only
  (E) I, II, and III

62. Diabetes mellitus, which results from the body's inability to convert glucose to glycogen, and from the inability of cells to absorb glucose, is due to a deficiency of which of the following hormones?

    (A) Glucogon
    (B) Adrenalin
    (C) Secretin
    (D) Parathormone
    (E) Insulin

63. Among married couples, complications can arise in pregnancy due to the Rh factor. These complications, however, can be avoided by taking a simple precautionary measure. This involves

    (A) treating Rh negative mothers of Rh positive babies with an injection of antibodies against the Rh positive antigen after the first pregnancy
    (B) treating Rh positive mothers of Rh negative babies with an injection of antibodies against the Rh negative antigen before the first pregnancy
    (C) treating Rh negative mothers of Rh positive babies with an injection of antibodies against the Rh negative antigen after the first pregnancy
    (D) treating Rh positive mothers of Rh negative babies with an injection of antibodies against the Rh negative antigen after the first pregnancy
    (E) treating Rh positive fathers with an injection of antibodies against the Rh positive antigen

Questions 64 to 66 refer to the following cladogram.

The nodes represent a hypothetical or known ancestor and the arrows represent the derived characteristics.

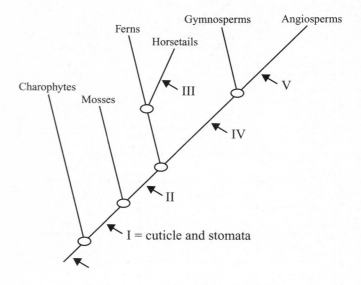

64. The charophytes are best described as which of the following?

    (A) Closest relatives of the plants
    (B) Primitive mosses
    (C) Liverworts
    (D) First plants
    (E) Seedless plants

65. What is the derived characteristic at IV?

    (A) Xylem
    (B) Phloem
    (C) Seeds
    (D) Flowers and fruit
    (E) Spores with cell walls

66. In the cladogram, which of the following groups represent a monophyletic clade?

    (A) Charophytes and mosses
    (B) Mosses and ferns
    (C) Mosses, ferns, and horsetails
    (D) Horsetails and gymnosperms
    (E) Gymnosperms and angiosperms

67. An enzyme facilitates a reaction by

    (A) increasing the free energy difference between reactants and products
    (B) decreasing the free energy difference between reactants and products
    (C) lowering the activation energy of the reaction
    (D) raising the activation energy of the reaction
    (E) increasing the time required for the experiment

68. Due to the dominant fauna in this period, the Mesozoic Era is often referred to as the

    (A) "Age of Reptiles"
    (B) "Age of Fishes"
    (C) "Age of Amphibians"
    (D) "Age of Birds"
    (E) "Age of Mammals"

69. Neutrophils in the human body are

    (A) phagocytes that engulf and digest bacteria
    (B) leukocytes that become specialized as B cells
    (C) leukocytes that become specialized as T cells
    (D) cells without nuclei that are fragments of large cells in the bone marrow
    (E) cells that develop from lymphoid stem cells

70. In 1928, bacteriologist Fred Griffith performed what is now considered a classic experiment. He had been studying the virulence of two strains of Pneumonococcus, the bacteria that cause pneumonia. One strain was dangerous and one was harmless. When grown on agar in petri dishes, the virulent strain produced "smooth" colonies; the harmless strain produced "rough" colonies.

    When the virulent strain was injected into mice, the mice died. When the rough strain was injected into mice, the mice lived. When the smooth strain was boiled first and then injected into mice, the mice lived. However, when boiled smooth strain and live rough strain, both harmless, were mixed together and then injected into mice, the mice died.

    The rough strain had somehow been converted to smooth. In subsequent experiments, various materials from the dead smooth bacteria were isolated

and purified and then injected to see if they were the transforming substance. Which of the following has since been determined to be the substance responsible for the transformation?

    (A) Ribonucleic acid (RNA)
    (B) Deoxyribonucleic acid (DNA)
    (C) Twenty different amino acids
    (D) Glucose polymers
    (E) Adenosine triphosphate (ATP)

71. Which of the following is true of both photosynthetic phosphorylation and oxidative phosphorylation?

    (A) The primary electron donor is a sugar molecule.
    (B) The ultimate electron acceptor is oxygen.
    (C) Both take place in the mitochondria.
    (D) Both produce chemical energy in the form of ATP.
    (E) Both use all of the same electron carriers.

72. The light reactions of photosynthesis include cyclic photophosphorylation and non-cyclic photophosphorylation. Of these two kinds of reactions, it can be said that

    (A) cyclic and noncyclic photophosphorylation each produce both ATP and NADPH necessary for the dark reactions of photosynthesis
    (B) both cyclic and noncyclic photophosphorylation involve photosystem I and photosystem II
    (C) cyclic photophosphorylation involves only photosystem I and produces only ATP; noncyclic photophosphorylation involves photosystem I and photosystem II and produces both ATP and NADPH
    (D) cyclic photophosphorylation involves the reduction of NADP and the liberation of oxygen; noncyclic photophosphorylation involves the reduction of NADP but not the liberation of oxygen
    (E) cyclic photophosphorylation involves the splitting of water, whereas noncyclic photophosphorylation does not involve the splitting of water

Question 73 is based on the diagram below.

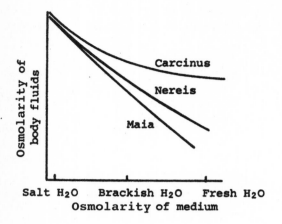

73. Pick the most probable conclusion from the data shown in the graph.

    (A) Maia, the spider crab, is an osmoconformer in salt water, but is capable of osmoregulation in fresh water.
    (B) Nereis, the clam worm, is an osmoconformer in freshwater and is capable of osmoregulation in brackish water.
    (C) Carinus, the shore crab, is an osmoconformer in brackish water, but is capable of osmoregulation in freshwater.
    (D) Carinus, the shore crab, is capable of osmoregulatin in brackish and fresh water.
    (E) Maia, the spider crab, is capable of osmoregulation in brackish and fresh water.

74. Most flowering plants generally meet their need for nitrogen by absorbing

    (A) amino acids and nucleotides
    (B) ammonium ions
    (C) nitrates
    (D) nitrites
    (E) diazonium salts

75. Fossil birds had feathers long before birds flew. This is an example of

    (A) an adaptation
    (B) homology
    (C) heterochrony

    (D) a preadaptation
    (E) polymorphism

76. Plant cells are able to withstand a much wider fluctuation in the osmotic pressure of the surrounding medium than animal cells. This is due to the plant cells'

    (A) stomata
    (B) lipid membrane
    (C) cell wall
    (D) chloroplasts
    (E) vacuole

77. Which of the following is concerned mainly with ATP production?

    (A) Cell membrane
    (B) Golgi apparatus
    (C) Ribosomes
    (D) Rough endoplasmic reticulum
    (E) Mitochondria

78. Prokaryotic cells differ from eukaryotic cells in that the former lack which of the following?
    I.   Ribosomes
    II.   Plasma membrane
    III. Endoplasmic reticulurn
    IV.   Nuclear membrane
    V.   Cell wall

    (A) I
    (B) I and IV
    (C) II and V
    (D) III and IV
    (E) V

79. The two strands of a DNA helix unwind and separate when treated by heating or by adding acids or alkalis to a solution of DNA. This unwinding and separation of the double helix is called melting. Single-stranded DNA has a higher absorbency at 260nm than double-stranded DNA. Below is a graph showing the DNA melting curves. Which of the following is or are correct conclusions?

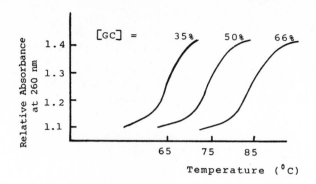

I. DNA double helix is a highly cooperative structure.
II. DNA double helix with a high GC content melts at a higher temperature.
III. The A-T rich regions of DNA melt earlier on heating.
IV. A-T base pairs are more stable than the G-C base pairs.

(A) I, II
(B) II, III
(C) III, IV
(D) I, II, III
(E) I, II, III, IV

80. The genetic code is described as "degenerate" because many amino acids are designated by more than one codon. Those codons that specify the same amino acid differ mainly in the third base of the triplet code. Which of the following is or are the biological significance of degeneracy of the genetic code?
I. The probability that a codon will mutate to a stop signal is minimized.
II. Some mutations at the third base of the triplet codon can be tolerated.
III. The same genetic information is allowed to vary in the DNA composition.

(A) I
(B) II
(C) III
(D) I, II
(E) I, II, III

81. In an enzyme purification experiment, you are provided with a whole cell extract, equipment, and chemicals required for the three kinds of chromatography below. Arrange these chromatographies in the appropriate order that the experiment should be conducted.
I. Affinity chromatography
II. Ion-exchange chromatography
III. Gel-filtration chromatography

(A) I, II, III
(B) II, III, I
(C) II, I, III
(D) III, I, II
(E) III, II, I

82. Which of the following statements concerning the X chromosomes in female mammalian cells are true?
I. Only one of the two X chromosomes is expressed.
II. The inactive X chromosome is faithfully inherited.
III. The inactive X chromosome exists in the form of euchromatin.

(A) I, II
(B) II, III
(C) I, III
(D) I, II, III
(E) III only

83. Insulin moves in the blood to hepatocytes where it is recognized by receptors in the plasma membrane. The result in the hepatocytes is decreased glycogenolysis. This is because

(A) insulin is recognized by an inhibitory G protein in the cell membrane
(B) there is an increase in the synthesis of cyclic adenosine monophosphate (cAMP)
(C) there is activation of the regulating protein phosphorylase kinase
(D) there is inactivation of glycogen synthetase
(E) there is an increase in glucose availability

84. One of the three major classes of hormones is the steroids. They are produced in the

    (A) adrenal cortex and gonads
    (B) hypothalamus
    (C) liver and pancreas
    (D) thyroid
    (E) adrenal medulla

85. First, second, and third order refer to what aspect of neuroendocrine hormonal arrangement?

    (A) The length of time that a certain hormone remains in the bloodstream after being released, first order being the shortest length of time, third order being the longest
    (B) The quantity of a particular hormone released into the bloodstream, first order being the smaller quantity, third order being the larger
    (C) The number of organs in the body directly affected by the release of a single hormone, first order meaning that one organ is affected, second order meaning that two organs are affected, third order meaning that three or more organs are affected
    (D) The number of hormones involved in a regulation process, first order meaning that one hormone is released, second order being when two hormones are used, third order being when three hormones are used
    (E) The areas of behavior affected: first order being those hormones which affect sexual behavior, second order being those which affect aggressive/submissive behavior, third order being those which affect feeding behavior

86. Many animals breed in spring in response to an environment stimulus. The environmental stimulus is increasing day length and breeding is due to increased secretions of sex hormone from the gonads. All of the following are part of the response pathway EXCEPT

    (A) the environmental stimulus is received by the brain
    (B) the hypothalamus secretes a neurotransmitter

(C) the anterior pituitary (adenohypophysis) is stimulated by releasing hormones
(D) the anterior hypophysis releases gonadotropic hormones
(E) the gonadotropic hormones stimulate the gonads to produce sex hormones

87. Oleander (*Nerium oleander*) is a xerophyte. Xerophytes live in very dry places and have which of the listed characteristics?
    I.   Very thick cuticle
    II. Single-layered epidermis on upper surface
    III. Sunken stomata on lower surface
    IV. Trichomes (epidermal hairs) surrounding the stomata

    (A) I and II
    (B) I, II, and III
    (C) II and IV
    (D) I, III and IV
    (E) II and III

88. Which of the following is the correct sequence for blood flow through the human circulatory system?

    (A) aorta, right atrium, right ventricle, pulmonary artery, pulmonary vein, left atrium, left ventricle, superior or inferior vena cava
    (B) superior or inferior vena cava, right atrium, right ventricle, pulmonary vein, pulmonary artery, left atrium, left ventricle, aorta
    (C) superior or inferior vena cava, left atrium, left ventricle, right atrium, right ventricle, pulmonary artery, pulmonary vein, aorta
    (D) superior or inferior vena cava, left atrium, left ventricle, pulmonary artery, pulmonary vein, right atrium, right ventricle, aorta
    (E) superior or inferior vena cava, right atrium, right ventricle, pulmonary artery, pulmonary vein, left atrium, left ventricle, aorta

89. In the transmission electron microscope (TEM)

    (A) structures coated with metals appear dark on the fluorescent screen as well as on the prints (positives) made by a photoplate
    (B) structures coated with metals appear light on the screen but dark on the prints

(C) structures coated with metals appear light on the screen and light on the positives

(D) metal-coated structures show up dark on the screen but light on the positives

(E) metal-coated structures appear dark or light on the screen depending on which metallic stain is employed

90. Rat liver cells homogenized in .25M sucrose and subsequently centrifuged for 10 minutes at 600 G. Which of the following would you expect to find in the pellet formed?

(A) Nuclei and whole cells
(B) Mitochondria and lysosomes
(C) Microsomes
(D) Synaptosomes
(E) Free ribosomes and soluble material

91. Which of the following are true of mutations that occur in humans?

(A) Only mutations that are dominant are passed to the offspring.
(B) None of the mutations are passed to the offspring.
(C) Only mutations occurring in the cells that form gametes are passed to the offspring.
(D) Only mutations that are recessive are passed to the offspring.
(E) All mutations occurring in parents are passed to their offspring.

92. When large numbers of roan cattle are interbred, percentages occur as follows: 25% red, 50% roan, and 25% white. These results illustrate

(A) independent assortment
(B) incomplete dominance
(C) dominance
(D) natural selection
(E) genetic mutation

Questions 93 to 95 refer to the following diagram.

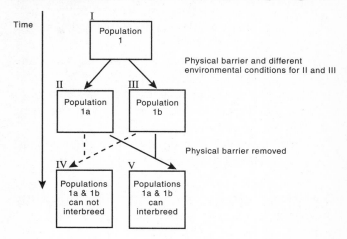

93. This scenario represents which evolutionary process?

(A) Temporal isolation
(B) Artificial selection
(C) Sympatric speciation
(D) Geographic isolation
(E) Behavioral isolation

94. In the given scenario, which population will have the greatest genetic variation?

(A) I
(B) II
(C) III
(D) IV
(E) V

95. Which of the following is a possible physical barrier that separated population 1 into 1a and 1b?

(A) A hurricane
(B) A glacier
(C) Geological formation of a new island
(D) A flash flood
(E) An invasion of a lethal virus into population 1

96. The rate of an enzyme-controlled reaction reaches a peak and then falls off after a certain temperature because of

(A) competition between the activation of enzymes and the inactivation of enzymes

(B) the antagonism of two opposing reactions, one whose rate increases with temperature, the other whose rate decreases with temperature

(C) the opposition caused by the increased temperature, which makes the reaction go faster, and on the other hand, inactivates enzymes

(D) a destruction of key vitamins with increasing temperature

(E) enzymatic degeneration into their original building block proteins

Questions 97–98 refer to the following paragraph.

Two plant species, A and B, from the same genus were used in an experiment. The pollen from A was collected and brushed on the stigma of plant B. The fertilized ova were treated with colchicine and a new plant C developed. Subsequent chromosome analyses showed that the vegetative cells of plants A and B had 18 chromosomes and plant C had 36 chromosomes.

97. The condition in plant C is called

(A) autopolyploidy
(B) polymorphism
(C) introgression
(D) allopolyploidy
(E) diploidy

98. What chromosomal change can result in the condition found in plant C?

(A) Nondisjunction
(B) Transduction
(C) Disjunction
(D) Natural variation
(E) Directional selection

99. Plant C was considered to be a new species. Which of the following crosses was/were required to demonstrate this conclusion?

Cross		Fertile Offspring (Yes or No)
I	A × A	Yes
II	B × B	Yes
III	A × C	No
IV	B × C	No
V	C × C	Yes

(A) I, II, III, IV, and V
(B) III and IV
(C) III, IV, and V
(D) V
(E) I, II, and III

100. The 14 species of Darwin's finches on the Galapagos Islands are believed to have evolved from competing populations of an ancestral finch that colonized the islands from the South American mainland. If so, this would be an example of

I. Convergent evolution
II. Adaptive radiation
III. Founder effect

(A) I
(B) II
(C) III
(D) I and II
(E) II and III

Questions 101–102 refer to the following.

Two species of flour beetles were grown together. In one case, they were grown in pure flour. In the second case, they were grown under the same conditions, including the same amount of flour, except that glass tubing was added.

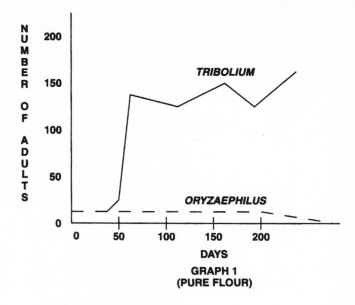

GRAPH 1
(PURE FLOUR)

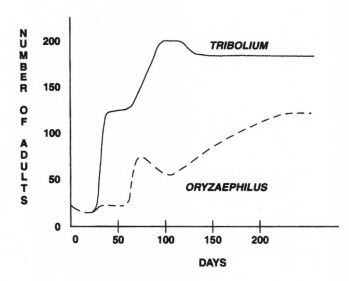

DAYS

101. When grown in pure flour, the number of *Tribolium* adults increases at the greatest rate between

   (A)   day 0 and 25 days
   (B)   25 and 50 days
   (C)   50 and 100 days
   (D)   100 and 150 days
   (E)   150 and 200 days

102. The presence of glass tubing

   (A)   inhibits growth in both species
   (B)   improves the growth of *Oryzaephilus*, but does not affect *Tribolium*
   (C)   improves the growth of *Tribolium*, but does not affect *Oryzaephilus*
   (D)   improves the growth of both species
   (E)   causes the extinction of *Oryzaephilus*

103. Echinoderms differ from arthropods, annelids, and mollusks because echinoderms undergo

   (A)   radial cleavage and the mouth develops from the blastopore
   (B)   determinate cleavage and the mouth develops from the blastopore
   (C)   radial and determinate cleavage and the mouth develops from the blastopore
   (D)   spiral and determinate cleavage and the anus develops from the blastopore
   (E)   spiral and indeterminate cleavage and the anus develops from the blastopore

104. Chemoheterotrophs use

   (A)   carbon dioxide as a carbon source and may use hydrogen sulfide as an energy source
   (B)   carbon dioxide as a carbon source and sunlight as an energy source
   (C)   inorganic molecules as both a carbon and an energy source
   (D)   organic molecules as both a carbon and an energy source
   (E)   organic molecules as a carbon source and light as an energy source

For questions 105–107, refer to the following diagram.

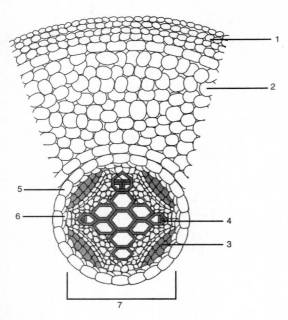

105. Which of the numbered structures corresponds (exclusively) to the phloem?

 (A) 3
 (B) 6
 (C) 4
 (D) 7
 (E) 5

106. This structure, often composed of a single layer of thin-walled parenchymatous cells, readily adopts meristematic activity to form lateral roots.

 (A) 5
 (B) 3
 (C) 6
 (D) 4
 (E) 2

107. Water entering the root may traverse large distances via the apoplast network without passing through a plasma membrane. However, water is compelled at some point to enter cells by the presence of the hydrophobic casparian strip, which is associated with cells indicated by

 (A) 3
 (B) 4
 (C) 2
 (D) 5
 (E) 7

108. A plant with no meristematic tissue will be unable to

 (A) photosynthesize
 (B) transport water
 (C) transport nutrients
 (D) produce fruit
 (E) respire

109. The following describes what category of plant?
 I.   Cambium adds a new ring each growing season
 II.   2 seed leaves
 III. Vascular bundles arranged in a single cylinder in stem
 IV.   Leaf vennation pinnate or palmate

 (A) II corresponds to dicots, while the others pertain to monocots
 (B) II and IV describe dicots, I and III describe monocots
 (C) dicots
 (D) I pertains to some but not all monocots, II, III, and IV are features of dicots
 (E) Monocots

110. John is injured and badly needs a blood transfusion. Carla volunteers, and the physician uses her blood. A few months later, Carla needs a transfusion. John volunteers but the physician turns him down. Given that the physician would accept any blood which was compatible, which of the following could be the blood types of John and Carla?

 (A) John is type B and Carla is type O
 (B) John is type B and Carla is type A
 (C) John is type AB and Carla is type O
 (D) Choices A and C only
 (E) Choices A, B, and C

111. The orderly change from one ecological community to another in an area is called

    (A) convergence
    (B) climax
    (C) succession
    (D) dispersal
    (E) progression

112. Which of the following molecules has an N group, O group, H, and a variable group all attached to the same carbon?

    (A) Nucleotides
    (B) Steroid
    (C) Amino acid
    (D) Fatty acid
    (E) Glycerol

113. When a young bird learns to follow the first large, moving object it sees and hears, this is an example of

    (A) trial-and-error learning
    (B) imprinting
    (C) habituation
    (D) taxis
    (E) kinesis

114. When a virus infects a bacterium, the material that enters the bacterium from the virus is

    (A) sulfur
    (B) a nucleic acid
    (C) a mutagen
    (D) a protein
    (E) a secondary messenger

115. The sequence of bases found in a strand of DNA which served as a template for the synthesis of mRNA is adenine – guanine – thymine. Which of the following sequences will be found on the tRNA?

    (A) cytosine – thymine – guanine
    (B) adenine – guanine – uracil
    (C) thymine – cytosine – adenine
    (D) uracil – cytosine – adenine
    (E) adenine – guanine – thymine

116. Which gland releases STH (growth hormone)?

    (A) Anterior pituitary
    (B) Posterior pituitary
    (C) Adrenal medulla
    (D) Adrenal cortex
    (E) Thyroid

Question 117 refers to the following passage.

According to the U.S. Geological Service, approximately one-third of the Earth's land surface is arid land with meager rainfall that supports only sparse vegetation and a limited population of people and animals. These arid regions may be hot or cold. They may be regions of sand or vast areas of rocks and gravel peppered with occasional plants. In 1953, Peveril Meigs divided these regions on Earth into categories according to the amount of precipitation they received.

117. The biome described here is called a

    (A) taiga
    (B) tundra
    (C) chapparal
    (D) grassland
    (E) desert

118. Natural selection can occur because

    (A) fossils have been found
    (B) the limbs of amphibians, reptiles, birds, and mammals are similar in structure
    (C) all living things contain DNA
    (D) more organisms are produced than can survive
    (E) extinction decreases genetic variability

119. Evidence for evolution includes all of the following EXCEPT

    (A) fossil record
    (B) similarities of proteins in different organisms
    (C) homologous limb structures
    (D) similarities in chromosome banding patterns
    (E) differences in physical appearance of individuals within a species

120. Which of the following depends upon a high degree of communication?

 I.   Social behavior
 II. Echolocation
 III. Circadian rhythms

 (A) I only
 (B) II only
 (C) III only
 (D) I and II
 (E) I and III

121. In humans, blond hair is dominant to red hair, and tasting of PTC is dominant to nontasting. What would be the expected F_2 phenotypic ratio when homozygous blond tasters marry homozygous blond nontasters?

 (A) all blond tasters
 (B) 75% blond tasters; 25% blond nontasters
 (C) 25% blond tasters; 25% blond nontasters; 25% red tasters; 25% red nontasters
 (D) 50% blond tasters; 50% red tasters
 (E) 50% blond tasters; 25% red tasters; 25% red nontasters

122. Ectrodactyly, an inherited human abnormality involving the lack of one or more fingers or toes or parts of them, occurs in homozygous-recessive individuals. If two normal parents have a daughter affected with this condition, and a normal son, what is the probability that the son will be a carrier of the recessive allele?

 (A) 25%
 (B) 33%
 (C) 50%
 (D) 67%
 (E) 75%

123. Some animal behavior recurs on a cyclical basis. Of the following, which three are the most common cycles of animal behavior?
 I.   Circadian (approximately 24 hours)
 II. Annual
 III. Monthly
 IV. Hourly
 V.   Weekly

 (A) I, II, and III
 (B) II, IV, and V
 (C) I, III, and V
 (D) II, III, and IV
 (E) I, II, and V

124. Which of the following general statements concerning biological succession are true?

 I.   The species composition changes continuously during the succession, but the change is usually more rapid in the earlier stages than in the later ones.
 II. The total number of species represented increases initially and then becomes fairly stabilized in the later stages.
 III. Both the total biomass in the ecosystem and the amount of non-living organic matter increase during the succession until a more stable stage is reached.
 IV. The food web becomes more complex, and the relations between species in them better defined.
 V.   Although the amount of new organic matter synthesized by the producers remains approximately the same, except at the beginning of succession, the percentage utilized at the various trophic levels rises.

 (A) I only
 (B) I and II only
 (C) III, IV, and V only
 (D) II, III, and IV only
 (E) I, II, III, IV, and V

125. All of the following statements are true about ecological efficiency EXCEPT

 (A) ecological efficiency is the percent of the energy transferred from one trophic level to the next
 (B) for two trophic levels, ecological efficiency is the ratio of net productivity of the higher trophic level to net productivity of the lower trophic level
 (C) in general, over 80 percent of the energy from one trophic level are not transferred to the next level

(D) ecological efficiency is a measure of the number of species in a particular trophic level

(E) ecological efficiency results in a decrease in energy available at higher trophic level

126. Gases and water enter and leave the leaves of plants through the stoma (singular: stomata). The extent of gases and water movement is controlled by the size of the stoma opening. All of the following explain the opening and closing of stoma EXCEPT

(A) guard cells on either side of the stoma expand and contract due to osmotic uptake and loss of water

(B) The stomata opens when the guard cells accumulate water and become more turgid

(C) The guard cells take up potassium, which results in osmotic uptake of water into the guard cells

(D) The guard cells release potassium, which results in the osmotic loss of water from the guard cells

(E) The guard cells have a red light receptor

127. Roots, in contrast to stems and leaves, do not possess structures specialized for gas exchange. Such structures are not needed because

(A) gas exchange does not occur in the roots

(B) most of the cells in the plant stem are dead, and the cells of the leaves obtain oxygen from photosynthesis; therefore, there is very little need for the roots to transport gases to the rest of the plant

(C) unlike the stem and the leaves, which are covered with a waxy cuticle to prevent dehydration, roots branch extensively into a tremendous surface area consisting of a myriad of porous root hairs. Gases can thus diffuse freely across the moist membranes of root hairs and other epidermal structures, making specialized structures unnecessary

(D) the plant obtains all of its needed CO_2 and other gases through the guard cells

(E) roots obtain oxygen with the water that is absorbed by them

128. An axolotl is a salamander in which larval morphs are sometimes capable of reproduction. This is an example of

(A) paedomorphosis
(B) metamorphosis
(C) polyploidy
(D) allometry
(E) genetic drift

129. The greatest number of mitotic divisions occurs in

(A) meristematic tissue
(B) xylem
(C) phloem
(D) guard cells
(E) cork

Questions 130–131 refer to the following.

The numbers of two different species in large tidal pools on the west coast of Oregon:

Observed Condition	Number of Species 1	Number of Species 2
a. Only species 1 present	60	0
b. Only species 2 present	0	45
c. Species 1 and 2 present	40	30

130. Which of the following best explains the result in condition c?

(A) Commensalism
(B) Interspecific competition
(C) Mutualism
(D) Parasitoidism
(E) Parasitism

131. If the species' numbers that were apparent after condition c were maintained, the outcome would best be described as

(A) competitive exclusion
(B) mutualism
(C) predation
(D) amensalism
(E) resource partitioning

Use the following for questions 132 and 133.

A species of water snake is eaten by birds. The snakes live on the mainland and are able to swim to an adjacent island. The mainland has a wide variety of habitats (rocks, sand, shrubs, marshes, and grasses), whereas the island is almost exclusively sand. A collection of the same numbers of snakes from the island and the mainland showed the following distribution of snake colors.

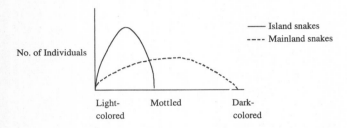

132. This is an example of

(A) stabilizing selection
(B) directional selection
(C) artificial selection
(D) diversifying selection
(E) sexual dimorphism

133. Which of the following situations would have the greatest likelihood of the snakes on the island eventually becoming a new species?

(A) Elimination of the movement of snakes from the mainland to the island
(B) Elimination of movement of snakes from the island to the mainland
(C) Removal of the rocks from the mainland habitat
(D) Addition of rocks to the island habitat
(E) Elimination of all movement of the snakes between the island and the mainland

134. Capillaries play a vital role in the exchange of substances between blood and intercellular tissue fluid because

(A) certain types of lymphocytes leave the bloodstream by way of the capillaries
(B) the blood is purified by the homeostatic action of capillaries

(C) materials leave and enter the bloodstream through the capillaries
(D) capillaries control blood pressure
(E) capillaries join arteries and veins

135. The human appendix, correctly called the vermiform appendix, is frequently described as

(A) on organ that is removed to avoid infection
(B) a vestigial organ
(C) analogous to the caecum of carnivores
(D) homologous to the colon in rabbits
(E) important in water re-absorption

DIRECTIONS: For each group of questions below, match the numbered word, phrase, or sentence to the most closely related lettered heading and mark the letter of your selection on the corresponding answer sheet. A lettered heading may be chosen as the answer once, more than once, or not at all for the questions in each group.

Questions 136–140

(A) 5-Bromouracil
(B) Nitrous acid
(C) Acridines
(D) Ultraviolet light
(E) N-nitroso compounds

136. Intercalating agents; give rise to frameshift mutations

137. Base analogue; primarily gives $GC \rightleftarrows AT$ transitions

138. Strong alkylating agents; result in $GC \rightleftarrows AT$ transitions, transversions

139. The damage done to this system is sometimes repaired by a photoreactivation system

140. Deamination, $GC \rightleftarrows AT$ transitions

Questions 141–144

    (A) Nonsense mutation
    (B) Missense mutation
    (C) Wobble
    (D) Competence
    (E) Genetic mosaics

141. Amber, ochre and opal (UAG, UAA and UGA, respectively)

142. Requirement for bacterial transformation

143. Specification of alternate amino acid by a mutant codon

144. AAG
     UUU

Questions 145–146 refer to the following.

Consider the following diagram of a nephron:

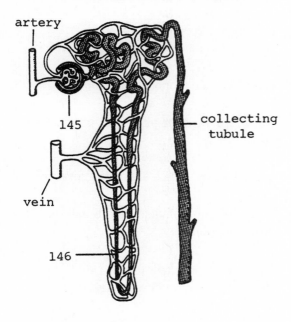

145. The name of this structure is the

    (A) loop of Henle
    (B) glomerulus
    (C) Bowman's capsule
    (D) inner renal medulla
    (E) outer renal cortex

146. This structure is the

    (A) loop of Henle
    (B) glomerulus
    (C) nephric capsule
    (D) inner renal medulla
    (E) outer renal cortex

Questions 147–150

    (A) Instinct
    (B) Insight learning
    (C) Habit
    (D) Tropism
    (E) Taxis

147. Flight pattern of bees gathering pollen

148. The "fight or flight" response

149. A hungry child stands on a chair to reach an apple resting on the table

150. Direction of plant growth toward the sun

Questions 151–155 refer to the following graph, which depicts the logarithmic growth rate of *E. coli* in hospitable medium.

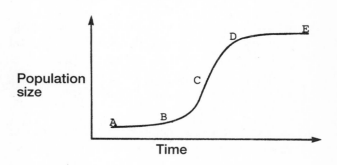

151. Point at which the population of *E. coli* has reached its carrying capacity

152. Point at which the population of *E. coli* has reached its optimal yield

153. Area of lag phase of growth

154. Area of stationary phase of growth

155. Area of logarithmic phase of growth

Questions 156–160

   (A) Competition
   (B) Mimicry
   (C) Mutualism
   (D) Parasitism
   (E) Allelopathy

156. Two species of aquatic life both feed off of salmon, although each species occupies a distinct ecological niche.

157. Two species of plant appear equally uninviting to predators; one species is toxic and the other is not.

158. A worm called *Schistosoma mansoni* is found in the Nile Valley of Egypt. The extensive irrigation canals fed by the Aswan Dam have allowed aquatic snails to flourish; these snails serve as intermediate hosts to the worm. *Schistosoma* mansoni enters the human body right through the skin when people wade in the water. The worm then feeds on the nutrients in the human blood.

159. A certain relationship exists between rhinoceri and a species of small birds. The birds perch on the rhino's back and feed on the small insects which infect the rhino's skin. This relationship is beneficial both to the rhinoceri and to the birds.

160. Yeast stop growing when their own waste product, ethanol, reaches 12%.

Questions 161–165

   (A) Mitochondria
   (B) Endoplasmic reticulum
   (C) Golgi bodies
   (D) Nucleus
   (E) Plasma membrane

161. Contains the chromosomes, and controls protein synthesis and transmission of hereditary information

162. The site for nearly all ATP synthesis in the cell

163. A complex of membranes that traverses the cytoplasm, providing a means of transport for lipids and proteins and containing enzymes that play an important role in the metabolic process

164. Involved in the condensation and concentration of protein, and also contains enzymes involved in the synthesis of complex carbohydrates and in the coupling of these carbohydrates to protein

165. Important for its discriminating permeability and its role in cell adherence

Questions 166–170

   (A) Bryophyta
   (B) Spermophyta
   (C) Angiosperms
   (D) Pterophyta
   (E) Chlorophyta

166. Is divided into monocots and diocots

167. Advanced vascular plants, all produce seeds and pollen

168. Spore formers which alternate generations

169. Contains chlorophylls a and b and has little cell differentiation

170. Contains chlorophylls a and b, are all multicellular with considerable cell specialization

Questions 171–175

    (A) Blastula
    (B) Zygote
    (C) Amnion
    (D) Corona radiata
    (E) Transpiration

171. The mechanism used by plants to rapidly evaporate the water it absorbs

172. During this stage the embryo resembles a hollow fluid-filled ball that is only one cell layer thick

173. An extraembryonic membrane that covers the embryo in its fluid-filled sad

174. A one-cell fertilized ovum

175. Dense covering of follicle cells

Questions 176–179

    (A) Dinoflagellates
    (B) Slime molds
    (C) Red algae
    (D) Brown algae
    (E) Diatoms

176. Agar is extracted from the cell wall and used as microbiological culture medium

177. The feeding stage is an amoeboid mass called a plasmodium

178. The cell wall is in two pieces and contains silica

179. Some members produce neurotoxin that accumulate in shellfish and cause paralytic shellfish poisoning

Questions 180–181

    (A) Active site
    (B) Apoenzyme
    (C) Coenzyme
    (D) Activation
    (E) Regulatory enzymes

180. During the enzymatic activity, the substrate binds to a portion of the enzyme

181. It has been determined that these enzymes have two binding sites

Questions 182–183

    (A) Convergent evolution
    (B) Parallel evolution
    (C) Character displacement
    (D) Allopolyploidy
    (E) Allopatric

182. Exemplified by relationship between fish and whales

183. Having different ranges

Questions 184–185

    (A) 2,3-diphosphoglycerate
    (B) T quaternary structure ("tense" structure)
    (C) Carbonic anhydrase
    (D) R quaternary structure ("relaxed" structure)
    (E) Parathyroid hormone (PTH)

184. Binds to deoxyhemoglobin but not oxyhemoglobin

185. Contains intersubunit salt linkages

DIRECTIONS: The following groups of questions are based on laboratory or experimental situations. Choose the best answer for each question and mark the letter of your selection on the corresponding answer sheet.

Questions 186–189

A series of experiments were performed as follows:

Experiment	Procedure	Observation
I	A frog limb was exposed to radiation of 6000R of x-rays and then amputated.	No nerve regeneration occurred
II	A frog limb was denervated and then amputated before the nerves regenerated.	No limb regeneration occurred
III	A frog limb was exposed to radiation and denervated and the limb was amputated before the nerves regenerated.	No nerve regeneration occurred
IV	A frog limb was exposed to radiation and denervated. The nerves are then allowed to grow back from the brachial plexus into the limb. The limb is then amputated.	Nerve regeneration occurs after amputation

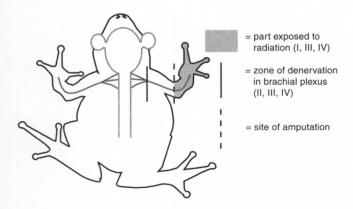

= part exposed to radiation (I, III, IV)

= zone of denervation in brachial plexus (II, III, IV)

= site of amputation

186. With reference to Experiment I, which of the following hypotheses could most likely be correct?

(A) Frog regeneration does not require the interaction of nerves.
(B) Limb regeneration is stimulated by X-rays.
(C) Frog tissue when X-rayed does not regenerate.
(D) Unless the nerves are permanently removed, X-rayed frog tissue will continue to regenerate.
(E) Limb regeneration is obviously stimulated by nerves.

187. Experiments I, II, and III were done because

(A) they aided in the determination of whether denervation or exposure to radiation was a more efficient way to prevent regeneration
(B) they proved that the results of experiment IV were most definitely false
(C) they were controls for experiment IV
(D) they demonstrated that the effect of radiation on limb tissues is opposite the effect of X-rays on nerves
(E) they revealed the effectiveness of denervation

188. It was possible to do experiment IV and obtain favorable results because

(A) the cell bodies of the limb were irradiated
(B) irradiated nerves were not likely to regenerate
(C) the myelinated nerves are unaffected
(D) the cells bodies of the limb nerves were not irradiated when the limb was exposed to X-rays
(E) irradiated and unirradiated nerves regenerate at the same rate

189. In another experiment, Experiment V, a segment of pigmented myelinated nerve which had not been irradiated was implanted into a limb which had been saturated with radiation. This segment was introduced into an albino host. After allowing some time for growth, the limb was severed at the site of the implant. The resulting regenerate was found to be pigmented. Such results imply that

(A) pigment will not be produced by axons of implanted segments of nerves

(B) the myelinated nerve cells were instrumental in this regeneration

(C) the irradiated areas did not support growth of the implanted nerve segment

(D) cells of the Schwann sheath failed to reproduce

(E) the implant had nothing to do with the limb's regeneration

Questions 190–191 refer to the following.

In raccoons the following genotypes of two independently assorting autosomal genes determine coat color:

A–B–   =   (gray)
A–bb   =   (yellow)
aaB–   =   (black)
aabb   =   (cream)

A complementary gene pair on a separate autosome determines whether any color will be produced. CC and Cc allow color expression according to the characteristics of the A and B alleles. The cc genotype results in albino raccoons regardless of the presence of the A and B alleles.

190. If a homozygous dominant raccoon for coat color but heterozygous for its complementary C gene was crossed with a raccoon heterozygous for both coat color and its complementary C gene the most likely phenotypic ratio would be:

(A) $\frac{4}{16}$ gray : $\frac{8}{16}$ yellow : $\frac{4}{16}$ albino

(B) $\frac{9}{16}$ gray : $\frac{3}{16}$ yellow : $\frac{4}{16}$ cream

(C) $\frac{1}{2}$ yellow : $\frac{1}{2}$ albino

(D) $\frac{12}{16}$ gray : $\frac{4}{16}$ albino

(E) $\frac{3}{8}$ black : $\frac{4}{8}$ yellow : $\frac{1}{8}$ cream

191. Given the inheritance pattern of coat color in raccoons, predict the genotype of the parents who produced the F_2 offspring

$\frac{3}{4}$ gray : $\frac{1}{4}$ albino

(A) AAbbCc × AabbCc
(B) aaBBCC × AaBbCc
(C) AABBCc × AABBCc
(D) AaBBcc × AaBbcc
(E) All produced the F_2 phenotypic ratio.

Questions 192–193 refer to the following.

The gonadal cortical tissue and gonadal medullary tissue were removed from the gonads of a group of male rats. They were removed before full sexual development was completed. These extracted gonadal tissues were then randomly replaced in the gonad of the rats and sexual development proceeded.

192. The result expected from such an experiment would be that

(A) all the embryos will be abnormal

(B) the rudimentary region of the gonads will be intact

(C) sperm cells will not be produced in the rats

(D) some of the rats will receive an original of the part of the gonad that was the same sex as their own, and some will receive one that was of a different sex

(E) the rats' sexes will all be changed

193. Such an experiment was done to determine

(A) if gonadal cortical tissue affected reproduction and growth

(B) whether a new species of rat could be formed

(C) if cortical dominance produces ovarian tissue while medullary dominance results in the development of testes

(D) whether ovarian tissue was produced by medullary dominance while cortical dominance causes the production of testes

(E) if the Müllerian duct will not be present in 50% of the embryos

Questions 194–196

194. Consider the following experiment: cysteine was incubated with its cognate tRNA and cysteine tRNA synthetase. The aminoacyl tRNA produced was then reacted with Raney nickel (which catalyses the removal of the third group from the primary carbon, ultimately giving a methyl group). The modified aminoacyl tRNA was then used in a cell-free protein synthesizing system with a template comprised of a random copolymer of U and G in 5:1 ratio. Note that such templates normally lead to incorporation of cysteine (UGU), but not alanine (GCU, GCC, GCA, or GCG). The peptides obtained from the above scenario

(A) incorporate alanine but not cysteine
(B) incorporate cysteine but not alanine
(C) incorporate both cysteine and alanine
(D) incorporate neither cysteine nor alanine
(E) there is not enough information to properly answer the question

195. Without considering the details of the catalytic process involved, the elongation of mRNA is most correctly referred to as which of the following reaction types?

(A) E_1
(B) E_2
(C) SN_1
(D) SN_2
(E) Disproportionation

196. During protein synthesis, information is taken from the DNA to the ribosomes. Which of the following is responsible for this transfer?

(A) Ribosomal RNA (rRNA)
(B) Messenger RNA (mRNA)
(C) Transfer RNA (tRNA)
(D) Nucleotides
(E) Endoplasmic reticulum

Questions 197–200 refer to the metabolic process given below:

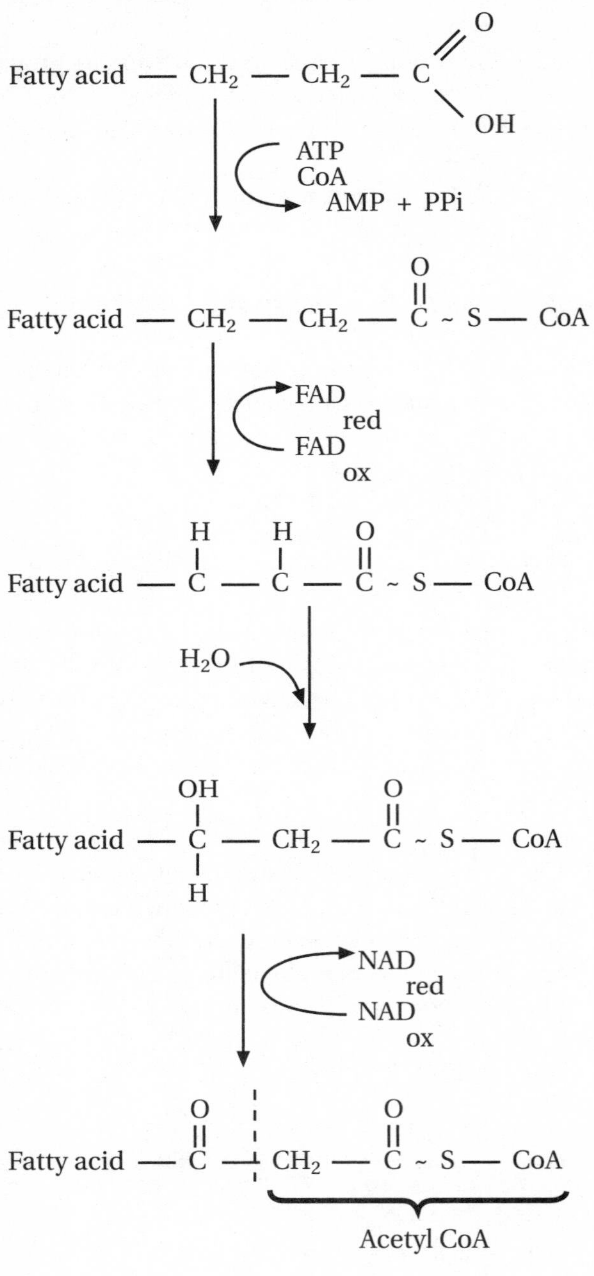

197. If the initial fatty acid chain is 16 carbons long, how much ATP is produced by the reduced products of this process?

    (A) 32
    (B) 35
    (C) 64
    (D) 80
    (E) 96

198. What would the total number of ATP produced be, if each Acetyl-CoA molecule produced then entered the Krebs cycle?

    (A) 96
    (B) 128
    (C) 119
    (D) 160
    (E) 192

199. The condition in which the body would resort to this metabolic pathway is

    (A) lactic acid buildup
    (B) muscle development
    (C) ketosis
    (D) acidosis
    (E) hypoglycemia

200. This pathway is commonly known as

    (A) the citric acid cycle
    (B) the fatty acid cycle
    (C) β-oxidation
    (D) gluconeogenesis
    (E) glucogenesis

Answer Key

1. (C)	35. (A)	69. (A)	102. (D)	135. (B)	168. (D)
2. (D)	36. (C)	70. (B)	103. (E)	136. (C)	169. (E)
3. (D)	37. (E)	71. (D)	104. (D)	137. (A)	170. (A)
4. (B)	38. (A)	72. (C)	105. (A)	138. (E)	171. (E)
5. (A)	39. (E)	73. (D)	106. (C)	139. (D)	172. (A)
6. (A)	40. (D)	74. (C)	107. (D)	140. (B)	173. (C)
7. (A)	41. (A)	75. (D)	108. (D)	141. (A)	174. (B)
8. (C)	42. (A)	76. (C)	109. (C)	142. (D)	175. (D)
9. (D)	43. (A)	77. (E)	110. (D)	143. (B)	176. (C)
10. (A)	44. (B)	78. (D)	111. (C)	144. (C)	177. (B)
11. (A)	45. (B)	79. (D)	112. (C)	145. (C)	178. (E)
12. (D)	46. (D)	80. (E)	113. (B)	146. (A)	179. (A)
13. (B)	47. (A)	81. (B)	114. (B)	147. (E)	180. (A)
14. (A)	48. (E)	82. (A)	115. (B)	148. (A)	181. (E)
15. (B)	49. (C)	83. (A)	116. (A)	149. (B)	182. (A)
16. (B)	50. (B)	84. (A)	117. (E)	150. (D)	183. (E)
17. (C)	51. (C)	85. (D)	118. (D)	151. (E)	184. (A)
18. (A)	52. (D)	86. (B)	119. (E)	152. (C)	185. (B)
19. (C)	53. (D)	87. (D)	120. (A)	153. (A)	186. (C)
20. (D)	54. (B)	88. (E)	121. (B)	154. (D)	187. (C)
21. (C)	55. (A)	89. (A)	122. (D)	155. (B)	188. (D)
22. (D)	56. (A)	90. (A)	123. (A)	156. (A)	189. (B)
23. (C)	57. (B)	91. (C)	124. (E)	157. (B)	190. (D)
24. (D)	58. (C)	92. (B)	125. (D)	158. (D)	191. (C)
25. (D)	59. (D)	93. (D)	126. (E)	159. (C)	192. (D)
26. (C)	60. (E)	94. (E)	127. (C)	160. (E)	193. (C)
27. (E)	61. (E)	95. (B)	128. (A)	161. (D)	194. (A)
28. (D)	62. (E)	96. (C)	129. (A)	162. (A)	195. (D)
29. (D)	63. (A)	97. (D)	130. (B)	163. (B)	196. (B)
30. (B)	64. (A)	98. (A)	131. (E)	164. (C)	197. (B)
31. (A)	65. (C)	99. (C)	132. (B)	165. (E)	198. (C)
32. (D)	66. (D)	100. (E)	133. (E)	166. (C)	199. (E)
33. (D)	67. (C)	101. (B)	134. (C)	167. (B)	200. (C)
34. (E)	68. (A)				

Practice Test 3
Detailed Explanations of Answers

1. (C)

There is a division of labor among the various members in which each member has a particular task to perform. For example, in the honey bee society, which is known as a hive, there is one reproductive female, the queen; several hundred males, the drones, whose main functions are reproductive; and many thousands of immature females, the workers. The workers build the hive, collect food, feed the queen and the drones, nurse the young and protect the colony from strange bees and enemies.

When a hive becomes overcrowded, the queen, some drones, and several thousand workers form a "swarm" and migrate to a new location to establish a new hive. In the old hive a new queen takes over.

2. (D)

The Law of Independent Assortment is Mendel's second law. Phrased in modern terms, this law states: the inheritance of a gene pair located on a given chromosome pair is unaffected by simultaneous inheritance of other gene pairs located on other chromosome pairs.

In other words, two or more traits produced by genes located on two or more different chromosomes pairs assort independently, each trait being expressed independently as if no other traits were present.

3. (D)

Heterotrophic organisms are unable to manufacture their own food because they are unable to synthesize the complex organic molecules they need from the simple ones found in nature. Hence they must obtain these compounds in an already manufactured form.

4. (B)

The R-group is always attached to a carbon atom. Attached to the same carbon atom are three other carbon groups: an amino group which has a positive charge, a carboxyl group which has a negative charge, and a hydrogen.

$$\boxed{R} - \underset{\underset{\underset{H}{|}}{\overset{\oplus}{\underset{|}{N}}\underset{H}{\overset{H}{|}}}{\overset{\overset{H}{|}}{C}} - C \underset{O^{\ominus}}{\overset{O}{\diagup}}$$

5. (A)

Meiosis is the process by which the number of chromosomes in the diploid cell or nucleus is reduced by half.

6. (A)

Gyrase (a form of topoisomerase) relaxes the supercoiled DNA. Initiator proteins (E) bind to the replication origin, then DNA helicase (D) binds to the initiator proteins and is loaded onto the DNA. The DNA helicase untwists the DNA. DNA primase (c) binds to helicase forming a complex called a primosome. The primosome synthesizes an RNA primer (B) to which new nucleotides can be added.

7. (A)

Evergreen trees have adapted to their environments by reducing their leaves to needles which are covered by a waxy secretion. This results in very low transpiration rates, thereby conserving water.

8. (C)

The female gametophyte is the embryo sac.

9. (D)

The endosperm nucleus indicated is the result of the combination of the diploid fusion nucleus (which arises from the joining of the 2 polar nuclei) with the (second) sperm. The result is thus a triploid nucleus, and after a series of divisions, a triploid endosperm tissue. The other sperm, of course, fertilizes the egg (double fertilization).

10. (A)

Many flowering plants illustrate the phenomenon of sporophyte dominance—it is the diploid stage that we are customarily aware of as the flowering plant, while the male and female gametophytes are closer to being reproductive intermediates with true independent stages. Sporophyll is derived from the Greek phyllon (leaf) and refers to a leaf modified to bear spores; sporangium is a more general term for any spore producing plant structure.

11. (A)

The structure indicated is the ovule. If named as a sporangium it would correspond to a megasporangium and the term meiosis on the figure should indicate a diploid structure.

12. (D)

The scanning electron microscope was developed in the 1940s. While its resolving power was not as wide as the standard electron microscope, it has the distinct advantage of producing the illusion of a three-dimensional image, with extraordinarily great depth of field.

13. (B)

If plants such as lettuce, celery, or carrots lose water by evaporation, their cells shrink. The turgor pressure within the cell diminishes, and the plants wilt. These plants if they have not died can be restored if they are placed in a solution of water. The water on the surface of the lettuce is at a higher concentration than that inside the cytoplasm and vacuole of the plant cell. The water molecules therefore diffuse inwardly, from a higher concentration to a lower concentration until the turgor pressure reaches a point where it balances the diffusion of water. If the produce man then sprayed the lettuce with a solution of salt in which the (number of salt molecules) equaled that in the cytoplasm of the lettuce cells, nothing will happen. The water concentration would be the same inside as outside and the spray would be called isotonic to the lettuce cytoplasm.

14. (A)

One group of Archaebacteria is the extreme halophiles which tolerate very high salt concentrations. Cyanobacteria, Clamydias, and Proteobacteria are eubacteria which are found in diverse habitats (anaerobic mud, freshwater, digestive tract) but not in extremely saline conditions. Rhodophyta are red algae, which cannot tolerate extreme salt concentrations.

15. (B)

Nucleic acid contains the genetic information. When a virus infects a bacterium, it infuses its own nucleic acid into the bacterium. Then, the bacterium, using the genetic code of the virus, begins to produce hundreds of additional viruses.

16. (B)

The plants differed because of the differences in amount of sunlight—that is, the environmental conditions. These characteristics were acquired during the life of the plant and are not hereditary. Plants require sunlight but not necessarily direct sunlight. All of the new plants were grown from cuttings and thus sexual reproduction and the expression of recessive, dominant, or polymorphic genotypes was not involved.

17. (C)

This question tests your ability to understand what a territory is, as defined in the question, and to apply that knowledge to understand its ecological functions. A territory is a particular area that is defended by its occupant, usually from others of the same species. Since the individual defends the borders of the territory, the territory actually decreases physical contact among species members. It, therefore, decreases the time and energy devoted to aggression.

A territory channels resources available to specific individuals, maximizing the individuals' chances of surviving and reproducing. It may insure a food supply for the individual, while limiting food available to those who have not been successful in establishing one. It also increases the reproductive success of the individual in it, by increasing the chance of mating. Therefore, it increases the reproductive success of the entire species. Equal distribution of resources among too many individuals could compromise the ability of any of them to survive. Therefore, the territory is a way of regulating population size.

18. (A)

Microtubules are small hollow tubes that, in cross-section, appear as tiny circles. Each microtubule is made of a spiral arrangement of spherical bodies of tubulin protein.

19. (C)

The Urochordate (lancelets), cephalochordate (e.g., amphioxus), and vertebrates are chordates and have the four chordate characteristics (pharyngeal gill slits, notochord, dorsal hollow nerve cord, and post anal tail) at some stage in their life. Hemichordates (acorn worm) have only two chordate characteristics.

20. (D)

Chloroplast and mitochondria, even though they are both cell organelles which are independent of the cells they inhabit, are different from each other in their electron transport system.

The mitochondrion system utilizes stalked F1 particles which project inwardly from the inner membrane surface. The chloroplast system utilizes CF1 particles which are on the outer surface of the thylakoid membrane.

The Z scheme of photosynthesis began with the splitting of water into hydrogen ions, electrons, and molecular oxygen while the electron transport of mitochondria ends with the combining of hydrogen ions and molecular oxygen into water.

Another difference is that the photosynthetic electron transport chain ended with the electrons being passed to NADP, while the mitochondrial system begins with electrons being passed from NADH to the first membrane bound carrier.

21. (C)

An ecosystem is the interaction of biotic and abiotic components of an area or habitat. A biogeochemical cycle involves the movement of mineral ions or molecules in and out of ecosystems. A food chain is not an ecosystem though it exists in one. A niche is the sum of all resources used by an organism in the environment. A population is all of the individuals of a particular species within a defined area.

22. (D)

The kangaroo rat, which is an inhabitant of the Southwestern deserts of the USA, has kidneys which are remarkably efficient at concentrating urine so that little water is lost. This rodent is also remarkable in that it obtains water from its own body metabolism. All of this makes the kangaroo rat adaptable to its environment.

23. (C)

Organisms which break down the compounds of dead organisms are called saprophytes (saprobes).

24. (D)

Hypertonic blood is caused by a lack of sufficient water in the body. Under these conditions, water must be conserved. To this end, the hypothalamus secretes antidiuretic hormone (ADH), which makes the cells of the collecting duct more permeable to water. There is also inhibition of the release of aldosterone. These two effects enable the kidney to excrete more sodium and less water—urine of greater concentration but lesser volume.

25—30

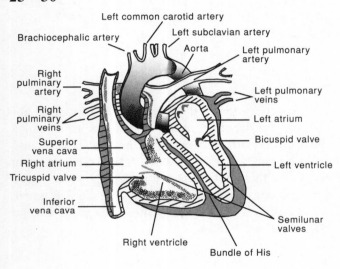

Left common carotid artery
Left subclavian artery
Brachiocephalic artery
Aorta
Left pulmonary artery
Right pulminary artery
Left pulmonary veins
Right pulminary veins
Left atrium
Superior vena cava
Bicuspid valve
Right atrium
Tricuspid valve
Left ventricle
Inferior vena cava
Semilunar valves
Right ventricle
Bundle of His

25. (D)

The pacemaker tissue of the heart consists of special fibers having their own automaticity. These fibers are located in the walls of the right atrium and it is from them that the heartbeat originates.

26. (C)

The bicuspid (mitral) valve consists of two flaps which separate the left atrium from the left ventricle. It prevents the backflow of blood from the left ventricle to the left atrium during ventricular systole.

27. (E)

Deoxygenated blood from the head, neck, and upper limbs returns to the heart through the superior vena cava. Blood from the rest of the body returns through the inferior vena cava. Both venae cavae open into the right atrium.

28. (D)

As the aorta arches around the pulmonary artery, it gives off its first set of branches. On the right, it gives off the brachiocephalic artery, which then branches into the right common carotid and subclavian arteries. The left common carotid and subclavian arteries arise directly from the aorta.

29. (D)

The arteries generally carry oxygenated blood except for the pulmonary arteries which pump deoxygenated blood to the lungs, where it is oxygenated.

30. (B)

In a normal individual, when the ventricles contract, there is a tremendous build-up of pressure, which reaches a maximum of about 120 mmHg. This pressure is transmitted to the walls of the aorta.

31. (A)

Corpus luteum, which is formed from ruptured follicle cells, secretes progesterone and estrogen. Progesterone has the effect of suppressing luteinizing hormone (LH) production.

32. (D)

Interactions among species is a characteristic of communities. Niches, ecosystems, and biomes include abiotic components of the environment. Populations include only one species.

33. (D)

Vitamin C is abundant in fresh fruits, especially citrus, tomatoes, and potatoes. A deficiency of this vitamin leads to scurvy, which is characterized by poor bone growth, hemorrhages, and slow healing of wounds.

34. (E)

By the beginning of anaphase, the centromeres have uncoupled and have begun moving toward opposite poles. In late prophase the chromosomes are not yet lined up along the equator and the nuclear membrane is still visible. The onset of metaphase occurs with the complete disappearance of the nuclear envelope and the alignment of chromosomes along the metaphase plate.

35. (A)

The larval wasp eats the insect, an animal, and is thus a secondary consumer. In addition, it kills its host and is a parasitoid. The fly, roundworm, and mink are secondary consumers because they consume animals. However, even if the human is considered a host for the blackfly, the blackfly does not kill the human. Similarly, the roundworm does not necessarily kill its host. The mink kills the fish but the fish is not a host. The bird is a primary consumer and the seed is not a host.

36. (C)

To answer this question you must know something about ecological pyramids. Energy flow, numbers of organisms, and biomass pyramids may be constructed for a particular ecosystem. These show the primary producers (plants) at the bottom, then primary consumers (herbivores), then secondary consumers, and finally tertiary consumers at the top. Members of each level feed on the members of the level below.

Productivity (photosynthesis) is only measured at the first level, since that is where the plants are. As one moves up the pyramid, the amount of available useful energy decreases, because energy transfers between trophic levels are inefficient and result in heat loss to the environment. Since energy levels decrease, usually the biomass would also decrease and the number of organisms would also decrease (less energy supports fewer organisms). The exception to the biomass pyramid occurs when the producers have a very high rate of reproduction. The exception to the population pyramid occurs when the primary producer is very large (for example, a tree).

37. (E)

Root tips show positive geotropism because they grow downward (i.e., with gravity). On the other hand, they show negative phototropism as they tend to grow away from light.

38. (A)

In a mutualistic relationship, both species benefit. The algae interspersed among fungi to obtain protection and moisture while providing photosynthetic nutrients to the heterotrophic fungus.

39. (E)

Testosterone belongs to the group of male sex hormones known as androgens. It is produced by the interstitial cells that lie outside the seminiferous tubules of the testes. It is vital to the development of the male's secondary sex characteristics, including body hair and baritone voice.

40. (D)

Two species of organism that occupy the same or similar ecologic niche in different geographical locations are termed ecological equivalents.

41. (A)

By definition an integral protein cannot be extracted from cell membrane without dissolving the membrane. The fact that the sample is undigested by proteolytic enzymes in an intact cell or sealed membrane but hydrolysed by the same enzymes when in leaky membranes indicates that they are exposed on the cytoplasmic face of the membrane but not on the outer surface.

42. (A)

In the alteration of generations in plants, the sporophyte begins with the fertilized egg cell. The egg cells mature and grow into the diploid sporophyte. The sporophyte generation ends with meiosis and the production of haploid spores, marking the beginning of the gametophyte generation.

43. (A)

The first priority of succulent plants adapting to arid environments is conservation of water. Since most water loss will occur above ground, the plant must adapt by reducing the surface exposed to hot, dry air. The less surface area exposed, the less water the plant will lose.

44. (B)

Sharks secrete urea into their blood and this makes the body fluids slightly hypertonic to the environment (water). Increased salt from drinking or absorption of salts from ionocytes does not occur in sharks. Salts are excreted (not absorbed) from the rectal gland in sharks. ADH is found in sharks.

45. (B)

The cDNA is a single stranded DNA synthesized by reverse transcriptase in vitro using mRNA as template. Eukaryotic genes containing noncoding intervening sequences are unsuitable for putting into bacteria for expression since bacteria do not have the machinery to remove intervening sequences. Though the other choices contribute to some extent to the success of having eukaryotic genes expressed in bacteria, they are not the most important ones.

46. (D)

All known bacterial mRNA has start and stop signals to define the length of polypeptide to be synthesized. The start signal is read by the initiator tRNA, which brings formylmethionine to the first amino acid position of a polypeptide chain. Initiator tRNA is different from the tRNA that puts methionine to internal positions on a polypeptide chain even though methionine is linked to these two kinds of tRNAs by the same aminoacyl-tRNA synthetase. Transformylase is the enzyme that transfers an activated formyl group from N10-formytetrehydrofolate to methionylinititor tRNA.

47. (A)

Maximum growth occurs on the side of the sporangiophore that faces away from the light, causing the sporangiophore to grow towards the light. This is positive phototropism. Negative phototropism is growth away from light. Thigmotropism is a response to touch. Gravitropism is a response to gravity. Thigmomorphogenesis is the developmental response to mechanical perturbation (e.g., short twisted trees in windy areas).

48. (E)

Turgor pressure is the pressure against the cell wall of the vacuole (B) and it increases due to the increased, not decreased, input of water into the vacuole (A), not out of it (E). The growth of the sporangiophore (C) is independent of the change in turgor pressure in the swelling. The water moves due to an increased, not decreased, osmotic pressure in the subsporangial swelling.

49. (C)

Plasmogeny (fusion of the cytoplasm) and karyogamy (fusion of the nuclei) are two stages in sexual union (syngamy) of fungal cells. This results in the formation of a diploid zygosporangium (A and B). The diploid zygosporangium undergoes meiosis, producing a sporangia that releases haploid spores (C). The sporangia is produced sexually, not asexually (E). The spores can grow mycelia and reproduce sexually or asexually. Between plasmogeny and karyogeny, the zygosporangium is karyotic (D) with two nuclei.

50. (B)

B (Bursa of fabricus) lymphocytes produce antibodies (protein produced by cells) that incapacitates pathogens and are responsible for humoral, or blood-born, immunity.

51. (C)

The nitrogenous bases of opposite chains of nucleotides are electrostatically attracted to each other by the formation of hydrogen bonds. This hydrogen bonding serves to maintain stability in the helical structure.

52. (D)

The process described is exocytosis. Endocytosis is just the opposite—moving material from outside the cell to the inside of the cell. This is accomplished by buckling the cell membrane inward, forming a depression which is eventually surrounded by the membrane and then becomes pinched off to form a new vacuole inside the cell, holding contents which were previously outside the cell. These contents can then be digested and used for energy.

Endocytosis is the name given to the general process; phagocytosis and pinocytosis are specific forms of endocytosis. Phagocytosis occurs when visible, solid material is engulfed; pinocytosis occurs when only dissolved materials, such as proteins, are consumed. Facilitated diffusion is the movement across membranes with the help of proteins.

53. (D)

Acetylcholine is found at neuromuscular junctions and central nervous system junctions. Norepinephrine, serotonin and dopamine are found at central nervous system junctions. Epinephrine is a hormone.

54. (B)

Water, like all other substances, diffuses from regions of greater concentration to regions of lower concentration. The diffusion of water through a semipermeable membrane is called osmosis. In this case, the concentration of water is higher inside the cell than out, so water will diffuse out of the cell, causing the cell to shrink.

55. (A)

The First Law of Thermodynamics says that energy can be transferred and transformed, but in a closed system it can neither be created nor destroyed.

56. (A)

January has a low mean temperature and high precipitation, making snow highly likely.

57. (B)

The temperatures show warm and cold seasons which are indicative of temperate climates. The mean temperatures are above freezing for all months and thus it is not arctic tundra. A tropical alpine biome has cyclic daily temperatures; the yearly variation is low. Tropical areas cycle primarily with respect to wet and dry conditions. The given area does have a wet and dry season but the temperature change is also significant. Therefore, it is not a tropical grassland or forest.

58. (C)

The bending of a plant shoot toward light in phototropism takes place as a result of unequal rates of growth between the side facing the light and the side shielded from the light. Since growth of an organism involves cell division and enlargement, the difference in the rate of growth in the two sides of the plant reflects a difference in the rate of cell division or enlargement or both. Auxins are known to accelerate cell growth in actively growing regions, such as stem tips and vascular cambium. In response to unidirectional light, it is found that there is a differential distribution of auxin in the stem; the side of the stem facing light receives a lower concentration of auxin than the side away from the light. The shaded side elongates faster than the illuminated side, and the effect is that the tip and the top part of the stem curve toward the source of the light.

59. (D)

Tissue damage releases thromboplastin, which stimulates the series of reactions; the result is the production of prothrombin activator. Prothrombin activator forms thrombin from prothrombin. Thrombin converts fibrinogen to the smaller insoluble protein fibrin.

60. (E)

The separated charges of a resting neuron are said to compose an electric potential. These electric charges are carried by ions that are part of, or are attached to, the two sides of the cell membrane of the neuron. In the rest state the positive and negative charges are prevented from coming together, and the membrane is said to be polarized electrically. When an impulse sweeps along a nerve fiber, the permeability of the membrane charges at successive points along the fiber. As this happens at any one point, an avenue is created through which the positive and negative ions of an adjacent point can pass, thus depolarizing that region. In this manner, the impulse is propagated wave-like along the fiber. Shortly after an impulse has passed a given point, the membrane at that point regains both its original state of permeability and polarization.

61. (E)

All three characteristics play an important role in dominance hierarchy. Physical condition is important because much of dominance position is attained through physical confrontation. Sex is a factor because females are usually weaker than males, and also are somewhat vulnerable and incapacitated during pregnancy and birth. Therefore, females tend to be subordinate to males. Experience is a factor because it can contribute favorably to an individual's survival: an experienced individual may be more adept at finding food and avoiding danger than a less experienced individual.

62. (E)

Insulin has two methods of regulating blood glucose. First, it promotes absorption of glucose by the cells, possibly by making the plasma membranes more permeable to glucose. Second, insulin promotes the absorption of glucose by the liver and the conversion of glucose to glycogen within the liver. Both of these actions tend to reduce the glucose level of the blood to normal.

63. (A)

When an Rh-negative woman is made pregnant by an Rh-positive man, the fetus she carries may be Rh-positive. This can result in serious and even fatal complications, not in the first pregnancy, but in subsequent pregnancies.

The incompatibility does not occur in the first pregnancy, because the Rh-negative woman does not build up antibodies against the Rh-positive antigen until Rh-positive red blood cells enter her bloodstream. If this happens, it takes place during birth; therefore, the first infant is unaffected.

If, at the time of delivery, small quantities of the baby's blood cells enter the mother's bloodstream, the mother may build up antibodies against the Rh-positive antigen.

A subsequent pregnancy with an Rh-positive baby could then be disastrous, because the mother's anti-Rh-positive antibodies could enter the baby's bloodstream and destroy its Rh-positive red blood cells.

To prevent this, the mother is given an injection of antibodies against the Rh-positive antigen shortly after her first pregnancy. This destroys the Rh-positive cells before they can trigger the immune response. Subsequent pregnancies are as normal as the first, because the mother has no anti-Rh-positive antibodies.

Treating the father would be a serious mistake as the antibodies would affect his Rh positive antigen production.

64. (A)

In the cladogram, the derived characteristics of plants (cuticle and stomata) give rise to the first plants (mosses). Charophytes are the closest relatives of the plants.

65. (C)

Seeds are first found in gymnosperms. Xylem (A) and phloem (B) are two main types of vascular tissue, the derived characteristic at II. Flowers and fruit (D) are the derived characteristics of angiosperms. Spores with cell walls (E) are found in all plants and are thus a derived characteristic at I.

66. (D)

A monophyletic group is each of the terminal groups or a common ancestor and all of its descendants. No individual groups are given as choices. The only combination of groups given that have a common ancestor and all of the descendents is gymnosperms and angiosperms have a common ancestor and no other members.

67. (C)

The following is a graph showing the function of an enzyme:

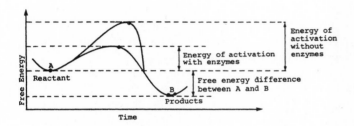

The enzyme decreases the energy barrier, which must be overcome in order for the reaction to proceed, thus allowing the reaction to take place more readily. The difference in free energy between reactants and products does not change.

68. (A)

The Mesozoic Era began some 230 million years ago and was characterized by a wide variety of reptiles. In fact, the Mesozoic Era is commonly referred to as the "Age of Reptiles"; common reptiles of this era were the primitive lizards: snakes, turtles, alligators, crocodiles, pterosaurs (flying reptiles), and later in the era, the dinosaurs. All of these, and also the mammals that came later, evolved from an important paleozoic group called the stem or root reptiles (cotylosaurs).

69. (A)

Neutrophils are phagocytes that engulf and digest bacteria (A). Lymphocytes develop from lymphoid stem cells (E) and develop into B cells (B) or T cells (C). Platelets are cells without nuclei that are fragments of large cells in the bone marrow (D). Lymphocytes and neutrophils are two types of leukocytes, and leukocytes and platelets are two types of cell elements.

70. (B)

After years of research, we know that DNA is the genetic material. In Griffith's experiment, the harmless rough cells took in smooth-cell DNA and were converted.

The demonstration that pure DNA extracted from smooth strain pneumonococcus could transform rough strain bacteria was made in 1944. However, for many years thereafter a good number of scientists remained unconvinced. They had thought that proteins were the genetic material, and old theories often die hard.

71. (D)

Both photosynthetic phosphorylation and oxidative phosphorylation produce ATP as the final energy-rich product. They differ, however, in their source of energy, site of reaction, in the nature of their electron donor and acceptor and electron carriers.

Photosynthetic phosphorylation uses light energy to synthesize ATP from ADP and inorganic phosphate. When light strikes a green plant, it is absorbed by the chlorophyll. The chlorophyll molecule then ejects a high-energy electron which passes down an electron transport chain. During this passage, the electron returns to its original energy level, producing ATP molecules in the process. The chlorophyll molecule serves as both the electron donor and electron acceptor, and the entire process takes place in the chloroplasts.

Unlike photosynthetic phosphorylation, which uses neither oxygen nor organic substrates, oxidative phosphorylation relies on both, because animals are incapable of utilizing sunlight as their energy source.

Oxidative phosphorylation, which takes place in the mitochondria, uses the energy released when certain reduced substrates are oxidized. The electrons released are passed in a series of redox reactions down the electron transport chain in a system of electron acceptors of decreasing reduction potential. In oxidative phosphorylation, the ultimate electron acceptor is oxygen and the primary electron donor is sugar or some other organic substrates. As in photosynthetic phosphorylation, the final product is energy-rich ATP.

72. (C)

In cyclic photophosphorylation, light striking a chlorophyll a molecule excites one of the electrons to an energy level high enough to allow it to leave the molecule. The chlorophyll$^+$ molecule, having lost an electron, is now ready to serve as an electron acceptor because of its net positive charge. However, the ejected electron does not return to its ground state and the chlorophyll$^+$ molecule directly; instead it is taken up by ferredoxin and passed along an electric transport chain. As the electron passes from ferredoxin to the cytochromes and finally back to chlorophyll$^+$, two ATP molecules are produced. No oxygen is liberated, since water is not split; no NADP is reduced since it does not receive electrons. ATP is formed during the electron flow, and light energy is thus converted into chemical energy in the ATP molecules. However, since NADPH$_2$ is not formed, cyclic photophosphorylation is not adequate to bring about CO$_2$ reduction and sugar formation, processes which require the energy of NADPH$_2$.

Non-cyclic photophosphorylation produces both ATP and NADPH$_2$ molecules necessary for the dark reactions of photosynthesis. In this process, electrons from excited chlorophyll a molecules are trapped by NADP in the formation of NADPH$_2$ and do not cycle back to chlorophyll a. Electrons are ejected from chlorophyll b to be donated to chlorophyll a through a series of electron carriers. To restore chlorophyll b to its ground level, water is split into protons, electrons, and oxygen. The electrons are picked up by chlorophyll b, the hydrogens are used to form NADPH$_2$, and oxygen escapes to the atmosphere in its molecular form. Both photosystems I and II are involved in the process of non-cyclic photophosphorylation.

73. (D)

In an osmoconformer, the body fluids have the same osmolarity as the external environment. In osmoregulators, the body fluids have a different osmolarity than the external environment. The diagram shows a direct relationship of osmolarity between the body fluids and the external environment for Maia—thus, Maia is an osmoconformer in all conditions. It is not an osoregulatory in freshwater or in brackish water. The relationship for Nereis is not a perfect straight line suggesting that Nereis may show some osmoregulation; it is not a complete osmoconformer in fresh water. Carinus, the shore crab, is an osmoregulator in both brackish and fresh water.

74. (C)

A few higher plants, especially grasses and trees, pick up quantities of ammonium ion and subsequently incorporate the nitrogen into more complex molecules. However, most flowering plants absorb required nitrogen in the form of nitrates.

75. (D)

The presence of structures in an organism before they are used for a particular function is called a pre-adaptation. Such was the case with feathers. An adaptation (A) is the change resulting from selection. Homology (B) is the condition in which different structures have a common origin. Heterochrony (C) is the rate of development of different parts of an organism. Polymorphism (E) is the existence of several alleles for a given trait.

76. (C)

Stomata are pores in plant stems and leaves and are not structures of a cell. The lipid membrane is common both to animals and plants, and chloroplasts, though structures in plant cells, are involved in the process of photosynthesis, not in the regulation of osmotic pressure. A vacuole is a sac inside the cell that stores materials including water; loss of water from the vacuole makes the plant wilt.

The osmotic pressure of a plant cell is regulated by the cell wall. Consider the separate fates of an animal cell and a plant cell placed into very hypotonic so-lutions. The animal cell will take in water, causing it to swell, and if the original difference in osmotic pressure is great enough, the cell may take in more water and build up more pressure than its membrane can with-stand, in which case the cell would burst (this is called lysis).

This lysis would not occur in a plant cell. A plant cell placed in a hypotonic medium would have water enter it, causing it to swell; however, an upper limit as to how much water can enter is imposed by the cell wall. As the cell swells, its plasma membrane exerts what is called turgor pressure on the cell wall. The wall exerts an equal and opposing pressure on the swollen mem-brane. Mature cell walls can be stretched only to a min-ute amount. When the pressure exerted by the cell wall is so great that further increase in cell size is not pos-sible, water will cease to enter the cell. Thus, plant cells will only absorb a certain amount of water, even in an extremely dilute medium.

77. (E)

The mitochondria are responsible for 95 percent of all ATP produced in the cell. For this reason, the mitochondria are commonly referred to as the "power-house" of the cell.

78. (D)

Prokaryotic cells lack the internal membranous structure characteristic of eukaryotic cells, and gener-ally belong to the simple life forms, such as bacteria. Prokaryotic cells lack endoplasmic reticulum, Golgi apparati, lysosomes, vacuoles, and a distinct nucleus surrounded by a nuclear membrane. They are generally less organized and less advanced than eukaryotic cells.

79. (D)

During heating or ionization by acids or alkalis, the hydrogen bonding of base pairs is disrupted and the double helix is melted into single strands. Refer-ring to the graph, a relative absorbance of below 1.1 corresponds to a double-helical structure, while a value of above 1.4 corresponds to a single-stranded DNA.

The suddenness of the transition from a double-helical structure to a single-stranded one as evidenced from the sigmoid curve suggests that the formation of a double helix is a cooperative process.

The graph clearly demonstrates that DNA with a higher GC content melts at a higher temperature. This is because each GC base pair is held by three hydro-gen bonds which are more stable than the two hydrogen bonds in each AT base pair.

80. (E)

Among the 64 codons, only 3, UAA, UAG and UGA, code for chain termination. The probability of having the other 61 codons mutate to these 3 stop co-dons is much lower than having 20 codons (one for each amino acid for a non-degenerate genetic code) mutate to 44 stop codons. Mutation at the third base of a triplet codon which changes one codon synonym to another does not change the amino acid sequence of a protein. For the same reason, one protein can be coded by many different DNA sequences.

81. (B)

Chromatography is a commonly used technique for protein fractionation. A mixture of proteins in solu-tion is applied to a chromatography column. Different proteins in the mixture interact with the column matrix at different degrees, and hence, their movement through the column is slowed down to different extents. Proteins can therefore be separately collected as they flow out of the column.

Ion-exchange chromatography utilizes the ionic association between proteins and the charged insoluble matrix. The commonly used matrices are diethyla-mi-noethyl-cellulose (DEAE-cellulose), which is positively charged, and carboxymethyl-cellulose (CM-cellulose), which is negatively charged. Proteins flow out separately as the column is washed with a solution of increasing salt concentration. The tightest binding protein requires the highest concentration of salt in order to dissociate from the charged matrix. Fractions of protein collected are tested for enzyme activity. Those fractions containing the desirable proteins are pooled and applied to a gel-filtration chromatography column. The matrix in the column is inert but porous so that proteins of different sizes can be separated. Small molecules enter through the spaces in the beads while large molecules stay between the beads. Large molecules are therefore found to flow out more rapidly than small molecules. Again, fractions are collected, tested for activity, pooled, and applied to the next chromatography column.

Affinity chromatography makes use of the biological interaction between an enzyme and its substrate. An enzyme substrate is immobilized onto an inert matrix such as a polysaccharide bead. Only enzymes with configurations complementary to their substrate remain in the column.

82. (A)

One of the two X chromosomes in each mammalian female cell is permanently inactivated. In mice, it has been found that one or the other of the two X chromosomes is condensed into heterochromatin during the period between the third and the sixth day of development. Heterochromatin is a highly condensed chromatin and is inactive in transcription. Euchromatin is the opposite of heterochromatin. It consists of all of the genome in the interphase nucleus excluding heterochromatin.

The selection of either the paternally inherited X chromosome or the maternally inherited X chromsome to be inactivated is a random process. Once a decision has been made, a clonal group of cells inherits the same inactive X chromosome. This explains why every female is a mosaic, with some groups of cells having the paternally inherited X chromosome expressed and other groups of cells having the maternally inherited X chromosome expressed.

83. (A)

Hepatocytes are liver cells, and glycogenolysis is the breakdown of glycogen to glucose. Therefore, in the liver cells there is decreased breakdown of glycogen to glycose. Insulin is important because it is recognized by the inhibitory G protein in the cell membrane. Increased levels of cAMP (B), activation of the regulating protein phosphorylase kinase (C), inactivation of glycogen synthetase (D), and an increase in glucose availability (E) are all results of increased glycogenolysis.

84. (A)

The steroid hormones consist of the glucocorticoids, the mineralcorticoids, and the sex hormones. They are produced in the adrenal cortex and gonads. The two other major classes of hormones are the protein hormones and the fatty acid hormones. The protein hormones consist of five groups:

(1) amino acids

(2) small peptides—less than 30 amino acid groups

(3) large peptides—approximately 30 amino acid groups

(4) polypeptides—very long chains of amino acids (e.g., STH with 190 amino acid groups)

(5) glycoproteins—very high molecular weights (e.g., TSH, LH, FSH)

The fatty acid hormones have twenty carbon groups. The protein hormones and the fatty acid hormones are produced in various tissues throughout the body.

85. (D)

The first, second, and third order of neuroendocrine hormonal arrangement refer to the number of hormones involved in a particular regulation process. The first order means that one hormone is used. An example is the regulation of the kidney by the hypothalamus, accomplished by the release of a single hormone, ADH. The second order means that two hormones are used. For example, a releasing hormone might be released from the pituitary, travel to another gland, and stimulate it to release another hormone. Third order involves three hormones. For example, a releasing hormone stimulates a gland to release a hormone, which in turn stimulates another gland to release yet another hormone.

86. (B)

The environmental stimulus is received by the brain. These higher centers in the brain secrete neurotransmitters that are passed to the hypothalamus. The hypothalamus secretes gonadotropic hormones that are received by the adenohypophysis. The adenohypophysis secretes gonadotropic-releasing hormones that move via the blood to the gonads, where the sex hormones are produced. Therefore, only (B) is untrue and therefore the correct response.

87. (D)

Xerophytes have structures that reduce water loss. These include a very thick cuticle (I), an epidermis that is several layers thick, sunken stomata (III), and hairs surrounding the stomata (IV). In some xerophytic plants there are hairs on the top and bottom of the leaves.

88. (E)

The right side of the heart receives deoxygenated blood from the inferior and superior vena cava (veins). This deoxygenated blood is pumped by the right ventricle through the pulmonary artery to the lungs. There, the blood is oxygenated and is returned to the heart via the pulmonary vein. The pulmonary system is the opposite of the rest of the circulatory system: the pulmonary artery, unlike other arteries, carries deoxygenated blood, and the pulmonary vein, unlike other veins, carries oxygen-rich blood.

The pulmonary vein empties into the left atrium, which in turn pumps blood into the left ventricle. The left ventricle pumps oxygenated blood into the aorta, which branches into the many arteries that distribute blood to cells throughout the body.

89. (A)

The metal-coated structures are impermeable to electrons, while the uncoated areas allow electrons to pass through them. These electrons make light spots on the screen, and dark ones on a photoplate. When positives are printed from the plate, these structures appear as they do on the screen—darker than their surroundings.

90. (A)

In centrifugation, the more massive structures pellet out before lighter or less dense structures. The diagram below shows the order of centrifugates obtained from rat liver cells. Note that rat liver cells do not contain synaptosomes.

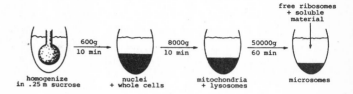

91. (C)

In humans, only mutations occurring in the gametes (sperm and eggs) have the potential of being passed to offspring. These mutations can be either dominant or recessive, since such attributes do not affect whether or not a mutant gene will actually be present in a gamete; dominance and recessiveness only come into play in the determination of phenotype.

92. (B)

Roan cattle result from breeding red cows and white bulls or vice versa. When roan cattle are interbred the red and white strains segregate as indicated in the Punnett square.

		R	W	Roan bull
Roan cow	R	RR	RW	RR = red
	W	RW	WW	RW = roan
				WW = white

The phenotype of the roan cattle is a result of incomplete dominance of the red and white alleles. The cattle have an appearance between the phenotypes of the two parental varieties.

Independent assortment means that genes that are not linked on the same chromosome are inherited independently of each other. *Dominance* indicates a trait which is able to mask another; in this case, neither trait (red or white) is dominant. *Natural selection* is the condition in which only the most fit of a species survive.

93. (D)

The physical barrier initially established that separated population 1 into populations 1a and 1b is an example of a geographic isolation. Temporal and behavioral isolations prevent interbreeding of individuals even though the organisms are in the same location. Artificial selection is not an evolutionary process; it is controlled. Sympatric speciation results when a new species arises within an existing population.

94. (E)

Population 1 (I) was divided by a physical barrier into populations 1a and 1b. Populations 1a (II) and 1b (III) were separated and lived under different environmental conditions. Microevolution would have affected the gene and genotype frequencies in these two populations differently because the conditions were different. If, after a period of time, populations 1a and 1b were reunited and were still able to interbreed, there would be greater genetic diversity in population V. If populations 1a and 1b were sufficiently different and could not interbreed (IV), they would only have the allele composition of 1a and 1b, not a mix as in (V).

95. (B)

Glaciers divide populations. A hurricane, a flash flood, and a viral invasion would reduce the size of the original population enhancing the opportunity for genetic drift (bottleneck effect). The geological formation of a new island would provide a new habitat and the conditions for adaptive radiation.

96. (C)

Increasing the temperature of the environment of a reaction speeds up the reaction, but after reaching a certain peak, the reaction rate tapers off and drops due to the denaturation of enzymes at high temperature. There are other control mechanisms that keep reactions from becoming incontrollable. Two processes of control are induction and repression. *Induction* means that certain enzymes are synthesized only when substrate is available; through repression, the accumulation of product molecules turns off the synthesis of the enzyme responsible for forming the product. Both enzyme induction and repression are highly selective regulatory processes

that act upon specific enzymes rather than affecting all of the enzymes in a cell.

The third factor determining the rate of an enzyme-mediated reaction is the catalytic activity of the individual enzyme in which the activity of certain enzymes is directly affected by the concentration of cofactors.

97. (D)

The vegetative cells of Plants A and B had two sets of chromosomes (2n = 18), whereas plant C had 4 sets of chromosomes (2n = 36). The condition in which organisms have more than two complete sets of chromosomes is called polyploidy. If the two parent plants are different species, the condition is called allopolyploidy. If the parent plants are the same species, it is called autopolyploidy. The diploid (2n) condition of the parent plants is called diploidy. Introgression is the input of alleles—not the addition of chromosomes—from one individual to a second. Polymorphism is the existence of several alleles for one trait.

98. (A)

Nondisjunction is a chromosomal mishap in which homologous chromosomes or sister chromatids do not separate during meiosis, or replicated chromosomes do not separate during mitosis. The result is polyploidy. Disjunction is the expected separation of chromosomes or chromatids. Transduction is the transfer of genes between bacteria and viruses. Natural variation occurs within and among alleles and genotypes of organisms. It is a consequence, not a cause, of allele and chromosome conditions. Directional selection is a process acting on chromosomal differences; it is not a condition.

99. (C)

Plant C is a separate species if it can produce fertile offspring when crossed with itself but cannot produce fertile offspring when crossed with other plant types. Crossing A × A and B × B determine that A and B can produce fertile offspring themselves; it does not affect the information about C as a species because C cannot produce fertile offspring with either A or B.

100. (E)

The dispersal of one species to many locations with different habitats can result in several new species, each descended from the original species. This is adaptive radiation (II) that occurred with the finches on the Galapagos Islands. The initial movement to other islands where there were no finches is founder effect (III). *Convergent evolution* (I) is the term sometimes used for condition homoplasy in which organisms have characteristics with similar functions but different origins. The important traits of the finches were all derived from the same origin and thus they are not convergent.

101. (B)

You are asked to determine which portion of Graph I shows the greatest rate of increase of *Tribolium* adults. Days 0 to 25 show no increase. Days 50 to 100, 100 to 150, and 150 to 200 show minor fluctuations, but no overall increase in numbers of adults. The largest increase (from 10 to 120 adults) occurs from day 25 to day 50. Compare the various slopes of the lines in the graphs to assess growth rate.

102. (D)

In order to answer this question, you need to compare the difference in growth of both species when grown in pure flour (Graph I) vs. growth in the same amount of flour with glass tubing added (Graph II). Graph II shows more *Tribolium* adults and more *Oryzaephilus* adults than Graph I. Therefore, the presence of glass tubing has resulted in an increase in growth of both species. The glass tubing probably offered a habitat choice to reduce competition.

103. (E)

Echinoderms are deuterostomes. Arthropods, annelids, and mollusks are protostomes. The development differs between protostomes and deuterostomes. Deuterostomes undergo radial and indeterminate cleavage and the anus develops from the blastopore. In protostomes, there is spiral and determinate cleavage and the mouth develops from the blastopore.

104. (D)

The term *chemoheterotroph* is a combination of two words: *chemotroph* and *heterotroph*. Heterotrophs obtain carbon from organic molecules and chemotrophs obtain energy from substances (chemicals) in the environment. Only (D) obtains carbon from organic molecules and energy from the environment. Inorganic carbon dioxide or an inorganic molecule is the carbon source for (A), (B) and (C). In (E), the energy source is light.

105–107

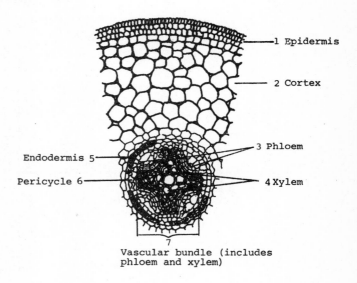

Vascular bundle (includes phloem and xylem)

105. (A)

The use of the term "exclusively" eliminates D as a possibility; the phloem cells intercalate between the arms of the x-shaped core of xylem.

106. (C)

The description corresponds to the pericycle. The pericycle is found within the endodermis, and forms the outside ring of the stele (a central cylinder of vascular tissue).

107. (D)

The Casparian strip is the waterproof band that comprises the radial and end walls of endodermal cells.

108. (D)

In a plant, only regions of meristematic tissue, also called embryonic plant tissue, continue to actively divide throughout the plant's life span. Fruit production requires active cell division. The removal of all of the meristematic tissue from a plant leaves it incapable of cell division and therefore incapable of producing fruits.

109. (C)

	Dicots	Monocots
No. of seed leaves (cotyledons)	2	1
Venation of leaves	Pinnate, palmate (i.e., branched network)	Parallel
Type of root system	Woody, taproot (i.e., large primary root with branch roots growing from it)	fibrous, no taproot, all roots approximately equal in size
Arrangement of vasculature in stem	Vascular bundles in a single ring	Vascular bundles irregularly distributed through pith tissue
Cambium/ growth rings	New growth ring each year or growing season	Stem and root devoid of cambium, no growth rings
Arrangement of flower parts	Arranged in twos, fours or most often fives	Arranged in threes

110. (D)

There are four blood types: A, B, AB, O. The basis of these blood types is the presence or absence of certain antigens, called agglutinogens, on the surface of the erythrocytes. The erythrocytes of type A blood carry agglutinogen A; those of type B carry agglutinogen B. Type O has neither of these agglutinogens, while type AB has both A and B agglutinogens. These agglutinogens react with certain antibodies, called agglutinins, that may be present in the plasma. An agglutinin-agglutinogen reaction causes the cells to adhere to each other. This clumping, or agglutination, would then block the small blood vessels in the body, causing death.

Agglutination is the clumping of erythrocytes by agglutinins. The blood clotting mechanism, or coagulation, is not involved. Type A blood, which has the A agglutinogen, does not have the anti-A agglutinin in its plasma. But it does have the anti-B agglutinin. A person with type B blood has only anti-A agglutinin. A person with type O blood has both anti-A and anti-B; type AB has neither type of agglutinin. A person does not have the agglutinin in his plasma which would clump his own red blood cells. A summary of the antibodies and antigens that each blood type contains is outlined in the following table:

Blood Type	Agglutinogens (antigens)	Agglutinins (antibodies)	Able to accept blood type
A	A	anti-B	A, O
B	B	anti-A	B, O
AB	A and B	none	A,B,AB,O
O	none	anti-A and anti-B	O

Blood typing is a critical factor in blood transfusions. Transfusions are usually between people of the same blood type; however, one may use another blood-type provided that it is compatible.

In this problem, we know that John can receive Carla's blood, but that Carla cannot receive John's. Choices (A) and (C) give blood types which fit this situation. In choice (B), John cannot receive Carla's blood, because his anti-A agglutinins would clump with Carla's A agglutinogens and agglutination would result.

111. (C)

Succession is a fairly orderly process of changes of communities in a region. The change in species composition is continuous, but is usually more rapid in the earlier stages than in the later ones. The number of species, the total biomass in the ecosystem, and the amount of non-living organic matter all increase during the succession until a more stable stage is reached. Toward the final stages, the food webs become more complex, and the relationships between species in them become better defined.

112. (C)

An amino acid molecule consists of an amine group, carboxyl group, and a variable group all attached to the same carbon.

113. (B)

For this question you must be familiar with animal behavior, and, in particular, different types of learning. A taxis is an oriented movement in response to an environmental stimulus, while a kinesis is a random response. Neither of these are learned behaviors.

Trial-and-error, habituation, and imprinting are all considered types of learning. Trial-and-error learning involves improvement in a response based upon previous experience. Habituation, on the other hand, is a decreased response to a stimulus after continued exposure to the stimulus. The behavior of the young bird is an example of imprinting. This type of learning occurs during a sensitive period in the development of the organism and usually is irreversible.

114. (B)

Nucleic acid contains the genetic information. When a virus infects a bacterium, it infuses its own nucleic acid into the bacterium, and the bacterium, using the genetic code of the virus, begins to produce hundreds of more viruses.

115. (B)

The DNA provides the template for mRNA. In mRNA, uracil replaces thymine. The mRNA then moves into the cytoplasm where the tRNA brings amino acids to the mRNA. The anticodon on the tRNA is opposite that of the codon on the the mRNA. Therefore, for the given DNA sequence, the mRNA codon would be uracil-cytosine-adenine, and the tRNA anticodon would be adenine-guanine-uracil.

116. (A)

STH—somatotropin—is the growth hormone which is species-specific for humans. It is released from the anterior pituitary during periods of fasting and during sleep. It is released in spurts during sleep; in children we see the characteristic growth "spurts" due to this periodic release.

117. (E)

The desert receives little rainfall and vegetation is sparse, making it difficult for large numbers of herbivores to live in this biome. (Peveril Meigs [1903-1979], incidentally, was a geographer whose writings include "World distribution of coastal deserts." Ch. 1 in Coastal Deserts: Their Natural and Human Environments, D.H.K. Amiran and A.W. Wilson (eds.). Univ. of Arizona Press, pp. 313, 1973.) The taiga (A) is characterized by largely evergreen forest vegetation. This eliminates this choice, as the biome described here cannot support trees. The tundra (B) has low temperatures unsuitable for trees or grasses, thereby eliminating this choice. The chapparal (C) is an incorrect choice because of its abundant rain in the winter and its ability to support evergreen trees. Like the desert, the grassland biome (D) is characterized by meager rainfall, but unlike the desert, it provides natural pastures for grazing animals and vegetation to support the large number of herbivores that can be found there.

118. (D)

This question asks you to determine the conditions under which natural selection occurs rather than the experimental evidence for natural selection. The fossil record and similarities in limb structure among organisms is evidence for the existence of natural selection. However, natural selection can only occur when more organisms are produced than can survive, allowing the best-adapted to survive and reproduce. The fact that all living things contain DNA is another piece of evidence that organisms may have a common heritage, but is not a precondition for evolution. Natural selection may lead to extinction, but extinction does not cause natural selection.

119. (E)

This question requires some knowledge of the experimental data that support the concept of evolution. The more traditional evidence includes the fossil record and similarities in limb structures in different organisms. More recent evidence comes from molecular studies that include similarities in proteins and chromosome banding patterns. Differences in physical appearance within a species merely reflect individual variation with no evidence of whether these traits are acquired or inherited, or whether changes have occurred over time.

120. (A)

To answer this question you need to know the functions of the three different forms of behavior. Social behavior involves the interactions of organisms within a species and includes, among others, mating behavior, caring for young, food procurement, and protection from predators. These all depend upon a high degree of communication. Echolocation is a mechanism used by bats for finding their way when they are flying. They emit sound waves that bounce off objects, and these sound waves are, in turn, perceived by the bats. Circadian rhythms are regular cycles of activity that occur

about every 24 hours. Neither echolocation nor circadian rhythms depend upon interaction with other organisms, and, therefore, they do not depend upon social communication.

121. (B)

R = blond

r = red

N = tasting

n = nontasting

Homozygous Blond Tasters: RRNN

Homozygous Blond Nontasters: RRnn

RRNN × RRnn = RRNn F_1

RRNn × RRNn → F_2:

	RN	Rn	RN	Rn
RN	RRNN blond taster	RRNn blond taster	RRNN blond taster	RRNn blond taster
Rn	RRNn blond taster	RRnn blond non–taster	RRNn blond taster	RRnn blond non–taster
RN	RRNN blond taster	RRNn blond taster	RRNN blond taster	RRNn blond taster
Rn	RRNn blond taster	RRnn blond non–taster	RRNn blond taster	RRnn blond non–taster

RR ⎧ 1NN → RRNN
⎨ 2Nn → 2RRNn
⎩ 1nn → RRnn

25% blond tasters
50% blond tasters
25% blond nontasters
75% blond tasters:
25% blond nontasters

122. (D)

The affected daughter must be homozygous recessive; therefore, each of the normal parents must be heterozygous.

	N	n
N	NN	Nn
n	Nn	nn

N = normal
n = lobster claw

Consider the possibilities for a normal son's genotype, as seen in a Punnet square. The son cannot be homozygous because he is normal. Of the three remaining possibilities, two contain the recessive allele, which would make him a carrier. Thus, the odds that he is a carrier are 2/3, or 67%.

123. (A)

Circadian, annual, and monthly cycles are the most common. The most obvious circadian rhythm is sleep/wakefulness, which for many animals tends to parallel the 24-hour cycle of day (sunlight) and night (darkness). Annual cycles include migration, hibernation, and mating. Monthly patterns, for many animals, include the ovulation cycle.

124. (E)

All of the statements are true regarding biological succession. If no disruptive factors interfere, most successions eventually reach a mature stage. The community of this stage is called the climax community. It has much less tendency than earlier successional communities to alter its environment in a manner injurious to itself. In fact, its more complex organization, larger organic structure, and more balanced metabolism enable it to control its own environment to such an extent that it can be self-perpetuating. Consequently, it may persist for long periods of time, provided that major environmental factors remain essentially the same.

In general, the trend of most successions is toward a more complex ecosystem in which less energy is wasted and a greater biomass is supported.

125. (D)

Ecological efficiency is the percent of the energy transferred from one trophic level to the next (A), and it is measured by the net productivity at successive trophic levels (B). In general, less than 20 percent of the energy from one trophic level is passed to the next. Thus, over 80 percent of the energy from one trophic level are not transferred to the next level (C) and there is a decrease in energy available at higher trophic levels (E). There is less biomass at succeeding trophic levels, but not necessarily fewer species (D).

126. (E)

The guard cells have blue-light receptors, which stimulate potassium and thus water uptake. The guard cells become more turgid, and this acts to open the stomata. Loss of potassium decreases the water content and the turgidity and thus the stomata closes. All of the statements are true except (E).

127. (C)

Roots play a vital role in the absorption and transport of not only gases, but water and nutrients as well. The roots' surfaces are readily permeable; therefore, no specialized structures are needed.

128. (A)

Paedomorphosis is a special case of heterochrony and is the retention of childlike (larval) morphology in adults. Metamorphosis is the transformation from a larval to adult morph. Allometry is the differential growth of different body parts. Polyploidy is a chromosomal change, and genetic drift is one cause of altered gene frequencies in populations.

129. (A)

To answer this question you need to recognize the different types of tissues in plants. Meristematic tissue is the undifferentiated, actively dividing tissue that is responsible for growth. Therefore, the meristem would show the greatest number of mitotic divisions. Cork (thick, water impermeable cells in bark) and xylem (hollow vessels that conduct water and minerals from roots to leaves) are both made of dead cells. Phloem and guard cells are highly differentiated. Phloem transports sap and guard cells regulate the size of stomata.

130. (B)

Interspecific competition is an interaction in which there is a negative effect for both species. In this example, the negative effect is the reduced numbers of each of the species. Commensalism and mutualism have no negative effects. Both parasitoids and parasites benefit and would increase in numbers. Their hosts would decrease in numbers.

131. (E)

Interspecific competition can have two main consequences. One of these, resource partitioning, is the one in condition c. The second, competitive exclusion, results in the elimination of one species. Mutualism is beneficial to both species so the numbers would not decrease. Predation results in a decrease in the number of only one of the species. Amensalism has a negative effect on one species but no effect on the second.

132. (B)

Birds prey upon the snakes and thus must see them from above. On the island, light-colored snakes would be less visible than dark ones to the birds. Therefore, on the island, there is directional selection for the light colored snakes. Stabilizing selection changes the range of values not the mean and diversifying selection results in a bimodal distribution. This is not artificial selection because there is no influence by humans. Sexual dimorphism relates to sexual election, and in this case there is no mention of differences between male and females.

133. (E)

Geographically isolating the island and mainland snakes could result in a new species on the island because the habitat on the island is different. As long as snakes can move either to or from the island, the alleles will continually mix in the island and mainland snakes. Adding rocks to the island habitat or the removal of rocks from the mainland habitat will reduce the differences between the habitats and minimize selection.

134. (C)

The capillaries are extremely narrow, so much so that their diameter can accommodate only one red blood cell. They consequently have enormous surface area, and materials can readily diffuse through their thin walls. Materials are exchanged between the capillaries and the cells outside the bloodstream, and between the capillaries and the intercellular, tissue fluids. The main function of these microscopic blood vessels is to provide a means for exchange, via diffusion, between the bloodstream and the rest of the cells in the body.

135. (B)

The human vermiform appendix is frequently described as a vestigial organ, one that was better developed in the ancestor. It is homolog with caecum, not the colon (D). It does not have the same function as the caecum of carnivores and therefore, it is not analogous (C). One function of the colon is water reabsorption, but the human vermiform appendix is not involved in water reabsorption (E). It can get infected but removal to avoid infection (A) is not recommended.

136. (C)   137. (A)   138. (E)   139. (D)   140. (B)

5-Bromouracil is a thymine analogue; it mispairs with guanine and can cause transitions in both directions between GC and AT base pairs. Nitrous acid is a deaminating agent, transforming cytosine to uracil (and adenine to hypoxanthine). The predominant mutagenic effects are again $GC \rightleftarrows AT$ AT transitions. Acridines, of the type illustrated below, function as intercalating agents. These complex with a portion of a DNA chain and mask it, thus giving rise to frameshift mutations.

Proflavin, a simple acridine

Ultraviolet light primarily causes dimerization of adjacent pyrimidines; since the dimers formed cannot fit into the double helix, replication and/or transcription are blocked until the defect is repaired.

This may occur through excision repair, which involves the following: sequential nicking by a UV-specific endonuclease, DNA synthesis by a polymerase, and $5' \rightarrow 3'$ excision by that same polymerase followed by DNA ligase-catalysed sealing of the nick. Alternatively, most cells contain a photoreactivating enzyme which, upon absorption of blue light, can recognize a pyrimidine dimer and catalyse its cleavage back to monomers.

N-nitroso compounds are among the most potent alkylating agents known. Guanine is particularly susceptible to these reagents, whose predominant mutagenic effects are $GC \rightleftarrows AT$ transitions.

141. (A)   142. (D)   143. (B)   144. (C)

A nonsense mutation refers to an alteration of the base sequence such that a terminator codon (UAG, UAA, or UGA) appears at an internal point in a structural gene. This is a much more serious lesion than a missense mutation because the chances of obtaining a functional polypeptide in spite of premature termination are minimal. Several strains of *E. coli* have been isolated with amber suppressor mutations (i.e., a tRNA is specified that misreads the termination codon and inserts an amino acid, resulting in a translated product with at least partial activity).

As the above suggests, the converse type of point mutation is a missense mutation which causes an alternative amino acid to be incorporated into the peptide. If the residue is not at a critical site in the product, and if the substituted amino acid is similar to the original (such as Asp incorporation for Glu) then protein function should show little or no impairment.

Wobble refers to the flexibility that exists in codon/anticodon recognition at the third codon base and results in the capacity of some tRNA molecules to bind several distinct codons. The pattern of degeneracies in the genetic code is such that XYU and XYC always specify the same amino acid, whereas XYA and XYG are not as consistent; this led Crick to hypothesize that the steric constraints on pairing of the third (3′ codon, 5′ anticodon) base would be less rigorous than for the first two. This is illustrated in the diagram:

anticodon   3′ ← 5′

A   A   G
⋮   ⋮   ⋮
U   U   U

codon   5′ → 3′

Competence refers to the ability of cells to take up fragments of naked DNA from the medium. If, for example, linkage or mapping in *E. coli* is to be studied via transformation experiments the cells are treated with concentrated $CaCl_2$ to make their membranes permeable to DNA (the cells must also be devoid of exonuclease I and V to function).

An organism is a genetic mosaic if it contains cells of more than one genotype. *Lyarization* is the facultative heterochromatization undergone by one X chromosome in female mammalian cells. The process is universal insofar as it occurs in all cells aside from those making up the germ line and, more importantly for the purposes of this question, the choice of whether the maternal or paternal X chromosome is to be inactivated is made at random.

145. (C)   146. (A)

The glomerulus is a tiny ball of blood capillaries surrounded by the Bowman's capsule. In each nephron, blood is filtered from the glomerulus to the interior space of the Bowman's capsule, where wastes are removed from the blood. Initial urine then flows through the nephric tubule, which is composed of the descending and ascending loop of Henle. The nephric tubule is highly convoluted and therefore extremely long and has enormous surface area, which is extremely important for the reabsorption of water. Water is in fact the most abundant of the substances that are filtered out from blood and later returned to it.

147. (E)   148. (A)   149. (B)   150. (D)

An instinct is a pattern of unlearned, automatic behavior that is usually beneficial to the species. Insight learning is the ability to use reason in order to solve a problem at hand. It is most common among the primates, especially chimpanzees, monkeys, and humans. A habit is a learned response. It is usually activity requiring little thought, thus leaving the conscious free for more complicated thought processes. A tropism is a growth pattern such as the phototropism of a plant toward a light source. A taxis refers to an orientation pattern such as the photoaxis of bees following the sun in order to maintain a straight flight pattern.

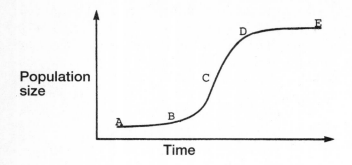

151. (E)   152. (C)   153. (A)
154. (D)   155. (B)

A = Lag phase of growth: that period of time during which the culture is first reproducing.

B = Logarithmic phase of growth: that period of time during which there is a sufficient number of members reproducing exponentially.

C = Point of optimal yield: point of largest rule of increase within a given environment.

D = Stationary phase of growth: that period of time during which the birth rate is equal to the death rate.

E = Point of carrying capacity: point in time at which the largest number of organisms of a particular species that may be maintained in a given environment.

156. (A)   157. (B)   158. (D)   159. (C)   160. (E)

Competition occurs when individuals or species vie for land or resources. When space and resources are limited, one individual or species will usually be defeated, and eliminated, by another.

Mimicry occurs because it is selectively favorable for one or more species. One species can avoid predators by mimicking the appearance of another species which is repulsive to predators.

Parasitism is a relationship in which one organism feeds off another organism. Most parasites live inside their hosts and suck up nutrients from various organs. The majority of parasitic relationships are harmful to the host, and some parasites can even be fatal to their hosts. Parasites would benefit were they to evolve into a form less damaging to their hosts. Killing their hosts can put the parasites at a great disadvantage.

Mutualism describes a relationship between two organisms in which both organisms benefit.

Allelopathy is the limiting of a population by the waste products which that population produces. In recent decades, much attention has been devoted to human allelopathy. Many environmentalists fear that the human race may poison itself to the point of extinction. Oil spills in the oceans, acid rain, deforestation, and toxic wastes are only a few of the threats to the oxygen and food supply. It is not inconceivable that humans could destroy their environment to the point that it could no longer support them.

161. (D)

The nucleus is the control center of the cell. The nucleus contains the chromosomes, which bear the genes, the ultimate regulator of life.

The genes contain the information that specifies the precise nature of each protein synthesized by the cell. Control of protein synthesis is the key to controlling the activities and responses of the cell, since a tremendous array of important biological and biochemical processes are regulated by enzymatic proteins. By switching particular genes on and off, the cell controls not only the kinds of enzymes that it produces, but also the amounts. Fine control of enzymatic proteins is crucial to the proper functioning of the cell.

162. (A)

Mitochondria are membrane-bounded organelles concerned principally with the generation of energy to support the various forms of chemical and mechanical work carried out by the cell. They are distributed throughout the cytoplasm, and are most abundant in regions of the cell that consume large amounts of energy.

Mitochondria are enclosed by two membranes. The outer one is a continuous delimiting membrane. The inner membrane features many folds, called cristae, that extend into the interior of the mitochondrion. Inside the inner membrane are various enzymes that catalyze a number of different reactions, all of which lead to the production of ATP.

163. (B)

The endoplasmic reticulum is responsible for transporting certain molecules to specific areas within the cytoplasm. In addition to transporting lipids and proteins, the endoplasmic reticulum contains enzymes important in cellular metabolism.

The membranes of the endoplasmic reticulum (ER) form interconnecting channels that can extend from the nuclear membrane to the plasma membrane. When the ER has ribosomes attached to its surface, it is called the rough endoplasmic reticulum (RER), and when there are no ribosomes attached, it is called the smooth endoplasmic reticulum (SER).

In cells actively engaged in protein synthesis, the protein, or polypeptide, chains are synthesized in the ribosomes and are then transported by the endoplasmic reticulum to other sites of the cell where they are needed.

164. (C)

The function of the Golgi apparatus is best understood in cells involved in protein synthesis and secretion. Protein synthesized on the rough endoplasmic reticulum is carried by vesicles to the Golgi bodies, where the protein is concentrated and condensed. The Golgi bodies release the protein in the form of secretory granules, which are then separated from the cytoplasm by a membrane that can fuse with the plasma membrane. When the secretory granule fuses with the plasma membrane, its content (protein in this case) is expelled from the cell, a process known as exocytosis.

The Golgi complex also has a synthetic role: it contains enzymes involved in the synthesis of complex carbohydrates and in the coupling of these carbohydrates to protein.

165. (E)

The plasma membrane consists primarily of lipid and protein. The lipid molecules are polar, with the two ends of each molecule having different electric properties. One end is hydrophobic, the other is hydrophilic. The lipid molecules arrange themselves in two layers in the plasma membrane so that the hydrophobic ends are near each other, and the hydrophilic ends face outside toward the water and are stabilized by water molecules. In this bilayer, individual lipid molecules can move laterally, so that the bilayer is actually fluid and flexible.

Protein molecules of the plasma membrane may be interspersed in various sites but embedded at different degrees in relationships to the bilayer. Some of them may be partially embedded in the lipid bilayer, some may be present only on the outer surfaces, others may be completely hidden in the interior, and still others may span the entire lipid bilayer from one surface to the other. The protein molecules, like the lipid molecules, are able to move laterally within the plane of the membrane.

The primary function of the membrane is to screen substances that would enter and leave the cell. The selectively permeable membrane allows the cell to regulate its internal environment.

The membrane is also important in cell adherence. The specificity of protein molecules on the membrane surface allows cells to recognize each other and to bind together through some interaction of the surface proteins. Communication between cells is important during cell division so that cells divide in an organized plane, rather than in random directions giving rise to an amorphous mass of cells as in cancer. Surface proteins play important roles in the immune response, hormonal communication, and conduction of impulses in nerve cells.

166. (C)

The angiosperms are flowering plants that can be divided taxonomically into two major groups. These are monocots and dicots.

167. (B)

The subdivision spermophyta are advanced vascular plants, which produce seeds and pollen.

168. (D)

Subdivision pterophyta contain the ferns that all have well-developed vascular tissue. The ferns alternate generations and are spore formers.

169. (E)

Chlorophyta or the green algae is a group which includes both unicellular and multicellular species. The algae contains both chlorophylls a and b. In this group there is little differentiation.

170. (A)

The division Bryophyta, which contains mosses, liverworts and hornworts, are multicellular plants with well-differentiated tissues and considerable cell specialization. They contain chlorophyll a and b.

171. (E)

Plants require a lot of water for their existence. This is due to their ability to rapidly evaporate absorbed water. The process used is transpiration.

172. (A)

During embryonic development, after numerous cell divisions, a stage is attained where the embryo resembles a hollow fluid-filled ball that is one cell layer thick—the blastula.

173. (C)

During embryonic development, the embryo is covered by four extraembryonic tissues: the chorion, amnion, yolk sac, and allantois.

The amnion is a tissue from the inner cell mass that develops into the "bag" or membrane of the "bag of waters;" the fluid in which the baby develops is called amnionic fluid or "waters."

174. (B)

During fertilization haploid male and female gametes join to form a diploid single-celled fertilized ovum called a zygote.

175. (D)

Corona Radiata is a dense covering of follicle cells that encompass the egg at ovulation. These cells are held by hyaluronic acid until penetrated by the enzyme hyaluronidase, which allows fusion of sperm and egg and thus fertilization.

176. (C)   177. (B)   178. (E)   179. (A)

Agar is contained in the cell wall of red algae. Algin is found in the cell walls of brown algae and used as an emulsifier.

The plasmodium is the feeding stage. It is a unicellular mass that can grow to be several centimeters.

Diatoms are known for walls that contain silica and fit together like a box.

Red tide is due to the presence of certain dinoflagellates. The dinoflagellates accumulate in shellfish. These shellfish can on occasion cause paralytic shellfish poisoning when consumed by humans.

180. (A)

During enzymatic activity, an enzyme combines with the substrate to form an intermediate enzyme—the substrate complex.

When the substrate binds to the enzyme, it binds specifically with a relatively small part of the enzyme molecule: the active site.

181. (E)

Some enzymes are capable of being inhibited by their own product. These enzymes are called regulatory enzymes. They have two binding sites: an active site and a regulatory site.

182. (A)   183. (E)

Convergent evolution denotes a situation whereby two groups that are not closely related develop similar characteristics in response to similar selection pressures. As an example, note the similarities between Australian marsupials and unrelated placental mammals. They are similar because their environments are similar.

Convergent evolution is equally well exemplified by the development of flippers by marine mammals such as the whale. Fish also employ flippers to maneuver through the water but are not closely related to any mammal.

Contrasting terms to convergent evolution are *divergent evolution* and *parallel evolution*. In the latter, two related species evolve in similar ways over an extended period of time. Divergent evolution denotes a splitting of one group into two groups that continually become more unlike over time.

Character displacement describes the tendency for closely related sympatric species to develop in such a fashion as to decrease both the competition between them and the possibilities of interbreeding. This occurs because of the adaptive advantages of reducing competition for a single niche and because of selective hybrid elimination: if two closely related sympatric species produce hybrids as well as or better adapted than themselves, then the original populations will soon become indistinct and will not properly represent different species. If, on the other hand, the hybrids are less well adapted than the parental populations, they (the hybrids) will die out, and any genetic factors in the parental population that promote correct mate selection will increase in frequency at the expense of those promoting incorrect selection.

Allopolyploidy denotes a multiplication in chromosome number in an interspecific hybrid. If, for example, two diploid individuals in distinct species produce a hybrid which (through a process of nondisjunction, for example) has its complement of chromosomes doubled, then the latter will have received a full diploid set from each parent. Thus, there is no requirement for pairing of chromosomes from different species in meiosis, and the hybrid is thus able to produce normal gametes. In all the above, there is no requirement for allopatry. Note that this is an important and frequent type of speciation among plants.

Remember that *allopatric* refers to groups with different ranges; *sympatric* denotes populations with the same range.

184. (A)   185. (B)

Briefly, the allosteric effects shown by hemoglobin are thought to depend on an equilibrium between two quaternary structures: tense and relaxed. The former is constrained by the presence of eight salt links between the four subunits and a low oxygen affinity; in the latter, the salt links are absent and the oxygen affinity is higher.

The two forms are thought to be in equilibrium; allosteric modulators exert their effect by shifting this equilibrium to one side or the other. The presence of 2,3-diphosphoglycerate tends to decrease the oxygen affinity of hemoglobin by binding only to deoxyhemoglobin (in fact oxygenation and 2,3-DPG binding are to a first approximation mutually exclusive, the latter being released from hemoglobin upon oxygenation). Protons and CO_2 are also involved in the formation of the salt links characterizing the T form; increased acidity and CO_2 concentrations favor decreased oxygen binding.

186. (C)

According to experiment I where the frog limb was exposed to radiation of 6,000R of X-rays and then amputated, regeneration did not occur. This result for experiment I shows that radiated frog tissue does not regenerate.

187. (C)

Experiments I, II, and III were done because they were controls for experiment IV.

In experiment I, the limb was exposed to radiation only and the limb amputated. In experiment II the limb was denervated and amputated before nerve regeneration and was not exposed to radiation. In experiment III the limb was exposed to radiation and denervated and the limb was amputated before nerve regeneration occurred.

In the fourth experiment the nerves were allowed to grow back from the brachial plexus.

188. (D)

In experiment IV the frog limb was not exposed up to the brachial plexus to radiation. The nerve cell bodies that were not radiated grew back into the limb. After the limb was amputated, they still continued to grow.

189. (B)

The resulting pigmented regenerate implies that the myelinated nerve cells were instrumental in the regeneration. The myelinated nerve cells that were pigmented were introduced into the albino host. If the resulting regenerate was pigmented it is obvious that the pigmented myelinated nerve cells had a great influence on the results.

190. (D)

The following genotypic and phenotypic configurations show the independent assorting autosomal genes that determine coat color.

A-	B-	=	gray
A-	bb	=	yellow
aa	B-	=	black
aa	bb	=	cream

CC and cc allow color expression according to the characteristic of the alleles. The cc genotype results in albino raccoons regardless of the presence of the A and B alleles.

Mating between AABBCc and AaBbCc:

$$AA \times Aa \qquad BB \times Bb \qquad Cc \times Cc$$
$$= \frac{1}{2}AA \times \frac{1}{2}Aa = \frac{1}{2}BB + \frac{1}{2}Bb = \frac{1}{4}CC + \frac{2}{4}Cc + \frac{1}{4}cc$$

When the three matings were performed 12/16 were dominated by the gray gene with either the CC or Cc as complementary gene and 4/16 were dominated by the cc gene that produces albino raccoons.

191. (C)

F_1 offspring are all gray.

$$AABBCc \times AABBCc$$

Step(1) $AA \times AA = 1\ AA$

$BB \times BB = 1\ BB$

$$Cc \times Cc = \frac{1}{4}CC + \frac{2}{4}Cc + \frac{1}{4}cc$$

Step (2) $AA \times BB = 1\ AB$

Step (3) $1AABB \times (\frac{1}{4}CC + \frac{2}{4}Cc + \frac{1}{4}cc)$

$$= \frac{1}{4}AABBCC + \frac{2}{4}AABBCc + \frac{1}{4}AABBcc$$

192. (D)

The result expected from the experiment would be that some of the rats would receive an original of the part of the gonad that was the same sex as their own, and some would receive one that was of a different sex. This would be because in the random replacement of the gonadal tissue, some of the gonadal cortical tissue and gonadal medullary tissue would be aligned in their correct positions.

193. (C)

This experiment was done to determine if cortical dominance produces ovarian tissue, while medullary dominance results in the development of testes.

There are two types of tissue which arise in the rudimentary region of the gonads: the gonadal cortical tissue and the gonadal medullary tissue. These two tissues are hypothesized to be antagonistic to each other. Cortical dominance produces ovarian tissue, while medullary dominance produces testes.

194. (A)

The polypeptide incorporates alanine in place of the usual cysteine. The experiment is important because it demonstrates that codon recognition is mediated exclusively by the anticodon rather than by some interaction involving the activated amino acid.

195. (D)

The reaction is obviously bimolecular, with the 3′ terminus of the growing chain reacting with a free nucleoside triphosphate. Furthermore, the reaction proceeds through nucleophilic attack by a 3′ oxygen on the α phosphorus atom of the triphosphate.

196. (B)

The DNA sub-unit contains all the genetic material for the synthesis of proteins. During protein synthesis, the sense strand of the DNA duplex acts as a template for the messenger RNA (mRNA). The mRNA then acts as a template for rRNA and arranges itself on the correct reading frame of ribosomes to synthesize the corresponding protein.

197. (B)

Each FAD yields two ATP and each NAD yields three ATP molecules in the subsequent process of electron transport. That leaves a total of five ATP molecules produced for each Acetyl-CoA. A 16-carbon fatty acid yields $[n/2] - 1$ Acetyl-CoA or 7.

$$7 \times 5 = 35 \text{ ATP}$$

198. (C)

We already know from question 197 that 35 ATP molecules have been produced. Each molecule of Acetyl-CoA may enter the Krebs cycle and yield an additional 12 molecules of ATP.

$$7 \times 12 = 84.$$
$$84 + 35 = 119.$$

199. (E)

Hypoglycemia—low blood sugar—depletes the cell's supply of carbohydrates, its prime source of energy. Lipid metabolism ensues. Lactic acid buildup can result from muscle development and both result in the conversion of lactic acid to pyruvic acid in the liver. Ketosis and acidosis result from the buildup of Acetyl-CoA as a result of lipolysis.

200. (C)

Although this is a form of fatty-acid metabolism, it is not a cycle. It is the oxidation of the β carbon of the fatty acid. The citric acid cycle is more commonly known as the Krebs cycle. Gluconeogenesis is the formation of glucose from non-carbohydrate molecules such as amino acids and lactic acid. Glucogenesis is the formation of glycogen from glucose.

GRE BIOLOGY

Index

Index